Lecture Notes in Mathematics 1501

Anna De Masi   Errico Presutti

Mathematical Methods for Hydrodynamic Limits

Springer-Verlag
Berlin Heidelberg New York
London Paris Tokyo
Hong Kong Barcelona
Budapest

Authors

Anna De Masi
Dipartimento di Matematica
Pura ed Applicata
Università di L'Aquila
67100 L'Aquila, Italy

Errico Presutti
Dipartimento di Matematica
Università di Roma Tor Vergata
Via Fontanile di Carcaricola
00133 Roma, Italy

Mathematics Subject Classification (1991): 60-02, 60K35, 82A05

ISBN 3-540-55004-6 Springer-Verlag Berlin Heidelberg New York
ISBN 0-387-55004-6 Springer-Verlag New York Berlin Heidelberg

Typesetting: Camera ready by author
Printing and binding: Druckhaus Beltz, Hemsbach/Bergstr.
46/3140-543210 - Printed on acid-free paper

To the memory of our friend Piero de Mottoni

Preface

The idea of writing this book came after a course given in Boulder, Colorado, in 1988, by one of us, (E.P.). At that time the paper of Guo, Papanicolaou and Varadhan had just appeared. Dave Wick, who had worked himself on entropy inequalities in interacting particle systems and was in Boulder at the same time, was very interested on these new results and we tried together to understand them. We were very much helped by the interest and participation of Dick Holley. This was the starting point, then the discussion went on, covering other methods used for studying the hydrodynamic behavior of interacting particle systems. In the end, we thought it worthwhile to save the material as an introductory course on the collective phenomena in particle systems. Unfortunately Dave did not want to take part to this last effort, and the task of actually writing down some notes was left to the present authors.

We used the notes in seminars and short courses, as well as in a School organized by Rolando Rebolledo in Santiago of Chile. The result was not what we were hoping for: too many details missing, at least for younger researchers without a solid background in probability and/or statistical mechanics. We therefore tried to add new topics, we made a preprint, [42], and waited for reactions. We have been lucky, many colleagues had the patience of reading our preprint and to suggest improvements. We are also very grateful to several students in L'Aquila and the University of Roma "La Sapienza" who read our work trying to understand it, they have been most helpful in rewriting the present version, which, in the meantime, has become a rather long text.

There are many colleagues who have helped us in this work. We have already mentioned Dave Wick, Richard Holley and Rolando Rebolledo. Most of the arguments that we have treated were explained to us by Joel Lebowitz and Herbert Spohn. Christian Maes and Ravi (Krishnamurthi Ravishankar) were the first to give us helpful criticism on the first draft of this work. Stefano Olla explained to us some of the tricks of the Guo-Papanicolaou-Varadhan method. We acknowledge many helpful discussions with Carlo Boldrighini and Sandro Pellegrinotti on the contents of Chapter IV. We understood together with Silvia Caprino and Mario Pulvirenti how to prove the validity of the kinetic limit in discrete velocity stochastic particle systems and with Joel Lebowitz and Lello Esposito how the theory applies to the study of the hydrodynamical limits for cellular automata. It has been a pleasure to work with Pablo Ferrari and Maria Eulalia Vares and we are certainly indebted to them for what we have understood on the exclusion process and its perturbations. In particular we are grateful to Maria Eulalia Vares for many suggestions and criticism on this book. We are indebted to Enza Orlandi, Piero de Mottoni, Enzo Olivieri and Giorgio Fusco for many helpful discussions on phase decomposition and interface dynamics and to Giambattista Giacomin and Luca Bonaventura for helpful comments on the last three chapters of this book.

CONTENTS

CHAPTER I

INTRODUCTION

The integral of a continuous function $f(r)$, $0 \le r \le 1$, with values in $[0,1]$, can be estimated, using the Monte Carlo method, by throwing at random a large number of points in the unit square and by computing the fraction of those which are below the graph of f. A way to do it is to introduce a mesh $\epsilon > 0$ and, for any integer x such that $0 \le \epsilon x \le 1$, to pick a variable with uniform distribution in $[0,1]$. If its value is smaller than $f(\epsilon x)$, a particle is put in x, which is otherwise left empty. The event is recorded by the variable $\eta(x)$, set equal to 1 if a particle is in x and 0 otherwise. Then

$$\epsilon \sum_{0 \le \epsilon x \le 1} \eta(x) \longrightarrow \int_0^1 dr f(r) \tag{1.1}$$

in probability when $\epsilon \to 0$. The random configurations $\eta = \{\eta(x), 0 \le \epsilon x \le 1\}$ introduced by this procedure allow to reconstruct the whole f and not only its integral. In fact we have that in probability

$$\epsilon \sum_{0 \le \epsilon x \le 1} \phi(\epsilon x)\eta(x) \longrightarrow \int dr \phi(r) f(r) \tag{1.2}$$

for all $\phi(r)$'s in a countably dense set in $C([0,1])$, f is then identified by the values of the integrals in (1.2).

Suppose now that there is a dynamics and let f_t, $t \ge 0$, be the solution of some equation with initial value $f_0 = f$. Is it possible to let the particles which approximate f move so that at any time t they also approximate f_t? Of course we would like the particles dynamics to depend only on the equation that f_t satisfies, so that if we vary f we only need to change the initial state of the particles and not the rules of their evolution. In fact we are even more exigent, pretending for instance that the updating rules can be economically implemented on a computer.

Consider as an example the case of the heat equation:

$$\frac{\partial f_t}{\partial t} = \frac{1}{2}\frac{\partial^2 f_t}{\partial r^2}, \qquad r \in [0,1] \tag{1.3a}$$

$$f_t(0) = f_t(1), \quad f_0(r) = f(r) \tag{1.3b}$$

Define the particles evolution so that at each unit step all the particles independently of each other jump to one of their two neighboring sites (with periodic conditions) with equal probability. Call η_t, $t \in \mathbb{N}$, the random particle configurations obtained in this way, the occupation numbers $\eta_t(x)$ being non negative integers, and not necessarily equal to 0 or 1, as in η. (In Chapter VI we will show that it is possible to simulate (1.3) by a dynamics where the exclusion condition, $\eta(x) = 0, 1$, is fulfilled at all t). The space-time scaling symmetry of (1.3) suggests that if the space mesh is ϵ then the time mesh should be ϵ^2. Indeed

$$\eta_{[\epsilon^{-2}t]} \longrightarrow f_t, \quad [a] = \text{integer part of } a$$

weakly, (in the sense of (1.2)), and in probability, as it will be shown in the next Chapter.

The analysis is not as simple if we replace the macroscopic equation (1.3) by a non linear PDE, in particular if it developes discontinuities and becomes degenerate. This is what happens in many applications and what one is usually more interested in. The simplest equation which exhibits formation and propagation of shock waves is the Burgers equation

$$\frac{\partial f_t}{\partial t} + a\frac{\partial}{\partial r}\left[f_t(1-f_t)\right] = 0, \quad a > 0 \tag{1.4a}$$

$$f_t(0) = f_t(1), \quad f_0(r) = f(r) \tag{1.4b}$$

To simulate (1.4) Boghosian and Levermore, [14], have introduced a particle model with two velocities, ± 1, and an exclusion rule which prevents particles with same velocity from being at the same site. The updating rule (times are discrete, integers) consists of two successive sub-updatings. The first one is called collisional: at all the sites where there is one particle the velocities are independently set equal to 1 with probability p and to -1 with probability $q = 1 - p$, independently of their previous values. All the other sites are left unchanged. The second sub-updating is streaming: each particle moves deterministically to its nearest neighbor site in the direction of its velocity, which is unchanged.

Since (1.4) is invariant when scaling space and time with the same parameter, for simulating (1.4) the ratio between the space and the time mesh is kept fixed. Chen, Lebowitz and Speer, [29], have done new experiments on the Boghosian-Levermore automaton to investigate the microscopic stability of the shock. They consider 2^{15} sites, hence a space mesh $\epsilon = 2^{-15}$, p is chosen larger than 1/2 (equal to 0.625). We show below pictures taken from [29].

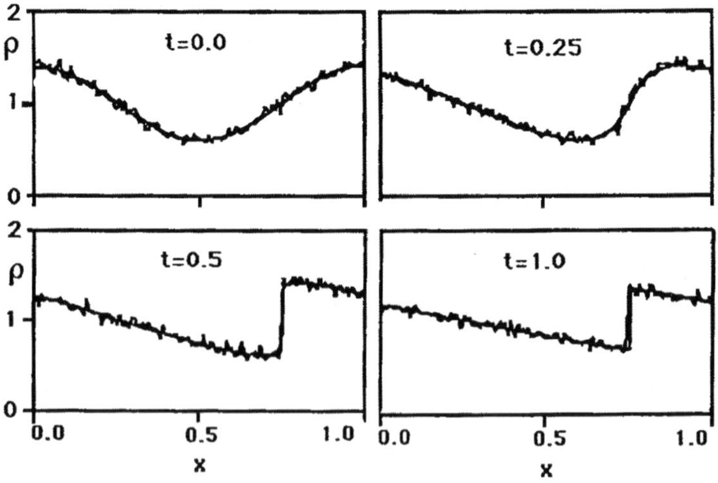

FIG. 1.1.

The continuous curve in Fig.1.1 represent the solution of (1.4) with $a = 1$ evolving from that at $t = 0$; the broken curves represent the observed density profiles at the indicated (macroscopic) times, measured by averaging the particles numbers in intervals containing 2^6 sites. The pictures clearly indicate that in the automaton a shock developes and that this happens as predicted by (1.4). The pictures below, Fig.1.2, taken from the same series, show the stability of the shock: the discontinuity decreases, as in (1.4), due to finite volume effects, yet, microscopically, the shock remains remarkably sharp:

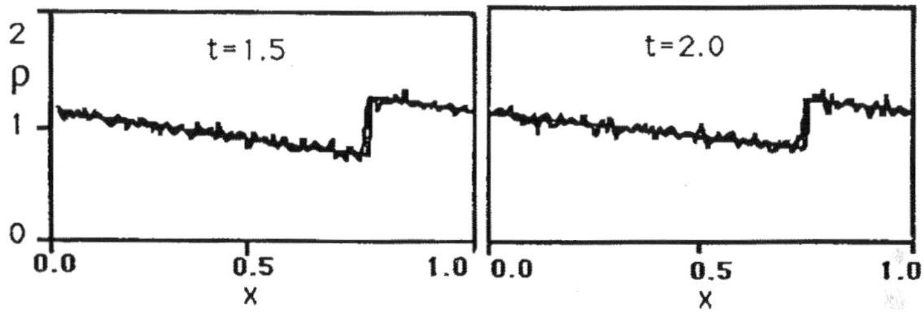

FIG. 1.2.

These results are well understood, mathematically: in this instance the theoretical analysis is even ahead of the experiments. Before the Boghosian and Levermore paper, a slightly different version of their model had already been solved, the microscopic structure of the shock being surprisingly well understood, to such a fine level that it is even difficult to observe it experimentally. Unfortunately it is not always like that, but in general we may say that the theory keeps the pace of the numerical experiments.

Even if the cellular automata, like the Boghosian-Levermore one, look extremely simple, they are nonetheless complex, many-component systems, as they should, if they have to simulate realistic equations describing physical macroscopic systems: it may seem therefore surprising that so much can be proven about them. *The key to the theoretical analysis is a separation of scales.* We have already seen it, in its more elementary form, when discussing the example of the Monte Carlo method. A small interval, where f looks essentially constant, almost a point in $[0,1]$, is represented by a huge number of lattice sites, when the mesh ϵ gets small. The lattice sites represent the microscopic world, the function f a macroscopic profile. Many different particles configurations are macroscopically the same, they produce in the limit the same value of f. Actually "most of them" have this property, as stated in (1.2). It is the law of large numbers which is responsible for such a behavior, but in a slightly unusual form. The distribution is in fact non homogeneous, if f is not a constant. But, for what said before, in a very small interval $[a,b] \subset [0,1]$ where the distribution is almost homogeneous, the number of variables $\eta(x)$ grows proportionally to ϵ^{-1}. Since they are independent, the law of large numbers holds and in each small interval, simultaneously, the particles density gets close to the value of f in that interval, with fluctuations which are depressed as $\epsilon \to 0$.

Something similar happens at later times, but the mechanism is now more subtle, the distribution of the particles at $t > 0$ is in fact no longer "prescribed by the outside", as at $t = 0$, but instead the result of all the interactions which have taken place during the evolution. In general in fact the motion of a particle has an influence on the others and even if initially they are independent, correlations will grow during the evolution. The validity of the law of large numbers is no longer trivial, since the variables may be even strongly correlated. Let us first discuss all that in the context of a real fluid, where the physical intuition may be of help. Suppose that the particles are molecules which interact via a Lennard-Jones potential, whose range typically is of the order of the angstrom. Macroscopic observations involve the measure of extensive quantities (like for instance the mass of the fluid) in regions whose size may be a few centimeters. Hence a distance which in macroscopic units is 1, has value 10^8, in microscopic units, thus ϵ is of the order of 10^{-8}. If the relevant macroscopic parameters, as the temperature, the pressure and so on, vary smoothly on the macroscopic scale, before a molecule understands that it is not in a homogeneous state, it undergoes so many collisions with the neighboring ones, that, in the meantime, it has thermalized. The equilibrium is only local, since in different regions the fluid parameters are in general different and such will be the equilibria that are reached. Since they are determined by the interactions between the molecules, they are described by Gibbs distributions which, in general, are not independent distributions.

There are therefore two steps in the transition from microscopics to macroscopics: the establishing of a local Gibbs structure, and the validity, for the Gibbs distributions, of the law of large numbers. While the second task is essentially solved by equilibrium statistical mechanics, the first one requires some stronger version of an ergodic theorem which seems beyond the reach of the present techniques. Thus the separation of scales appears also in the mechanical models of the real fluids, but the derivation of hydrodynamics requires the solution of a very hard dynamical problem as the proof of local equilibrium. If, on the other hand, this is assumed to hold, then it is possible to obtain the Euler equations from the Newton's law. A derivation along these lines was first presented, we think, by Morrey, [99], (see also the Introduction in [38] for some more details). If a suitable amount of stochasticity is added to the evolution rules, the difficulties can, sometimes, be overcome and, to some extent, the program of deriving the hydrodynamical equations can be completed.

This has opened a very active line of research on a subject which looked closed to mathematical investigations. Many models have been studied and even if they are often far from realistic, yet they seem to catch some of the features present in the real systems. There should be some universality which makes even quite different systems behave almost identically at a macroscopic level. Hopefully these stochastic particle models are in the same universality class as the real systems, and if this is so even when the hydrodynamic equations are no longer valid, they could be of help in investigating critical phenomena as turbulence, phase separation, shocks ...

Accordingly, the research on the hydrodynamical limit of stochastic particle system follows two main directions. One is devoted to the validity problem, namely to define suitable limiting procedures and, correspondingly, to prove, for possibly general classes of models, convergence to hydrodynamic or kinetic equations. The other is to go beyond that, if necessary by restricting the analysis to particular models, and to study, in such cases, the behavior of the system when it is no longer described by the macroscopic equation. Both directions are pursued using theoretical and numerical studies. The results are extremely encouraging, if you listen to people who work in this direction. More skeptical are others, you often hear people say "but how does it apply to hamiltonian system?" or, "would it not be better, numerically, to use conventional discretization methods directly on the Euler, or the Navier-Stokes equations?"

The only way to answer such difficult questions requires an analysis of what has been achieved and of its physical relevance and this is not an easy task, since really many papers have been published. A comprehensive analysis of the state of the art, both for deterministic and stochastic systems, is however available in a book recently written by Herbert Spohn, [125]. We definitely address the reader to this work, which is extremely helpful both to beginners and to more experienced readers, since it contains in an easily accessible way the most relevant results and many of the mathematical techniques and of the physical ideas in this field.

Our aim here is to present some of the mathematical methods which are more often used by examining in a self-contained way particular models. Even though the origin of the models is in the underlying physical phenomena, the structures which arise define beautiful mathematical problems interesting in themselves, as in the best tradition of mathematical physics. In particular probability has plaid and still plays a leading role in the development of the theory, as the theoretical articles are more or less equally shared by journals of probability and mathematical physics. We have been certainly influenced by that when writing this book.

The macroscopic equations that are derived in the models we consider are linear and non linear parabolic equations, reaction diffusion equations and discrete velocity Boltzmann equations. We discuss also critical phenomena by studying the appearence of hydrodynamic behavior in kinetic particle models and phase separation phenomena in systems which undergo phase transition. The particle models are zero range processes and their perturbations obtained by adding birth-death branchings and by giving velocities to the particles. We also consider lattice gases like the symmetric simple exclusion process and its perturbations.

Many of the methods that we present come from the literature in interacting particle systems, see [90] and [55]. The entropy inequalities which have been more recently introduced in this subject starting from the paper of Guo, Papanicolaou and Varadhan, [71], are also discussed, but the main part of the book deals with correlation functions techniques. We refer to the Sznitmann's survey, [128], for a discussion on these latter topics in a more probabilistic context. We conclude this paragraph presenting the contents of the single chapters.

In Chapter II we discuss independent particles. Our aim is to make the reader familiar with the various scalings and to introduce important concepts as correlation functions, density fields and local equilibrium. We also recall tightness criteria which are extensively used to derive the hydrodynamical equations.

In Chapter III we present, in the context of the symmetric zero range processes, the entropy methods introduced by Guo, Papanicolaou and Varadhan. We also prove, for these models, the Kipnis, Olla and Varadhan "superexponential estimate" which constitute a basic ingredient in the analysis of the large deviations from the hydrodynamical behavior.

In Chapter IV we introduce a system of random walks with birth-death processes, which simulate reaction diffusion equations. Here we use correlation function techniques which can be applied in a standard way, because the death mechanism provides a priori bounds on the particle densities.

In Chapter V we discuss a model for the Carleman equation, a particular discrete velocity Boltzmann equation. Again we use correlation functions techniques, but now the analysis is much more difficult due to the conservation of the particles number. This gives us the opportunity to introduce the "v-functions", a kind of truncated correlation functions, and, by their help, to prove convergence at all times. The usual local-time limitation in the validity of the Boltzmann equation is thus overcome.

Chapter VI is essentially devoted to the symmetric simple exclusion process, we discuss couplings and duality techniques and apply them to prove convergence to a reaction-diffusion equation for the "Glauber+Kawasaki" spin process.

Chapters VII and VIII deal with critical phenomena, they have a more qualitative structure, no proof is given, we just state the main results and discuss conjectures and open problems. In Chapter VII we study long time effects, as the appearence, at longer times, of the hydrodynamic equations, in models which, at shorter times, are described by kinetic equations. In Chapter VIII we discuss phenomena like phase decomposition and interface dynamics, which arise at macroscopic times when the system undergoes a phase transition.

The separation of phases is then studied in details in Chapters IX and X, in the context of the Glauber+Kawasaki process, using the v-functions technique, in a way similar to that of Chapter V.

On each argument we only quote some of the most pertinent works, and add a paragraph of Bibliographical Notes with additional references at the end of the Chapter, if necessary. A more comprehensive bibliography can be found in Spohn's book. Some of the paragraphs, or subparagraphs, are marked by a *: this means that they are of more elementary nature and that they may be skipped by the more experienced readers. Each chapter is essentially independent of the others, crossed references between chapters are meant to help the reader but, in general, they are not essential.

§1.1 Bibliographical notes.

A survey on macroscopic limits can be found in [124], and, for stochastic systems, in [38], but many progresses have been obtained in the field since then. As already mentioned in this chapter the most updated and comprehensive survey is Spohn's, [125]. Reference [111] collects two short surveys, more updated is [89]. In these last three papers most of the proofs are not reported.

There are many papers on the derivation of the Burgers equation and on the study of the shock. In the context of the asymmetric simple exclusion and zero range processes, see [3], [4], [5], [10], [37], [61], [119], [90], [131], [69], [59], [11], [118], [82]. For the weakly asymmetric simple exclusion process the limiting equation is the Burgers equation with viscosity, this has been shown in [43] and [68]. The structure of the shock waves in these systems is studied in [114], [50] and [49]. The Boghosian-Levermore model introduced in [14], was proven in [86] to behave, in its weakly asymmetric version, as the Burgers equation with viscosity. The completely asymmetric case has been studied in [63]. Other models for the Burgers equation are considered in [23] and [127]. Simulations on the asymmetric model are reported in [15] for the exclusion model and in [14] and in [29] for the Boghosian-Levermore automaton. Travelling waves appear also in the contact process, [67], and in a model for the KPP equation [18], [24].

CHAPTER II

HYDRODYNAMIC LIMITS FOR INDEPENDENT PARTICLES

We have chosen the system of independent particles for introducing in a simple, elementary way basic concepts in the theory of collective phenomena, as the various hydrodynamic scaling limits and properties like local equilibrium and propagation of chaos. Our aim in this chapter is to introduce the reader to these concepts and in particular to make him acquainted with the many space-time scalings, which are extensively used throughout this book. To avoid diversions from this objective our original attitude has been to skip most of the mathematical details, but since the audience was mixed, we had to compromise and, in a first version, we have added a few more proofs. The compromise was a failure: boring for the experts and incomplete for the beginners. We try again, this time by inserting new paragraphs, marked by a star, to fill some of the mathematical details at least to some reasonable extent, but in such a way that they may be skipped without loosing the continuity of the presentation.

§2.1 The independent process.

The configuration space is $\mathbb{N}^{\mathbb{Z}}$, its elements are the particles configurations, denoted by $\eta = \{\eta(x) \in \mathbb{N}, x \in \mathbb{Z}\}$. $\eta(x)$ is called the occupation number at x, and it is interpreted as the number of particles which are in x in the configuration η. One often considers configuration spaces where $\mathbb{Z}$ is replaced by $\mathbb{Z}^d$, $d > 1$, or, sometimes, by finite sets Λ.

The evolution is a Markov process with values in the state space $\mathbb{N}^{\mathbb{Z}}$. Its generator is denoted by L_0, it acts on the cylindrical functions f (f is cylindrical if it depends on η only through finitely many of its entries) as

$$(L_0 f)(\eta) = \sum_x \eta(x)[p\{f(\eta - \delta_x + \delta_{x+1}) - f(\eta)\} + q\{f(\eta - \delta_x + \delta_{x-1}) - f(\eta)\}] \tag{2.1}$$

where p and q are non negative reals and $p + q = 1$; $\delta_y \in \mathbb{N}^{\mathbb{Z}}$ is the configuration with just one particle in y and sums-differences of configurations are defined componentwise. Here we restrict to nearest neighbor jumps, but more general situations can be studied as well.

Since f is cylindrical the sum on the right hand side of (2.1) has finitely many terms hence it is well defined. There is a theorem which states that there is a unique Markov process on (a dense subset of) $\mathbb{N}^{\mathbb{Z}}$ whose generator agrees with the expression given in (2.1). We shall neither prove this theorem nor give its precise statement, but just refer to the literature, see for instance [53], Ch. VIII, Section 5. In the next paragraph we discuss the heuristic meaning of (2.1) and we give details of technichal nature on the structure of the independent process.

§2.2.* Construction of the process.

We start by constructing the process when only one particle is present. This extends easily to the case with finitely many particles and then, by a limiting procedure, to the general case.

2.2.1 The process of a single random walk.

The state space is the set of all the integers $\mathbb{Z}$. The process is realized by means of two auxiliary processes: the first one is a sequence $\{\tau_n,\ n \geq 1\}$, of independent identically distributed (i.i.d.) non negative variables with exponential distribution of intensity $\lambda > 0$: namely

$$\mathbb{P}(\{\tau_n \geq \tau\}) = e^{-\lambda\tau}, \quad \text{for any } n \geq 1$$

The point process $\{t_n = \tau_1 + \cdots + \tau_n,\ n \geq 1\}$ is then called the Poisson process with density λ.

The second auxiliary process is $\{\sigma_n,\ n \geq 1\}$ where again the variables are i.i.d. each value being ± 1, 1 with probability p and -1 with probability $q = 1 - p$. Let $\mathcal{P}$ be the probability on the direct product $\mathcal{X}$ of these two spaces, so that all the τ_n and σ_n are mutually independent. Given an initial state $x_0 \in \mathbb{Z}$, we define in $\mathbb{Z}$ the random walk $x(t)$, $t \geq 0$, as follows: take an element in $\mathcal{X}$, i.e. the two sequences $\{t_n, \sigma_n,\ n \geq 1\}$, then $x(t) = x_0$ for $0 \leq t < t_1$; $x(t) = x_1 = x_0 + \sigma_1$ for $t_1 \leq t < t_2$; $\dots$ $x(t) = x_{n+1} = x_n + \sigma_{n+1}$ for $t_{n+1} \leq t < t_{n+2}$, and so on. Notice that $x(t)$ is almost surely defined for all t, because with probability 1, $t_n \to \infty$.

The τ_n's are the time intervals between the successive rings of an "exponential clock of intensity λ". When the clock rings a coin is tossed, it indicates right ($\sigma_n = 1$) with probability p and left ($\sigma_n = -1$) with probability q; the random walk moves accordingly.

Denote by $P_t(x \to y)$ the probability of being in y at time t, starting from x at time 0. If the initial distribution is the probability μ, $\mu(x)$, $x \in \mathbb{Z}$, being the probability of being at x, the distribution at time $t > 0$ is μ_t, where

$$\mu_t(y) = \sum_x \mu(x) P_t(x \to y)$$

Given a bounded function f on $\mathbb{Z}$, we denote by $\mathbb{E}_{\mu_t}(f)$ its expectation:

$$\mathbb{E}_{\mu_t}(f) = \sum_x \mu_t(x) f(x)$$

After some simple computation one finds that:

$$\frac{d}{dt}\mathbb{E}_{\mu_t}(f) = \mathbb{E}_{\mu_t}(Lf), \qquad (Lf)(x) = \lambda p[f(x+1) - f(x)] + \lambda q[f(x-1) - f(x)]$$

Identifying x with the configuration δ_x, we see that when $\lambda = 1$, L is the same as the restriction of L_0, as defined in (2.1), to the configurations with just one particle.

We can reconstruct $P_t(x \to y)$ from L by the following formula, which we shall prove next:

$$(e^{Lt} 1_y)(x) = P_t(x \to y)$$

where 1_y is the characteristic function at y and L is thought of as an operator on the bounded functions on $\mathbb{Z}$: the left hand side in the above equation is then defined by power expanding the exponential, since L is bounded the series converges.

We have

$$\frac{d}{dt}(e^{Lt} 1_y)(x) = (e^{Lt}[L1_y])(x), \qquad L1_y = \lambda p(1_{y-1} - 1_y) + \lambda q(1_{y+1} - 1_y)$$

Think now of x as fixed and define

$$f_t(y) = (e^{Lt}1_y)(x)$$

then

$$\frac{df_t(y)}{dt} = \lambda p[f_t(y-1) - f_t(y)] + \lambda q[f_t(y+1) - f_t(y)] = \lambda[pf_t(y-1) + qf_t(y+1)] - \lambda f_t(y)$$

which can be rewritten in integral form as

$$f_t(y) = e^{-\lambda t} f_0(y) + \int_0^t ds e^{-\lambda(t-s)} \lambda [pf_s(y-1) + qf_s(y+1)]$$

By iteration we then express f_t as a series. The first term is $e^{-\lambda t} f_0(y)$. Notice that $e^{-\lambda t}$ is the probability that $t_1 > t$, t_1 being an exponential time of intensity λ. The second term is

$$[pf_0(y-1) + qf_0(y+1)] \int_0^t ds e^{-\lambda(t-s)} \lambda e^{-\lambda s}$$

The integral is the probability that $t_1 \leq t$ and $t_2 > t$; this is then multiplied by the sum of f_0 computed at $y-1$ with weight p and at $y+1$ with weight q. Since analogous interpretation can be given to the other terms, we have

$$f_t(y) = \sum_z Q_t(y \to z) f_0(z)$$

where Q_t is the transition probability of a random walk jumping after exponential times of intensity λ with probability q to the right and p to the left, just the opposite as in the original process:

$$Q_t(y \to z) = P_t(z \to y)$$

Then, because $f_0 = 1_x$, we get that $f_t(y) = P_t(x \to y)$.

The process with transition probabilities Q_t is the "dual process". With respect to the direct process (whose transition probabilities are P_t), the time runs "backwards". The transition probability of the dual process is in fact obtained by reversing the transition in the direct process. In the cases when the dual process exists, it provides useful and powerful tools of investigations as we shall extensively see in this and in the next chapters.

Before going on we discuss the existence of a dual process for general jump processes on countable spaces, (for a general reference see [116]), notation and concepts will be useful in the sequel.

2.2.2 Jump processes.

The state space is now a countable set S. The process is determined by a strictly positive, bounded, intensity function $c(x)$ on S and by a transition probability $\pi(x \to y)$ on $S \times S$, in the following way. Let $\{t_n, n \geq 1\}$ be the process considered earlier, where $t_n = \tau_1 + \cdots + \tau_n$, the τ_i being exponential times of intensity $\lambda = 1$. Let $\{x_n, n \geq 0\}$ be the Markov chain with transition probability π, assuming for instance that x_0 is given. Let $\mathcal{P}$ denote again the law of the direct product $\mathcal{X}$ of these two processes. The jump process $\{x(t), t \geq 0\}$ is then defined on $\mathcal{X}$ as follows: $x(t) = x_0$ for $0 \leq t < \tilde{t}_1 = c(x_0)\tau_1, \ldots, x(t) = x_n$ for $\tilde{t}_{n-1} \leq t < \tilde{t}_n = \tilde{t}_{n-1} + c(x_{n-1})\tau_n$, and so on. Notice that $\tilde{\tau}_n = c(x_{n-1})\tau_n$ has exponential distribution with intensity $c(x_{n-1})$.

By extending to this case the previous definitions, we define $P_t(x \to y)$, μ_t and L which has now the form:

$$(Lf)(x) = \sum_y c(x)\pi(x \to y)[f(y) - f(x)]$$

Again

$$(e^{Lt}1_y)(x) = P_t(x \to y)$$

in fact, let $f_t(y)$ denote the left hand side in the above equation. Then, proceeding like before, we have

$$f_t(y) = e^{-c(y)t} f_0(y) + \int_0^t ds e^{-c(y)(t-s)} \sum_z c(z)\pi(z \to y) f_s(z)$$

By iteration the right hand side reconstructs $P_t(x \to y)$, showing therefore that $f_t(y) = P_t(x \to y)$. However the right hand side cannot always be interpreted in terms of a dual process, for this we need that

$$\sum_z c(z)\pi(z \to y) = c(y)$$

namely that

$$\tilde{\pi}(y \to z) \equiv \frac{1}{c(y)} c(z)\pi(z \to y)$$

is a transition probability. In this case

$$f_t(y) = \sum_z Q_t(y \to z) f_0(z)$$

where Q_t is the transition probability of the jump process with same intensities $c(x)$ as in the original process, but with transition probability $\tilde{\pi}$. This is the dual process.

We conclude this sub-paragraph by a notational remark: one can realize the jump process with intensities $c(x)$ and transition probability $\pi(x \to y)$ in terms of the following *graphical representation of the process.* For each ordered pair x and y, consider a Poisson process $\{t_n(x,y),\ n \geq 1\}$, with intensity $c(x,y) = c(x)\pi(x \to y)$, (namely $\tau_n = t_n - t_{n-1}$, $t_0 = 0$, are i.i.d. exponential times with intensity $c(x,y)$). At each of these times draw an oriented arrow from x to y. Assume that all these Poisson processes are independent. Given x, the initial position of a "particle" and a realization of all these processes, define $x(t)$ as the site reached at time t by a particle which moves following the oriented arrows whenever they occur starting from the site where the particle is. Then the distribution of $x(t)$ is the same as that defined previously, as it can be checked for instance by computing the generator of this process. One often says that $c(x,y)$ is the intensity of the jump from x to y, and this jump process is sometimes defined directly in terms of the $c(x,y)$ alone, one can easily reconstruct, in fact, $c(x)$ and $\pi(x \to y)$ from $c(x,y)$.

Notice that in this way we can realize simultaneously processes starting from all the possible initial states. Moreover if two processes with different initial conditions are realized in this way, then if they are equal at some time, then they remain equal at all the later times. Such a property has been often used for proving ergodic theorems.

2.2.3 Finitely many independent random walks.

We consider n independent labelled random walks, namely a Markov process with state space $\mathbb{Z}^n = \{\underline{x} : \underline{x} = (x_1, \ldots, x_n)\}$ and transition probability $\tilde{P}_t(\underline{x} \to \underline{y})$

$$\tilde{P}_t(\underline{x} \to \underline{y}) = \prod_{i=1}^{n} P_t(x_i \to y_i)$$

where $P_t(x_i \to y_i)$ is the transition probability of a single random walk. Given a probability μ we define μ_t in terms of $\tilde{P}_t$, as before. We have

$$\frac{d}{dt}\mathbb{E}_{\mu_t}(f) = \mathbb{E}_{\mu_t}(\tilde{L}f), \qquad (\tilde{L}f)(x) = \lambda \sum_{i=1}^{n} \{p[f(\underline{x}^{(i,+)}) - f(\underline{x})] + q[f(\underline{x}^{(i,-)}) - f(\underline{x})]$$

where $\underline{x}^{(i,\pm)} = (x_1, \ldots, x_i \pm 1, \ldots, x_n)$.

In many cases the labels of the particles are unimportant, so that the relevant state space is $\Omega_n = \{\eta \in \mathbb{N}^{\mathbb{Z}} : \sum_{x \in \mathbb{Z}} \eta(x) = n\}$. On this space we define the transition probability

$$P_t(\eta \to \eta') = \sum_{\underline{y} \in \mathbb{Z}^n} \tilde{P}_t(\underline{x} \to \underline{y}) 1(\eta' = U(\underline{y})) \tag{2.2a}$$

where

$$U(\underline{z}) = \sum_{i=1}^{n} \delta_{z_i}, \quad \underline{z} \in \mathbb{Z}^n \tag{2.2b}$$

and $\underline{x}$, in (2.2a), is any configuration such that $\eta = U(\underline{x})$: notice in fact that the right hand side of (2.2a) is independent of $\underline{x} \in U^{-1}(\eta)$.

Let us now compute the generator L of the unlabelled process: this is defined by the identity

$$\frac{d}{dt} \sum_{\eta'} P_t(\eta \to \eta') f(\eta') = \sum_{\eta'} P_t(\eta \to \eta')(Lf)(\eta') \tag{2.3}$$

By using (2.2) the left hand side becomes:

$$\frac{d}{dt} \sum_{\underline{y}} \tilde{P}_t(\underline{x} \to \underline{y}) f(U(\underline{y})) = \sum_{\underline{y}} \tilde{P}_t(\underline{x} \to \underline{y})(\tilde{L} f \circ U)(\underline{y})$$

By direct inspection

$$(\tilde{L} f \circ U)(\underline{y}) = (L_0 f)(U(\underline{y}))$$

with L_0 as in (2.1) (we here suppose that $\lambda = 1$). In particular this shows that $(\tilde{L} f \circ U)(\underline{y})$ is a symmetric function of $\underline{y}$, hence

$$\sum_{\underline{y}} \tilde{P}_t(\underline{x} \to \underline{y})(\tilde{L} f \circ U)(\underline{y}) = \sum_{\eta'} P_t(\eta \to \eta')(L_0 f)(\eta')$$

and, from (2.3)

$$\sum_{\eta'} P_t(\eta \to \eta')(Lf)(\eta') = \sum_{\eta'} P_t(\eta \to \eta')(L_0 f)(\eta')$$

Take $t = 0$, then, since $P_0(\eta \to \eta') = 1(\eta = \eta')$ we have that $L = L_0$ restricted to Ω_n (if $\lambda = 1$).

We can now easily recognize that the unlabelled process we have defined is a jump process. Consider in fact the jump process on Ω_n with constant intensity $n\lambda$ and transition probabilities (assume below $\eta(x) > 0$)

$$\pi(\eta \to \eta + \delta_{x+b} - \delta_x) = \begin{cases} p\eta(x)/n & \text{if } b = 1 \\ q\eta(x)/n & \text{if } b = -1 \end{cases}$$

The generator associated to this process is easily seen to be L, the generator of the process of n unlabelled independent random walks.

2.2.4 The independent process with infinitely many particles.

The state space $\mathbb{N}^{\mathbb{Z}}$ has uncountably many configurations with infinitely many particles, no hope therefore that it is a jump process, (we have seen that the jump intensity increases proportionally to the number of particles). Given a configuration η and $N > 0$ we define $\eta^{(N)}$ as

$$\eta^{(N)}(x) = \begin{cases} \eta(x), & \text{if } |x| \le N \\ 0, & \text{otherwise} \end{cases} \tag{2.4}$$

Let f be a cylindrical function, then we denote by

$$A(f, t, N|\eta) \equiv \sum_{\eta'} \tilde{P}_t(\eta^{(N)} \to \eta') f(\eta')$$

which is well defined because $\eta^{(N)}$ has finitely many particles.

We want to define the transition probability starting from η by the limiting values as $N \to \infty$ of $A(f, t, N|\eta)$, when f varies in the set of all the bounded cylindrical functions.

We will show below that: (1) for $N > N'$

$$|A(f, t, N|\eta) - A(f, t, N'|\eta)| \le \|f\| \gamma(N', \ell|\eta) \tag{2.5}$$

where $\|\cdot\|$ denotes the sup norm; $\ell = \ell(f)$ is such that if f depend only on $\eta(x)$ with $x \in \Lambda$, then

$$\ell = \max_{x \in \Lambda} |x|$$

We will then show that (2) $\gamma(N', \ell|\eta)$ vanishes when $N' \to \infty$, if η is any configuration in the set defined in (2.6) below.

Proof of (2.5) Let and let $N > N' > \ell$. Then the difference in the left hand side of (2.5) is bounded by the sup norm of f times the probability that any particle starting from $|x| > N'$ is in Λ at time t. (It is often convenient, as above, to switch back to the labelled process, recalling that the unlabelled one is defined as its marginal over the symmetric functions). To make such a probability small, we must evidently impose some constraint on the way $\eta(x)$ grows when $|x| \to \infty$.

The probability that a random walk (with intensity λ and jumps on the neighboring sites) travels a distance larger than $d \in \mathbb{N}$ in a time t is bounded by:

$$\sum_{n \ge d} e^{-\lambda t} \frac{(\lambda t)^n}{n!} \le \frac{(\lambda t)^d}{d!}$$

(This bound only counts the number of jumps, assuming they are all unfavourable). Then the probability that any of the particles in $\eta^{(N)}$ at $|x| > N'$ is in Λ at time t is bounded by the sum of the probabilities for each individual particle starting at distances larger than N', namely

$$\gamma(N', \ell|\eta) \equiv \sum_{|x|>N'} \eta(x) \frac{(\lambda t)^{|x|-\ell}}{(|x|-\ell)!}$$

uniformly in N. We specify now the range of η as

$$X(c,n) = \{\eta \in \mathbf{N}^{\mathbf{Z}} : \eta(x) \le c(|x|^n + 1), \text{ for all } x \in \mathbf{Z}\} \tag{2.6}$$

Then

$$\lim_{N' \to \infty} \gamma(N', \ell|\eta) = 0$$

which proves (2.5). □

We have seen that for any $t > 0$, for any bounded, cilyndrical function f and for any $\eta \in X(c,n)$, the limit $A(f,t|\eta)$ of $A(f,t,N|\eta)$ exists. We shall now prove that $A(f,t|\eta)$ is the expectation of f with respect to a probability on $\mathbf{N}^{\mathbf{Z}}$.

Denote by $\mathbf{P}_{\eta^{(N)}}$ the probability of the process starting from $\eta^{(N)}$, and by $\mathbf{E}_{\eta^{(N)}}$ its expectation. Call $\nu^N_{\Lambda,t}$ the probability on $\mathbf{N}^\Lambda$ which is the marginal on this space of the law $\mathbf{P}_{\eta^{(N)}}$ at time t.

By the Prokhorov theorem (see 2.7.1 below) $\{\nu^N_{\Lambda,t}\}$ is a tight family if for any k,

$$\mathbf{E}_{\eta^{(N)}}([\sum_{x\in\Lambda} \eta(x,t)]^k)$$

is bounded independently of N. (We shall in fact prove that the expectation converges as $N \to \infty$ to a finite value). To prove this we first observe that, denoting by $\eta(\cdot,t)$ the occupation numbers at time t, we have for all integer $n \ge 1$ and for all $N' < N$,

$$\mathbf{P}_{\eta^{(N)}}(\{\sum_{x\in\Lambda} \eta(x,t) \ge n + \sum_{|x|\le N'} \eta(x)\}) \le \gamma(N', \ell|\eta)^n, \qquad \ell = \max_{x\in\Lambda} |x| \tag{2.7a}$$

Hence choosing N' so large that $\gamma(N', \ell|\eta) < 1/2$, for any k we have that

$$\mathbf{E}_{\eta^{(N)}}([\sum_{x\in\Lambda} \eta(x,t)]^k) \le \sum_{n=1}^{M} n^k + \sum_{n\ge 1} (M+n)^k (\frac{1}{2})^n, \qquad M = \sum_{|x|\le N'} \eta(x) \tag{2.7b}$$

Since the expectation on the left hand side of (2.7b) is a nondecreasing function of N, its limit as $N \to \infty$ exists and, by (2.7b) it is finite:

$$A_k(t,\eta) = \lim_{N\to\infty} \mathbf{E}_{\eta^{(N)}}([\sum_{x\in\Lambda} \eta(x,t)]^k) < \infty$$

By general arguments in measure theory, we can then prove that $A_k(t,\eta)$ is the expectation of $[\sum_{x\in\Lambda} \eta(x,t)]^k$ with respect to a probability on $\mathbf{N}^{\mathbf{Z}}$. For the sake of completeness we give below some more details and precise references.

By tightness, $\nu^L_{\Lambda,t}$ converges by subsequences. Let $\nu_{\Lambda,t}$ be a limiting point, ($\nu_{\Lambda,t}$ is therefore a probability on $\mathbb{N}^\Lambda$), then

$$\mathbb{E}_{\nu_{\Lambda,t}}(f) = A(f,t|\eta)$$

for any bounded cylindrical function depending only on $\eta(x)$, $x \in \Lambda$. This proves that $\nu^N_{\Lambda,t}$ converges weakly to $\nu_{\Lambda,t}$ and not only by subsequences. Then by varying Λ over all the finite subsets of $\mathbb{Z}$, we obtain a consistent family of probabilities, which, by the Kolmogorov theorem, Theorem V.7.1 in [107], determines a unique probability $P_t(\eta \to d\eta')$ on $\mathbb{N}^{\mathbb{Z}}$, $\eta \in X(c,n)$, such that

$$A(f,t|\eta) = \int P_t(\eta \to d\eta')f(\eta'), \quad \text{for all the bounded cylindrical functions } f$$

Notice, for later purposes, that the product of any finite number of occupation numbers is integrable with respect to $P_t(\eta \to d\eta')$. This latter is the transition probability of the independent process: one can easily check that for any bounded cylindrical function

$$\frac{d}{dt}\int P_t(\eta \to d\eta')f(\eta') = \int P_t(\eta \to d\eta')(L_0 f)(\eta')$$

with L_0 given by (2.1). By arguments similar to the above ones, one can verify that dP_t is supported by the union over c and n of the sets $X(c,n)$.

If the initial distribution is not an individual configuration but a probability μ supported by the union over c and n of the sets $X(c,n)$, we define, for any $t > 0$, μ_t as

$$\mathbb{E}_{\mu_t}(f) = \int d\mu(\eta) \int P_t(\eta \to d\eta')f(\eta') \tag{2.8a}$$

for any bounded measurable function f. (From the previous analysis one can deduce that the P-integral of f is a measurable function of η). If f is also cylindrical then

$$\mathbb{E}_{\mu_t}(f) = \lim_{N\to\infty} \int d\mu(\eta) \sum_{\eta'} P_t(\eta^{(N)} \to \eta')f(\eta') \tag{2.8b}$$

§2.3 The equation for the one-body correlation functions.

The fastest way to find the macroscopic equations for the independent process is to look at the one-body correlation functions.

2.3.1 Definition: The one body correlation functions.

The one-body correlation functions for a probability μ on $\mathbb{N}^{\mathbb{Z}}$ are defined if i) μ is supported by configurations η on which the independent process is defined, i.e. on the union over c and n of the sets $X(c,n)$ defined in (2.6), and ii), denoting by μ_t the law of the process at time t starting from μ, then the variables $\eta(x)$ are integrable with respect to μ_t, for all $x \in \mathbb{Z}$ and $t \geq 0$. The one-body correlation functions are then

$$u(x,t|\mu) = \mathbb{E}_{\mu_t}(\eta(x)) \equiv \mathbb{E}_{\mu}(\eta(x,t)), \quad \text{for all } x \in \mathbb{Z},\, t \geq 0 \tag{2.9}$$

If μ is supported by a single configuration η we shall simply write $u(x,t|\eta)$.

If there is a constant $c > 0$ such that

$$\sup_{x \in \mathbf{Z}} \mathbf{E}_\mu(\eta(x)) < c < \infty \tag{2.10}$$

then, see the next paragraph, μ is supported by the union of the sets $X(c, n)$ and, for any $t > 0$, μ_t satisfies the same inequality (2.10), so that the one body correlations are defined. Furthermore

$$\begin{aligned} \frac{d}{dt} u(x,t|\mu) &= \mathbf{E}_{\mu_t}(L_0 \eta) = p\big(u(x-1,t|\mu) - u(x,t|\mu)\big) + q\big(u(x+1,t|\mu) - u(x,t|\mu)\big) \\ &= \frac{1}{2}\big(u(x+1,t|\mu) + u(x-1,t|\mu) - 2u(x,t|\mu)\big) + \frac{(p-q)}{2}\big(u(x-1,t|\mu) - u(x+1,t|\mu)\big) \end{aligned} \tag{2.11a}$$

hence, see §2.4,

$$u(x,t|\mu) = \sum_y Q_t(x \to y) u(y,0|\mu) \tag{2.11b}$$

where Q_t is the transition probability of a single random walk which jumps after exponential times of intensity one to the right with probability q and to the left with probability p. The corresponding Markov process is called the dual process.

In the next paragraph we prove all the above statements, in the meantime notice that the last equality in (2.11a) defines a discretized PDE: by a suitable scaling procedure, we will prove that it converges to a true PDE, which is therefore the macroscopic equation for the independent process.

§2.4* Proofs.

Proof that: if μ satisfies (2.10) then it is supported by the union over c and n of $X(c,n)$, as defined in (2.6). Using the Chebyshev inequality

$$\mu(\{\eta(x) > |x|^n\}) \leq |x|^{-n} \mathbf{E}_\mu(\eta(x)) \leq c|x|^{-n}$$

hence the claim, by the Borel-Cantelli lemma, after choosing n, in the above inequality, large enough.

Proof that if μ_t is the law starting from a configuration with finitely many particles, then (2.11b) holds. We first compute the action of L_0 on the occupation variables and we find an expression as in (2.11a), recalling that all the expectations are bounded, because the total number of particles is finite. (2.11a) can be rewritten in integral form and solved by iteration as in §2.2, concluding the proof of the claim.

Proof of (2.11). We have that

$$\int d\mu(\eta) [\sum_{\eta'} P_t(\eta^{(N)} \to \eta') \eta'(x)] \leq c$$

for all N. Then for any $x \in \mathbf{Z}$ and $t \geq 0$, $\eta(x)$ is integrable with respect to μ_t, as follows from (2.8b), (the limit exists by monotonicity). Same arguments prove (2.11a) and (2.11b) for any μ which satisfies (2.10).

§2.5 Hydrodynamic limits.

The hydrodynamic equations for the independent process are the continuum version of (2.11a). To have true derivatives instead of lattice derivatives we introduce a scaling parameter ϵ, as in the following definition:

2.5.1 Definition: The lattice approximation.

The family $\rho_\epsilon(x)$, $x \in \mathbb{Z}$, $\epsilon > 0$, is a lattice approximation to the continuous function $\rho(r)$, $r \in \mathbb{R}$, if for any r

$$\lim_{\epsilon \to 0} \rho_\epsilon([\epsilon^{-1} r]) = \rho(r) \tag{2.12}$$

$[a]$ denoting the integer part of a. $\rho(r)$ is a ***density profile*** if it is a non negative, uniformly bounded, continuous function with continuous bounded derivatives. Finally the family of probability measures μ^ϵ approximates the density profile $\rho(r)$ if $\mathbf{E}_{\mu^\epsilon}(\eta(x))$ is a lattice approximation to $\rho(r)$.

This definition does not imply that $\mathbf{E}_{\mu^\epsilon}(\eta(x))$ is uniformly bounded. However, we shall henceforth consider initial measures μ^ϵ such that

$$u(x, 0|\mu^\epsilon) \equiv \mathbf{E}_{\mu^\epsilon}(\eta(x)) = \rho(\epsilon x), \quad \text{for all } x \in \mathbb{Z} \tag{2.13}$$

2.5.2 Proposition. *Given the density profile $\rho(r)$, let μ^ϵ be as in (2.13) and μ^ϵ_t, $t > 0$, the distribution of the process at time t. The following holds:*
(i) Let $p > q$, then for any t, $\mu^\epsilon_{\epsilon^{-1}t}$ approximates the density profile $\rho_t(r)$, where $\rho_t(r)$ solves

$$\frac{\partial \rho_t}{\partial t} = -(p-q)\frac{\partial \rho_t}{\partial r} \tag{2.14a}$$

$$\rho_0(r) = \rho(r) \tag{2.14b}$$

(ii) For $p = q$, (2.14) becomes trivial, i.e. $\rho_t(r) = \rho(r)$, for all t. However for any t the family $\mu^\epsilon_{\epsilon^{-2}t}$ approximates the solution $\rho_t(r)$ of

$$\frac{\partial \rho_t}{\partial t} = \frac{1}{2}\frac{\partial^2 \rho_t}{\partial r^2} \tag{2.15a}$$

$$\rho_0(r) = \rho(r) \tag{2.15b}$$

(iii) Let $p - q = \epsilon a$, $a > 0$, then for any t, $\mu^\epsilon_{\epsilon^{-2}t}$ approximates the solution $\rho_t(r)$ of

$$\frac{\partial \rho_t}{\partial t} = \frac{1}{2}\frac{\partial^2 \rho_t}{\partial r^2} - a\frac{\partial \rho_t}{\partial r} \tag{2.16a}$$

$$\rho_0(r) = \rho(r) \tag{2.16b}$$

Proof. By the local central limit theorem, [58] we have

$$\lim_{t \to \infty} \sum_y \left| Q_t(0 \to y) - \frac{1}{\sqrt{2\pi t}} e^{-(y-vt)^2/(2t)} \right| = 0, \quad v = q - p$$

recalling that $Q_t(x \to y) = Q_t(0 \to y - x)$. From this, (2.11) and (2.13) we obtain an expression for $\mu^\epsilon_t(\eta(x))$, which is easily seen to imply the statements contained in the Proposition. □

Notice that the equations (2.14) and (2.15) are a fixed point for the space-time scaling which has been used to derive them: it is the limiting procedure which forces such symmetries which are absent at the original particle level. On the other hand (2.16) has not any space-time symmetry and in fact, when deriving it, the generator changes with the scaling parameter. Such a limit has therefore a different nature than the previous ones, which were *pure hydrodynamic limits*.

2.5.3 Definition: Microscopic and macroscopic space-time variables.

The variables (x,t), $x \in \mathbf{Z}$ and $t \geq 0$, which appear in the definition of the independent process are the *microscopic space and time coordinates* while $(\epsilon x, \epsilon t)$ when $p > q$ [$(\epsilon x, \epsilon^2 t)$ in the other cases] are the corresponding *macroscopic coordinates.*

The transition from microscopic to macroscopic variables induces in the limit when $\epsilon \to 0$ a transition from the original microscopic evolution defined by (2.1) to the macroscopic evolution (2.14), (2.15), (2.16) according to the various scalings.

2.5.4 Definition: Macroscopic limits.

We call *hydrodynamic limits, respectively the Euler and the diffusive limits,* the limits used to derive (2.14) and (2.15). The one used to derive (2.16) is not purely hydrodynamical because the dynamics depends on ϵ, so that we shall refer to it simply as a macroscopic or a continuum limit. A physically relevant example is the kinetic, Boltzmann-Grad limit, under which the Boltzmann equation is derived: for hard spheres their diameter is varied with ϵ in such a way that the mean free path and the Knudsen number remain finite, we shall discuss such limits in Chapter VII, see in particular §7.1.

The linearity of the limiting PDE's (2.14)-(2.16), reflects the triviality of the original microscopic model. To obtain more interesting, e.g. non linear, macroscopic equations, we need to switch on an interaction among the particles, so that the jump intensities of the particles are influenced by the presence of the others, as in Chapter III where we present a zero range process whose macrosocpic behavior is described by a non linear diffusion equation. In the other chapters we study evolution laws which depend on ϵ, as in case iii) of Proposition 2.2. The limiting equations are of two types, non linear reaction diffusion equations and non linear discrete velocity Boltzmann equations.

Different formulations of the macroscopic limits are discussed below.

§2.6 The density fields.

In Proposition 2.5.2 we have derived the macroscopic equation by taking averages of the random variables $\eta(x,t)$ over the different realizations of the process: evaluated at a single realization, such variables have no reason to be related to the solution of the macroscopic equation, having non vanishing fluctuations around their averages.

On the other hand one would like to recognize the macroscopic equation even by looking at single realizations, as in a computer simulation. As discussed in the introduction, there is a natural way to achieve this, exploiting the ergodic theorem (law of large numbers) with respect to space shifts.

2.6.1 Definition: The density fields.

Let $S(\mathbf{R})$ be the set of "smooth functions with fast decay at infinity" (we shall be more specific when needed). The density fields $X_t^\epsilon(\phi)$, $\phi \in S(\mathbb{R})$, $t \geq 0$, are defined as

$$X_t^\epsilon(\phi) = \epsilon \sum_x \phi(\epsilon x)\eta(x, \epsilon^{-b}t), \qquad \begin{cases} b = 2 & \text{for diffusive scalings} \\ b = 1 & \text{for Euler scalings} \end{cases} \tag{2.17}$$

where $\eta(x,t)$ is the random variable which counts the number of particles in x at time t.

We denote by

$$\omega \equiv \{\omega_t, t \in \mathbb{R}_+\}$$

an element of $D(\mathbb{R}_+, S'(\mathbb{R}))$ (the definition and some properties of this space are given in §2.7 below) and by $\{X_t(\phi)\}$ the canonical coordinate process, i.e.

$$X_t(\phi)(\omega) = \omega_t(\phi)$$

We denote by $\mathcal{P}^\epsilon$ the measure on $D(\mathbb{R}_+, S'(\mathbb{R}))$ induced by the measure $\mathbb{P}^\epsilon_{\mu^\epsilon}$ on the independent process by means of the relation (2.17). Notice that the dependence of $\mathcal{P}^\epsilon$ on ϵ is due both to the presence of ϵ in the definition of the density fields, and in the initial measure μ^ϵ.

If ϕ approximates the characteristic function of an interval $\mathcal{I} \subset \mathbb{R}$, then $X_t^\epsilon(\phi)$ approximates $|\mathcal{I}|$ times the density of particles in $\epsilon^{-1}\mathcal{I}$, at time $\epsilon^{-1}t$. Therefore the knowledge of the values of the density field, in the limit as $\epsilon \to 0$, gives macroscopic information on the state of the system, the details of what is happening at the single sites is lost, and only the collective behavior is recorded, this is an aspect of the separation of scales mentioned in the introduction.

The density field (2.17) can be as well considered as the integral of ϕ over the measure

$$\epsilon \sum_x \eta(x, \epsilon^{-b}t)\delta_{\epsilon x}$$

where δ_r is the delta function at r. As a consequence $\mathcal{P}^\epsilon$ is supported for each ϵ by $D(\mathbb{R}_+, M)$, where M is the set of the σ-finite measures on $\mathbb{R}$. In several cases, as in ours, it is possible to specify further the range of the process to gain more regularity, see for instance the analysis of Spohn, [125], on the density fluctuations at equilibrium.

In the setup of the density field formalism, convergence to the macroscopic equation means convergence of $\mathcal{P}^\epsilon$ to a law $\mathcal{P}$ supported by a single trajectory in $D(\mathbb{R}_+, S'(\mathbb{R}))$, whose values at each time are specified by a density which solves the desired macroscopic equation. To attack this problem one proves first tightness of the $\mathcal{P}^\epsilon$. Once this is done there is convergence by subsequences and one need to identify the limiting law as supported by the solutions of some PDE. If uniqueness holds then one concludes the convergence problem.

We start by discussing tightness. Mitoma, [97], has proved that the tightness of $\mathcal{P}^\epsilon$ is implied by the tightness of the law of any single density fields $(X_t^\epsilon(\phi))_{t\geq 0}$. We shall not discuss this result further and just refer to [97]. We should also say that in all the cases we discuss we could also prove directly the tightness of $\mathcal{P}^\epsilon$ by establishing tightness of the density fields by means of estimates which depend on ϕ via suitable seminorms of ϕ.

Next we study the tightness of the laws of the single density fields. The criterion we use is given in the following theorem whose proof will be discussed in the next paragraph. (Some of the conditions appearing below may be relaxed, in particular the moment conditions in (2.20) below, but we will not enter into this).

2.6.2 Theorem. *Let $T > 0$, and, for each $0 < \epsilon \le 1$, let $\mathcal{P}^\epsilon$ be a probability on $D([0,T],\mathbf{R})$ and denote by $\mathcal{E}^\epsilon$ its expectation. Call $x(t)$ the coordinate variable in this space. Assume that $x(t) \in L^2(\mathcal{P}^\epsilon)$ for each ϵ and t as above; assume further that there are $L^2(\mathcal{P}^\epsilon)$ functions $\gamma_i^\epsilon(t)$, $i = 1,2$, for each ϵ and t, such that*

$$x(t) - \int_0^t ds \gamma_1^\epsilon(s) = M^\epsilon(t) \text{ is a } \mathcal{P}^\epsilon\text{-martingale} \tag{2.18}$$

$$M^\epsilon(t)^2 - \int_0^t ds \gamma_2^\epsilon(s) = N^\epsilon(t) \text{ is a } \mathcal{P}^\epsilon\text{-martingale} \tag{2.19}$$

If furthermore there is c such that

$$\sup_{0\le t\le T} \mathcal{E}^\epsilon\left(\gamma_1^\epsilon(t)^2 + \gamma_2^\epsilon(t)^2 + x(t)^2\right) \le c, \quad \text{for all } \epsilon \tag{2.20}$$

then $\mathcal{P}^\epsilon$ is tight in $D([0,T],\mathbb{R})$. The functions $\gamma_1^\epsilon(t)$ and $\gamma_2^\epsilon(t)$ are called the first and the second compensators of $x(t)$ and $x(t)$ and $M^\epsilon(t)^2$ are called semimartingales.

In our applications $x(t)$ is imbedded in a Markov process, in such a case we have a formula for its compensators:

2.6.3 Theorem. *Let $\xi(t)$ be a Markov process with state space X, generator L and law $\mathcal{P}$. Let $f(t)$, $0 \le t \le T$, be a family of real valued measurable functions on X and consider the process*

$$x(t) = f(\xi(t),t), \quad 0 \le t \le T$$

Assume that $Lf(\cdot,t)$ and $L[f(\cdot,t)^2]$ are well defined for all $0 \le t \le T$. Set

$$\gamma_1(t) = Lf(\xi(t),t) + f_t(\xi(t),t), \quad f_t(\cdot,t) = \frac{d}{dt}f(\cdot,t) \tag{2.21}$$

$$\gamma_2(t) = L[f(\xi(t),t)^2] - 2f(\xi(t),t)Lf(\xi(t),t) \tag{2.22}$$

Assume that $x(t)$ and both the $\gamma_i(t)$ are in $L^2(\mathcal{P})$ for all $0 \le t \le T$ and that

$$\sup_{0\le t\le T} \mathcal{E}\left(\gamma_1(t)^2 + \gamma_2(t)^2 + x(t)^2\right) \le c \tag{2.23}$$

Then $\gamma_1(t)$ and $\gamma_2(t)$ are the first and second compensators of $x(t)$.

The proof of this theorem is discussed in the next paragraph, but first let us see how it applies to our case. We are considering the variable $X_t^\epsilon(\phi)$: it does not depend explicitly on time, but the time t appearing as a subscript is actually time $\epsilon^{-b}t$ in the independent particle process, as before $b = 1$ for Euler scalings and $= 2$ for diffusive scalings. Therefore

$$\gamma_1^\epsilon(\phi,t) = \epsilon^{-b} L_0 X_t^\epsilon(\phi) \tag{2.24}$$

$$\gamma_2^\epsilon(\phi,t) = \epsilon^{-b}[L_0(X_t^\epsilon(\phi))^2 - 2X_t^\epsilon(\phi)L_0X_t^\epsilon(\phi)] \tag{2.25}$$

After some computation we get, writing for simplicity $x_\pm = x \pm 1$:

$$\gamma_1^\epsilon(\phi,t) = \epsilon \sum_x \epsilon^{-b}\{p[\phi(\epsilon x_+) - \phi(\epsilon x)] + q[\phi(\epsilon x_-) - \phi(\epsilon x)]\}\eta(x) \tag{2.26}$$

$$\gamma_2^\epsilon(\phi,t) = \epsilon^{-b+2} \sum_x \{p[\phi(\epsilon x_+) - \phi(\epsilon x)]^2 + q[\phi(\epsilon x_-) - \phi(\epsilon x)]^2\}\eta(x) \tag{2.27}$$

In γ_1^ϵ the divergent factor ϵ^{-b} is compensated by the ϵ's coming from the discrete derivatives. In γ_2^ϵ on the other hand the factors ϵ coming from the discrete derivatives overcome ϵ^{-b}, and, as we shall see, γ_2^ϵ vanishes as $\epsilon \to 0$. This will imply the desired result that the density field does not fluctuate in the limit.

One final remark: we take the expectation in the martingale relation (2.18) which relates $X_t^\epsilon(\phi)$ to γ_1^ϵ. By using the properties of the martingales, (we consider simply the case $p = q$ for notational simplicity)

$$\lim_{\epsilon\to 0} |\mathbb{E}_{\mu^\epsilon}(X_t^\epsilon(\phi)) - \mathbb{E}_{\mu^\epsilon}(X_0^\epsilon(\phi)) - \int_0^t ds \mathbb{E}_{\mu^\epsilon}(X_t^\epsilon(\frac{1}{2}\phi''))| = 0 \tag{2.28}$$

where ϕ'' denotes the second derivative of ϕ. By (2.11) and (2.13)

$$\begin{aligned} |\mathbb{E}_{\mu^\epsilon}(X_t^\epsilon(\phi))| &\le c\epsilon \sum_x |\phi(\epsilon x)| \le c \max_r\{(1+r^2)|\phi(r)|\} \epsilon \sum_x \frac{1}{1+(\epsilon x)^2} \\ &\le c' \max_r\{(1+r^2)|\phi(r)|\} \end{aligned} \tag{2.29}$$

We assume that $\phi(r)$ decays at infinity faster than any power of r, then the function $t \to \mathbb{E}_{\mu^\epsilon}(X_t^\epsilon(\phi))$ is equibounded. In order to show that it is also equicontinuous, we observe that from (2.28) it follows that

$$\lim_{\epsilon\to 0} |\mathbb{E}_{\mu^\epsilon}(X_t^\epsilon(\phi)) - E_{\mu^\epsilon}(X_s^\epsilon(\phi)) - \int_s^t ds' \frac{1}{2} \mathbb{E}_{\mu^\epsilon}(X_{s'}^\epsilon(\phi''))| = 0$$

The equicontinuity follows from this and (2.29). (We are assuming that all the derivatives of $\phi(r)$ decay at infinity faster than any power of r). For each ϕ, by the Ascoli-Arzelà theorem, $E_{\mu^\epsilon}(X_t^\epsilon(\phi))$ converges by subsequences as $\epsilon \to 0$ on the compacts. By a diagonalization procedure this holds for all ϕ in a dense set, hence, by continuity, for all ϕ's. All the limiting values $\omega_t(\phi)$ satisfy the equation

$$\omega_t(\phi) - \omega_0(\phi) = \frac{1}{2} \int_0^t ds \omega_s(\phi'') \tag{2.30}$$

which is (2.15) in weak form. As we shall see this identifies

$$\omega_t(\phi) = \int_{-\infty}^{\infty} dr \phi(r) \rho(r,t) \tag{2.31}$$

providing an alternative way for deriving (2.12): the proof of the convergence of $\mathcal{P}^\epsilon$ parallels the above one. Not always the convergence of the density fields is obtained as a byproduct of the analysis of the correlation functions, which in some cases might be quite harder, as for the zero range process which will be studied in the next chapter.

§2.7 Tightness.

We start by recalling the Prokhorov theorem:

2.7.1 Theorem. *Let X be a complete separable metric space and M the space of all the probabilities on X with the weak topology. Then a subset $\Gamma \subset M$ is tight if and only if for every $\epsilon > 0$ there is a compact set $K_\epsilon \subset X$ such that $\mu(K_\epsilon) > 1 - \epsilon$ for all $\mu \in \Gamma$.*

The proof of this theorem can be found for instance in [107], Theorem II.6.7. We apply this theorem when $X = D([0,T], \mathbb{R})$.

2.7.2 Definition: The space $D([0,T],\mathbb{R})$.*

A function $x(t)$, $0 \le t \le T$, is in $D([0,T],\mathbb{R})$ if and only if

i) for all $t \in [0,T)$ $x(t)$ equals its limit from the right

ii) for all $t \in (0,T]$ $x(\cdot)$ has limit from the left

The trajectories of a jump process are in $D([0,T],\mathbb{R})$, see §2.2. In such a process the jumps of $x(t)$ are determined by exponential clocks, so that it looks natural to consider close two realizations $x(t)$ and $y(t)$ if the clocks ring for them almost at the same time and they have almost the same jumps. This is not reflected by sup norms, the "right" topology, instead, is the Skorohod topology, corresponding to the Skorohod metric

$$s(x,y) = \inf_{\lambda \in H} [\sup_{0 \le t \le T} |x(t) - y(\lambda t)| + \sup_{0 \le t \le T} |t - \lambda t|] \tag{2.32}$$

where H is the set of all the homeomorphisms λ from $[0,T]$ onto itself such that $\lambda(0) = 0$ and $\lambda(T) = T$. For a discussion on this definition and its implications, see [107], Section VII.6.

2.7.3 The compact sets in $D([0,T],\mathbb{R})$.*

(In this and the next paragraphs we may use x to denote an element in $D([0,T],\mathbb{R})$, not to be confused with a site in $\mathbb{Z}$). The Ascoli-Arzelà theorem requires equiboundedness and equicontinuity and gives tightness in the sup norm. In the Skorohod topology the equicontinuity is replaced by a weaker condition, [107], Theorem VII.6.2,

The closed set $K \subset D([0,T],\mathbb{R})$ is compact in the Skorohod topology if and only if

$$\sup_{x \in K} \sup_{0 \le t \le T} |x(t)| < \infty, \qquad \lim_{\delta \to 0} \sup_{x \in K} \tilde{\omega}_x(\delta) = 0 \tag{2.33}$$

where

$$\tilde{\omega}_x(\delta) = \max \Bigg(\sup_{0 \le t \le T} \; \sup_{t-\delta \le t' \le t \le t'' \le t+\delta} \{\min[|x(t) - x(t')|, |x(t) - x(t'')|]\}; \\ \sup_{0 \le t \le \delta} |x(t) - x(0)|; \sup_{T-\delta \le t \le T} |x(T) - x(t)| \Bigg) \tag{2.34}$$

2.7.4 Theorem. *(Theorem* VII.7.2 in [107]). Let $\mathcal{P}^\epsilon$, $0 < \epsilon \le 1$, be a family of probabilities on $D([0,T],\mathbb{R})$. Then $\mathcal{P}^\epsilon$ is tight if and only if

i) for any $\zeta > 0$ there is c such that

$$\mathcal{P}^\epsilon(\{\sup_{0 \le t \le T} |x(t)| \le c\}) \ge 1 - \zeta, \quad \text{for all } 0 < \epsilon \le 1$$

ii) for any $\zeta > 0$ and c there is $\delta > 0$ such that

$$\mathcal{P}^\epsilon(\{\tilde{\omega}_x(\delta) \le c\}) \ge 1 - \zeta, \quad \text{for all } 0 < \epsilon \le 1$$

Sufficient conditions for i) and ii) are:

2.7.5 Theorem. *Condition i) of Theorem 2.7.4 is a consequence of*

$$\sup_{0<\epsilon\le 1} \mathcal{E}^\epsilon\big(\sup_{0\le t\le T} x(t)^2\big) < \infty \tag{2.35}$$

while condition ii) follows from the Censov criterion

$$\sup_{0\le t_1<t_2<t_3\le T} \mathcal{E}^\epsilon\big([x(t_2)-x(t_1)]^2[x(t_3)-x(t_2)]^2\big) \le c(t_3-t_1)^2 \tag{2.36}$$

which must hold for all ϵ and for some constant c independent of ϵ.

Using the Chebyshev inequality one readily sees that (2.35) implies i) of Theorem 2.7.3. We refer to the literature, Theorem 15.6 of Billingsley, [13], for the proof that (2.36) implies ii).

2.7.6. Proof of Theorem 2.6.2.

We first prove a weaker version containing the additional assumption that there is c so that for all ϵ

$$\mathcal{E}^\epsilon(\bar\gamma_2^\epsilon(T)) \le c, \quad \bar\gamma_2^\epsilon(t) = \sup_{0\le s\le t} |\gamma_2^\epsilon(s)| \tag{2.37}$$

We follow the proof of Holley and Stroock, [73], [74], [75], see also [38]. We have

$$\mathcal{E}^\epsilon\Big(\sup_{0\le t\le T} x(t)^2\Big) \le 2\mathcal{E}^\epsilon\Big(\sup_{0\le t\le T} M^\epsilon(t)^2\Big) + 2\mathcal{E}^\epsilon\Big(\sup_{0\le t\le T} \Big[\int_0^t ds\gamma_1^\epsilon(s)\Big]^2\Big) \tag{2.38}$$

Using the Doob's inequality on martingales and (2.19) we have

$$\begin{aligned}\mathcal{E}^\epsilon\Big(\sup_{0\le t\le T} M^\epsilon(t)^2\Big) &\le 4\mathcal{E}^\epsilon(M^\epsilon(T)^2) = 4\mathcal{E}^\epsilon(N^\epsilon(T) + \int_0^T dt\gamma_2^\epsilon(t))\\ &\le 4\mathcal{E}^\epsilon(x(0)^2) + 4T\sup_{0\le t\le T}[\mathcal{E}^\epsilon(|\gamma_2^\epsilon(t)|)]\end{aligned} \tag{2.39}$$

Since the last term in (2.38) can be bounded in a similar way, we thus conclude the proof of (2.35).

We still need to prove ii) of Theorem 2.7.4. We rewrite for brevity (2.18) as

$$x(t) = \Gamma_1^\epsilon(t) + M^\epsilon(t)$$

$\Gamma_1^\epsilon(t)$ being the second term on the left hand side of (2.18). We then need to prove condition ii) both for Γ_1^ϵ and M^ϵ. We start by the former. Since

$$\tilde\omega_x(\delta) \le \omega_x(\delta) \equiv \sup_{0\le s\le t\le s+\delta\le T} |x(t)-x(s)|$$

by the Chebyshev inequality:

$$\begin{aligned}\mathcal{P}^\epsilon(\{\tilde\omega_{\Gamma^\epsilon}(\delta) > c\}) &\le c^{-2}\mathcal{E}^\epsilon\Big(\sup_{0\le s\le t\le s+\delta\le T}\Big[\int_s^{s+\delta} dt\gamma_1^\epsilon(t)\Big]^2\Big)\\ &\le c^{-2}\mathcal{E}^\epsilon\Big(\delta\int_0^T dt\gamma_1^\epsilon(t)^2\Big)\\ &\le c^{-2}T\delta\sup_{0\le t\le T}\mathcal{E}^\epsilon(\gamma_1^\epsilon(t)^2)\end{aligned}$$

It remains to prove ii) for $M^\epsilon(t)$. We have already proven in (2.39) and below that

$$\mathcal{E}^\epsilon\big(\sup_{0\le t\le T} M^\epsilon(t)^2\big) < c, \quad \text{for all } 0<\epsilon\le 1$$

As a consequence, given any $\zeta > 0$ there is b so that

$$\mathcal{P}^\epsilon\big(\{\sup_{0\le t\le T} M^\epsilon(t)^2 \le b\}\big) > 1-\zeta, \qquad \mathcal{P}^\epsilon\big(\{\bar\gamma_2^\epsilon(T)\le b\}\big) > 1-\zeta$$

For the last inequality we have used (2.37). For any $b > 0$ we define the stopping time τ as the first time $t \le T$ when either $M^\epsilon(t)^2 \ge b$ or $\bar\gamma_2^\epsilon(t) \ge b$ or both ($\bar\gamma_2^\epsilon$ is defined in (2.37)), otherwise we set $\tau = \infty$. Therefore

$$\mathcal{P}^\epsilon(\{\tau \le T\} \ge 1-\zeta$$

It is enough to prove ii) for $\tilde M^\epsilon(t) = M^\epsilon(t\wedge\tau)$, $t\wedge\tau$ being the minimum between t and τ. We use at this point the criterion (2.36) applied to $\tilde M^\epsilon(t)$. We notice first that

$$\tilde M^\epsilon(t)^2 - \int_0^{t\wedge\tau} ds\gamma_2^\epsilon(s) = \tilde N^\epsilon(t) \text{ is a } \mathcal{P}^\epsilon \text{ martingale}$$

Then

$$\begin{aligned}
&\mathcal{E}^\epsilon\Big([\tilde M^\epsilon(t_2) - \tilde M^\epsilon(t_1)]^2[\tilde M^\epsilon(t_3) - \tilde M^\epsilon(t_2)]^2\Big)\\
&= \mathcal{E}^\epsilon\Big([\tilde M^\epsilon(t_2) - \tilde M^\epsilon(t_1)]^2 \int_{t_2\wedge\tau}^{t_3\wedge\tau} dt\gamma_2^\epsilon(t)\Big)\\
&\le b(t_3-t_2)\mathcal{E}^\epsilon\Big([\tilde M^\epsilon(t_2) - \tilde M^\epsilon(t_1)]^2\Big) \le b^2(t_3-t_2)(t_2-t_1)
\end{aligned}$$

We have therefore proven (2.36) for $\tilde M^\epsilon$ and this concludes the proof of ii) of Theorem 2.7.4.

The proof of Theorem 2.6.2 without the extra assumption (2.37) requires more sophisticated techniques. The key point is a new tightness criterion for semimartingales, i.e. for processes $x(t)$ for which the representation (2.18) holds:

2.7.7 Theorem. *Assume that (2.18) holds for $x(t)$ as well as condition i) of Theorem 2.7.4. Assume further that for any ζ and c positive there is $\delta > 0$ so that for any stopping time τ and any $0 < \epsilon \le 1$*

$$\mathcal{P}^\epsilon\big(\sup_{\tau\le t\le\tau+\delta} |x(t)-x(\tau)| \le c\big) \ge 1-\zeta$$

Then $\mathcal{P}^\epsilon$ is tight.

We refer to the literature for the proof of this theorem, see [78], [94], [66], [115].

To apply Theorem 2.7.7 to our case we write

$$\begin{aligned}
\mathcal{P}^\epsilon\big(\sup_{\tau\le t\le\tau+\delta} |x(t)-x(\tau)| > c\big) &\le c^{-2}\mathcal{E}^\epsilon\big(\sup_{\tau\le t\le\tau+\delta}[x(t)-x(\tau)]^2\big)\\
&\le 2c^{-2}\mathcal{E}^\epsilon\big(\sup_{\tau\le t\le\tau+\delta}[\int_\tau^t ds\gamma_1^\epsilon(s)]^2 + \sup_{\tau\le t\le\tau+\delta}[M^\epsilon(t)^2 - M^\epsilon(\tau)^2]\big)
\end{aligned}$$

We have already estimated analogous expressions before, we omit the details. The proof of Theorem 2.6.2 is thus concluded. □

Next we give a support property of the limiting law $\mathcal{P}$ which is useful when studying the fluctuation fields.

2.7.8 Theorem. *Let $\mathcal{P}^\epsilon$ be the law induced by the density field $X_t^\epsilon(\phi)$ and let $\mathcal{P}$ be a weak limit point of $\mathcal{P}^\epsilon$. Then*

$$\mathcal{P}\big(C(\mathbb{R}_+, S'(\mathbb{R}))\big) = 1$$

where $C(\mathbb{R}_+, S'(\mathbb{R}))$ is the subspace of continuous trajectories in $D(\mathbb{R}_+, S'(\mathbb{R}))$.

Proof. We follow Spohn, [125]. Let Δ be the function on $D([0,T], S'(\mathbb{R}))$

$$\Delta(x) = \sup_{0\le t\le T} |x(t_+) - x(t_-)|$$

namely $\Delta(x)$ is the sup of the jumps of $x(t)$.

We fix ϕ and call $x(t)$ the canonical variable $X_t(\phi)$ on $D([0,T], S'(\mathbb{R}))$ and denote by $x = \{x(t), 0 \le t \le T\}$. Since the values of countably many of such variables determine uniquely an element of $D([0,T], S'(\mathbb{R}))$, it will be enough to prove that

$$\mathcal{P}\big(\{\Delta(x) = 0\}\big) = 1$$

namely that

$$\mathcal{E}\big(\Delta(x)\big) = 0$$

because $\Delta(x) \ge 0$. On the other hand for $\epsilon > 0$

$$\mathcal{E}^\epsilon\big(\Delta(x)\big) \le 2\epsilon \sup |\phi(r)|$$

because, with probability 1, there is only one jump at a time, and when a particle jumps $X_t^\epsilon(\phi)$ varies by ϵ times a difference of ϕ's at different sites.

The proof is then completed if we show that

$$\lim_{\epsilon\to 0} \mathcal{E}^\epsilon\big(\Delta(x)\big) = \mathcal{E}\big(\Delta(x)\big)$$

This is a consequence of the definition of $\mathcal{P}$ as a weak limit of $\mathcal{P}^\epsilon$, once we prove that $\Delta(x)$ is a continuos function in $D([0,T], \mathbb{R})$. This is so because, by the definition of the Skorohod metric,

$$\Delta(x) \le \Delta(y) + s(x,y) \qquad \Delta(y) \le \Delta(x) + s(x,y)$$

We have therefore concluded the proof of Theorem 2.7.8. □

§2.8 Convergence of the density fields.

The tightness criteria established so far are an essential step in the proof of the following theorem.

2.8.1 Theorem. *Assume that μ^ϵ satisfy (2.9) and (2.13) and that there is c such that for all ϵ and all $x \in \mathbb{Z}$*

$$\mathbb{E}_{\mu^\epsilon}(\eta(x)^2) < c, \quad \sum_y \big|\mathbb{E}_{\mu^\epsilon}(\eta(x)\eta(y)) - \mathbb{E}_{\mu^\epsilon}(\eta(x))\mathbb{E}_{\mu^\epsilon}(\eta(y))\big| < c \tag{2.40}$$

Then the law $\mathcal{P}^\epsilon$ of the density fields $X_t^\epsilon(\phi)$ converges weakly to $\mathcal{P}$, the law on $D(\mathbb{R}_+, S'(\mathbb{R}))$ supported by the distribution valued trajectory ω_t such that

$$\omega_t(\phi) = \int dr \phi(r)\rho(r,t) \tag{2.41}$$

where $\rho(r,t)$ solves either (2.14) or (2.15) or (2.16) according to the values of p and q.

To prove the theorem we first verify the tightness conditions. To this end we use the following remarkable identity:

2.8.2 Lemma. *Let*

$$D((x,y),\eta) = \begin{cases} \eta(x)\eta(y) & \text{if } x \neq y \\ \eta(x)^2 - \eta(x) & \text{if } x = y \end{cases} \tag{2.42}$$

Then

$$\begin{aligned} LD((x,y),\eta) =& q[D((x_+,y),\eta) - D((x,y),\eta) + D((x,y_+),\eta) - D((x,y),\eta)] \\ &+ p[D((x_-,y),\eta) - D((x,y),\eta) + D((x,y_-),\eta) - D((x,y),\eta)] \end{aligned} \tag{2.43}$$

The proof of the lemma requires a little computation and a lot of patience, we leave it to the readers as an exercise. We will be back on this in §2.9, when generalizing the identity to cases with more than two sites.

Defining the two body correlation functions as

$$u((x,y),t|\mu^\epsilon) = \mathbb{E}_{\mu^\epsilon}(D((x,y),\eta_t))$$

and, using Lemma 2.8.2, we have

$$u((x,y),t|\mu^\epsilon) = \sum_{(z_1,z_2)} Q_t((x,y) \to (z_1,z_2))u((z_1,z_2),0|\mu^\epsilon) \tag{2.44}$$

hence, by (2.40),

$$u((x,y),t|\mu^\epsilon) \leq c \tag{2.45}$$

As a consequence, from (2.26) and (2.27) we verify that the conditions of Theorem 2.6.2 are fulfilled: $\mathcal{P}^\epsilon$ converges then by subsequences. We shall next identify the limiting laws of $\mathcal{P}^\epsilon$.

2.8.3 Lemma. *(Assume $p = q$). If $\mathcal{P}$ is a limiting point of $\mathcal{P}^\epsilon$, then for all $T > 0$ and all $\phi \in S(\mathbb{R})$*

$$\mathcal{P}(\{\sup_{0 \leq t \leq T} |X_t(\phi) - X_0(\phi) - \frac{1}{2}\int_0^t ds X_s(\phi'')| = 0\}) = 1 \tag{2.46}$$

Proof. Let ϵ_k be a sequence such that $\mathcal{P}^{\epsilon_k}$ converges to $\mathcal{P}$. Then for any $\delta > 0$

$$\begin{aligned} &\mathcal{P}(\{\sup_{0 \leq t \leq T} |X_t(\phi) - X_0(\phi) - \frac{1}{2}\int_0^t ds X_s(\phi'')| > \delta\}) \\ &\qquad = \lim_{k \to \infty} \mathcal{P}^{\epsilon_k}(\{\sup_{0 \leq t \leq T} |X_t(\phi) - X_0(\phi) - \frac{1}{2}\int_0^t ds X_s(\phi'')| > \delta\}) \end{aligned} \tag{2.47}$$

We now recall that the law $\mathcal{P}^\epsilon$ is the same as the law of the density field $X_t^\epsilon(\phi)$ under $\mathbb{P}_{\mu^\epsilon}$. We have by (2.26), recalling that $p = q = 1/2$,

$$K_t^\epsilon \equiv |\frac{1}{2}X_t^\epsilon(\phi'') - \gamma_1^\epsilon(\phi,t)| \leq \epsilon \sum_x \psi(\epsilon x,\epsilon)\eta(x,\epsilon^{-2}t)$$

$$\psi(\epsilon x,\epsilon) = \frac{\epsilon}{3!} \sup_{|r| \leq \epsilon} |\phi'''(\epsilon x + r)|$$

The right hand side in the last inequality is a semimartingale and using the Chebyshev and the Doob inequalities, as before, we prove that there is a constant c such that

$$\mathbb{P}_{\mu^\epsilon}(\{\sup_{0\le t\le T} K_t^\epsilon > \delta'\}) \le c\hat{\delta}^{-1}\,\epsilon$$

We choose $\hat{\delta} = \delta/2$, then the probability on the right hand side of (2.47) is bounded by (writing ϵ for ϵ_k)

$$\mathbb{P}_{\mu^\epsilon}(\{\sup_{0\le t\le T} |X_t^\epsilon(\phi) - X_0^\epsilon(\phi) - \int_0^t ds\gamma_1^\epsilon(\phi,s)| > \delta/2\}) + c\hat{\delta}^{-1}\,\epsilon$$

We use the representation (2.18), then again by the Chebyshev and the Doob inequalities, the last probability is bounded by

$$\begin{aligned}(\delta/2)^{-2}4\,\mathbb{E}_{\mu^\epsilon}(M^\epsilon(\phi,T)^2) &\le (\delta/2)^{-2}4\mathbb{E}_{\mu^\epsilon}(\int_0^T ds|\gamma_2^\epsilon(\phi,s)|)\\ &\le (\delta/2)^{-2}4T\sup_{0\le t\le T}[\mathbb{E}_{\mu^\epsilon}(\gamma_2^\epsilon(\phi,t)^2)]^{1/2}\end{aligned}$$

which, by (2.27) and (2.45), vanishes as $\epsilon \to 0$, we omit the details. From all these estimates the proof of the Lemma follows. □

We adapt to the present case the proof of Holley and Stroock, [73], [74], [75].
By Lemma 2.8.3 we know that any limiting law $\mathcal{P}$ is supported by distribution valued trajectories ω_t such that, for any ϕ

$$\omega_t(\phi) = \omega_0(\phi) + \int_0^t ds\,\omega_s(A\phi), \quad A\phi = \frac{1}{2}\phi''$$

On the other hand defining $\phi_t = e^{At}\phi_0$ we get for any distribution ω

$$\omega(\phi_t) - \omega(\phi_0) = \int_0^t ds\,\omega(A\phi_s)$$

and from the above two identities we get that $(t,s) \to \omega_t(\phi_s)$ is a continuous function of the two variables jointly, and from this it can be seen that

$$\omega_s(\phi_{t-s}) \text{ is constant for } 0\le s\le t\le T, \text{ hence } \omega_t(\phi_0) = \omega_0(\phi_t)$$

This proves that $\omega_t = \rho(r,t)dr$ if it is so at $t = 0$. Setting $\tilde{\eta}(x) = \eta(x) - \mathbb{E}_{\mu^\epsilon}(\eta(x))$, we have

$$\mathbb{P}_{\mu^\epsilon}(\{|\epsilon\sum_x \phi(\epsilon x)\tilde{\eta}(x)| > \delta\}) \le \|\phi\|\delta^{-2}\epsilon^2\sum_{x,y}|\phi(\epsilon x)||\mathbb{E}_{\mu^\epsilon}(\tilde{\eta}(x)\tilde{\eta}(y))|$$

where $\|\phi\|$ is the supremum of $|\phi(r)|$. By (2.40) the sum over y is bounded uniformly on x and ϵ so that the last expression vanishes when $\epsilon \to 0$. On the other hand, as we have already seen, the expectation of the density field at time 0 converges to the integral of the test function over the initial density, thus proving that $\mathcal{P}$ is supported by the distribution which at time 0 has density $\rho(r,0)$. This concludes the proof of Theorem 8.2.1. □

§2.9 Local Equilibrium.

We next discuss local equilibrium and propagation of chaos, properties which play a basic role in the transition from microscopics to macroscopics. Local equilibrium means that at the hydrodynamical times the local structure of the state is close to equilibrium, namely to the invariant measure at the corresponding density. For the independent process the extremal invariant measures are product of identical Poisson measures, so that local equilibrium implies that "approximately" the $\eta(x, \epsilon^{-2}t)$ are independent Poisson distributed variables. Usually, by propagation of chaos, people mean a less detailed condition, namely that the occupation numbers at macroscopic distances become independent. We first devote a subparagraph of more elementary nature on the invariant measures for the independent process.

2.9.1 Invariant measures.*

Consider first a finite volume $\Lambda = [-L, L]$ with periodic boundary conditions. Let N be the total number of particles and suppose for the moment being that they are labelled. We distribute them independently and with equal probability in Λ. Such a distribution is clearly invariant, because it is obviously so when there is just one particle, and the particles are put independently and move independently. By some combinatorics, it is easy to see that if we let $L \to \infty$ and $N \to \infty$ in such a way that $N/(2L) \to \rho$, the marginals of the above measures over the unlabelled particles converge to ν_ρ on $\mathbb{N}^{\mathbb{Z}}$, the product of Poisson measures with density ρ, namely

$$\nu_\rho(\{\eta(x) = k\}) = e^{-\rho}\frac{\rho^k}{k!}. \quad \forall x \in \mathbb{Z}\,, \ \forall k \in \mathbb{N} \tag{2.49}$$

From this it follows that ν_ρ is invariant, of course this could be checked right away from (2.1). We shall prove in the sequel that on a suitable set of $\mathbb{N}^{\mathbb{Z}}$ the ν_ρ are the only extremal invariant measures.

Notice that the parameter ρ in (2.49) has the physical interpretation of a density, since

$$\mathbb{E}_{\nu_\rho}(\eta(x)) = \sum_k k\nu_\rho(\{\eta(x) = k\}) = \rho \tag{2.50}$$

When $p = q = 1/2$, the generator L_0 is a selfadjoint operator on $L_2(\mathbb{N}^{\mathbb{Z}}, \nu_\rho)$, for any given ρ (as it is easy to check), in particular this proves that ν_ρ is invariant (actually reversible). If $p > q$, ν_ρ is again invariant but no longer reversible.

2.9.2 Duality for the independent particles.

We consider first $p = q = \frac{1}{2}$. As already seen

$$L_0\eta(x) = \frac{1}{2}[\eta(x+1) + \eta(x-1) - 2\eta(x)] \tag{2.51}$$

so that the right hand side can be interpreted as the action of the generator of a random walk on a function of x. More precisely let

$$\Omega_1 = \{\xi \in \mathbb{N}^{\mathbb{Z}} : |\xi| \equiv \sum_x \xi(x) = 1\}$$

then we define the function

$$D : \Omega_1 \times \mathbb{N}^{\mathbb{Z}} \to \mathbb{R}$$

$$(\xi, \eta) \to D(\xi, \eta) \tag{2.52}$$

$$D(\xi, \eta) = \prod_x D_{\xi(x)}(\eta(x)) \tag{2.53}$$

where $D_0(n) = 1$ while $D_1(n) = n$. Then $L_0 D(\xi, \eta)$ is the same both when the variable on which L_0 acts is ξ and when it is η. This happens also when $|\xi| = 2$, see Lemma 2.8.2, but it holds also in

$$\Omega = \{\xi \in \mathbb{N}^{\mathbb{Z}} : |\xi| \equiv \sum_x \xi(x) < \infty\} \tag{2.54}$$

if we define D on $\Omega \times \mathbb{N}^{\mathbb{Z}}$ by means of (2.53) choosing $D_k(\cdot)$ as the *Poisson polynomial of order k*, namely

$$D_k(n) = \begin{cases} 1 & \text{if } k = 0; \\ n(n-1)\cdots(n-k+1) & \text{otherwise} \end{cases} \tag{2.55}$$

2.9.3 Correlation functions.

The measure μ has n-body correlation functions if $D(\xi, \eta) \in L_1(\mu)$ for any $|\xi| \leq n$. Then the n-body correlation functions are

$$u(\xi, t|\mu) = \mathbb{E}_\mu(D(\xi, \eta_t)), \quad |\xi| = n$$

If μ is supported by a single configuration η, we simply write $u(\xi, t|\eta)$.

This definition generalizes the notion of one and two body correlation functions given earlier.

2.9.4 Proposition. *Let $\eta \in X(c, n)$, see (2.6), for some c and n, and let $\xi \in \Omega$. Then for any $t \geq 0$*

$$u(\xi, t|\eta) = \mathbb{E}_\xi(u(\xi_t, 0|\eta)) \tag{2.56a}$$

If we consider an initial measure μ such that for some $k > 0$

$$\mathbb{E}_\mu\Big(\prod_{i=1}^k \eta(x_i)\Big) \leq c < \infty \quad \text{for all } x_1, \ldots x_k \in \mathbb{Z}$$

then, for all $t \geq 0$ and ξ such that $|\xi| = k$:

$$u(\xi, t|\mu) = \mathbb{E}_\xi(u(\xi_t, 0|\mu)) = \sum_{\xi'} P_t(\xi \to \xi') u(\xi', 0|\mu) \tag{2.56b}$$

where $P_t(\xi \to \xi')$ is the transition probability for the independent process when the jumps to the right have probability q and those to the left p. D is called the dual function and (2.56) the duality relation (for the independent particles).

Proof. For simplicity we only consider the case $p = q$ for which we present two proofs of (2.56).

Proof 1). After some computation, it is possible (but not so easy) to see that $L_0 D(\xi, \eta)$ is the same both if L_0 is considered as an operator acting on functions of ξ or of η. (In the second proof below we avoid this point). We then call ξ-process and η-process the processes with generator L_0 starting respectively from $\xi \in \Omega$ and $\eta \in \mathbb{N}^{\mathbb{Z}}$. Consider then the process which is the direct product of the ξ and the η processes and call $\mathcal{E}$ the expectation with respect to its law starting from (ξ, η). Then, for any s and t, $0 \le s \le t$,

$$\frac{d}{ds}\mathcal{E}(D(\xi_{t-s}, \eta_s)) = 0 \tag{2.57}$$

Integrating (2.57) over s we have

$$\mathcal{E}(D(\xi_0, \eta_t)) = \mathcal{E}(D(\xi_t, \eta_0)) \tag{2.58}$$

This completes the first proof of the Proposition.

Proof 2). Given η let $\eta^{(L)}$ be as in (2.4) and let N be the total number of particles in $\eta^{(L)}$. Given a labelled configuration $\underline{x} \in \mathbb{Z}^N$, call $U(\underline{x})$ the corresponding unlabelled configuration in $\mathbb{N}^{\mathbb{Z}}$. Let λ_0 be the probability on $\mathbb{Z}^N$ which gives weight $1/N!$ to any $\underline{x}$ such that $U(\underline{x}) = \eta^{(L)}$, namely λ_0 is obtained by labelling with equal probability the particles in $\eta^{(L)}$. Let then $\xi \in \Omega$ and let $(y_1, \ldots, y_k)$ be such that $U((y_1, \ldots, y_k)) = \xi$. Then

$$D(\xi, \eta^{(L)}) = \sum_{(i_1,\ldots,i_k)} \lambda_0(\{x_{i_1} = y_1, \ldots, x_{i_k} = y_k\})$$

where the sum is over all the possible choices of $(i_1, \ldots, i_k)$.

Denote by $\mathbb{P}_{\eta^{(L)}}$ the probability of the unlabelled process starting from $\eta^{(L)}$ and by $\mathcal{P}_{\lambda_0}$ that of the labelled process starting from λ_0. Then

$$\mathbb{P}_{\eta^{(L)}}(\{\eta_t = \eta'\}) = \mathcal{P}_{\lambda_0}(\{U(\underline{x}(t)) = \eta'\})$$

and all the configurations $\underline{x}(t)$ such that $U(\underline{x}(t)) = \eta'$ have same probability. Then

$$u(\xi, t|\eta^{(L)}) = \sum_{(i_1,\ldots,i_k)} \mathcal{P}_{\lambda_0}(\{x_{i_1}(t) = y_1, \ldots, x_{i_k}(t) = y_k\})$$

$$= \sum_{(i_1,\ldots,i_k)} \sum_{(z_1,\ldots,z_k)} [\prod_{h=1}^{k} P_t(y_h \to z_h)] \lambda_0(\{x_{i_h} = z_h,\ h = 1, \ldots, k\})$$

and calling $\xi_t = U((z_1, \ldots, z_k))$,

$$u(\xi, t|\eta^{(L)}) = \mathbb{E}_\xi(D(\xi_t, \eta^{(L)}))$$

By taking the limit as $L \to \infty$ and using the assumption on the growth of $\eta(x)$ as $|x| \to \infty$, see §2.2, we then obtain (2.56a). The proof of (2.56b) easily follows from (2.56a), so the Proposition is proven. □

2.9.5 Theorem. *(Doob's Theorem) If μ is a product of Poisson measures with bounded averages, then, for any $t \ge 0$, μ_t, the distribution at time t, is also a product of Poisson measures with averages satisfying (2.11).*

Proof. Since μ is a product of Poisson measures, then for all ξ and η,

$$u(\xi, 0|\mu) = \prod_x u(x, 0|\mu)^{\xi(x)} \tag{2.59}$$

The converse is also true, namely if (2.59) holds for all ξ then μ is a product of Poisson measures, (this is the reason why the $D_k(n)$'s are called Poisson polynomials).

From (2.56) and (2.59) we have that

$$u(\xi,t|\mu) = \sum_{\xi'} P_t(\xi \to \xi') \prod_x u(x,0|\mu)^{\xi'(x)} \tag{2.60}$$

Therefore

$$u(\xi,t|\mu) = \prod_x \Big[\sum_{x'} P_t(x \to x')u(x',0|\mu)\Big]^{\xi(x)} \tag{2.61}$$

□

2.9.6 Definition: Local equilibrium.

The family μ^ϵ, $0 < \epsilon \le 1$, of measures on $\mathbb{N}^{\mathbb{Z}}$ is a local equilibrium distribution for the density profile $\rho(r)$ if for any $r \in \mathbb{R}$ and for any bounded cylindrical function f,

$$\lim_{\epsilon \to 0} \mathbb{E}_{\mu^\epsilon}(d_{[\epsilon^{-1}r]}f) = \mathbb{E}_{\nu_{\rho_0(r)}}(f) \tag{2.62}$$

where $d_x f(\eta) = f(\tilde{d}_x \eta)$, $\tilde{d}_x \eta(y) = \eta(y+x)$ for all y.

2.9.7 Proposition. *Let μ^ϵ, for each $0 < \epsilon \le 1$, be a product of Poisson measures satisfying (2.13) and (2.9). Then if $p > q$ [$p = q$] $\mu^\epsilon_{\epsilon^{-1}t}$ [$\mu^\epsilon_{\epsilon^{-2}t}$] is a local equilibrium distribution for $\rho_t(r)$ solution of (2.14), [(2.15)]. Analogous statement holds for $p - q = \epsilon a$ and $\rho_t(r)$ solution of (2.16).*

Proof. By a density argument it is enough to show that for any fixed $\xi \in \Omega$ and denoting by $t_\epsilon = \epsilon^{-1}t$ or $t_\epsilon = \epsilon^{-2}t$, according to the values of p,

$$\lim_{\epsilon \to 0} u(\tilde{d}_{[\epsilon^{-1}r]}\xi, t_\epsilon|\mu^\epsilon) - u([\epsilon^{-1}r], t_\epsilon|\mu^\epsilon)^{|\xi|} \tag{2.63}$$

but this is a consequence of (2.61) and Proposition 2.5.2. □

§2.10 Basin of attraction of ν_ρ.

We shall only consider the case $p = q$. For c and n positive and $\rho \ge 0$ set

$$\mathcal{H}(c,n,\rho) = \{\eta \in \mathbb{N}^{\mathbb{Z}} : \eta(x) \le c|x|^n, \text{ for all } x, \lim_{L\to\infty} A_L(\eta) = \rho\} \tag{2.64}$$

where

$$A_L(\eta) = \lim_{L\to\infty} \frac{1}{2L+1} \sum_{|x|\le L} \eta(x)$$

2.10.1 Theorem. *Given n, c and ρ, let $\eta \in \mathcal{H}(c,n,\rho)$. Then the transition probability $P(\eta \to \eta')$ converges weakly to ν_ρ.*

*Proof**. Recall that

$$u(x,t|\eta) = \sum_y \pi_t(|y-x|)\eta(y) \tag{2.65}$$

where $\pi_t(|x|)$ is the probability that a single random walk which starts from the origin is at time t in x (since the walk is symmetric, the probability does not depend on the sign of x).

We first prove that $u(0,t|\eta) \to \rho$. This comes from the identity:

$$u(0,t|\eta) = \sum_{\ell=1}^{\infty} h_t(\ell)A_\ell(\eta) - 2\pi_t(1)\eta(0) + \pi_t(0)\eta(0)$$
$$h_t(\ell) = [\pi_t(\ell) - \pi_t(\ell+1)](2\ell+1)$$

obtained by writing for $\ell > 0$

$$\eta(\ell) + \eta(-\ell) = \zeta(\ell) - \zeta(\ell-1), \quad \zeta(\ell) = \sum_{y=0}^{\ell}[\eta(y) - \eta(-y)]$$

and by "integrating by parts" (2.65) with such substitutions. By assumption $A_\ell(\eta)$ converges to ρ. By the local central limit theorem, see [58],

$$\lim_{t\to\infty} \sum_x |\pi_t(x) - G(x/\sqrt{t})|(|x|/\sqrt{t})^n = 0, \qquad G(r) = \frac{1}{2\pi}e^{-r^2/2}$$

hence

$$\limsup_{t\to\infty} \sum_x |h_t(x)| \le c$$

and

$$\lim_{t\to\infty} \sup_x |h_t(x)| = 0$$

These properties prove that $u(0,t) \to \rho$.

It is also easy to see that if $\eta \in \mathcal{H}(c,n,\rho)$ then also the translates of η are in the same space (maybe with a different c): this shows that also $u(x,t|\eta) \to \rho$, for any fixed x.

For a finite configuration ξ let

$$u(\xi,t|\eta) = \mathbf{E}_\eta(D(\xi,\eta_t))$$

We want to show that this converges to ρ^k, if ξ has k particles. Assume first that they are all at $x = 0$. Then by duality:

$$u(\xi,t|\eta) = \sum_{x_1,\dots x_k} \prod_{i=1}^{k} \pi_t(x_i)u((x_1,\dots,x_k),0|\eta)$$

The function $u(\underline{x},0|\eta)$, $\underline{x} = (x_1,\dots,x_k)$, equals the product of the $\eta(x_i)$ if these are all different, otherwise if there are ℓ x_i equal we get instead of $\eta(x_i)^\ell$ the Poisson polynomial of order ℓ computed at $\eta(x_i)$. Consider all the cases where there are just h different sites with multiple occupancy, and fix in each of them the labels of the particles which are there, call such sets of labels $C_1,\dots,C_h$. Let particle i be in C_1 then the factor coming from x_i is $\eta(x_i)\dots[\eta(x_i) - k_1 + 1]$, calling k_1 the cardinality of C_1. There is a which only depends on k, such that

$$|\eta(x_i)\dots[\eta(x_i) - k_1 + 1] - \eta(x_i)^{k_1}| \le a\eta^{k_1-1}$$

Hence the contribution due to C_1 is the sum of the "right factor" $\eta(x_i)^{k_1}$ plus a term bounded by $a\eta^{k_1-1}$. Consider the contribution of the latter. The factors due to all the other clusters can be bounded by constants times the "right factors" (products of η). We bound $\pi_t(x_i)$ by a constant times $1/\sqrt{t}$, by the local central limit theorem. Hence these terms are a part of the contribution due to just $k-1$ particle (in the dual process) assuming the correlation function at time 0 equals the product of the η's even if they are at the same site. This contribution is therefore bounded by a constant times $1/\sqrt{t}$.

Proceeding iteratively we prove that the factors due to the various clusters can be changed into the "right ones" with an error which vanishe like $1/\sqrt{t}$. This concludes the proof that $u(\xi, t|\eta) \to \rho^k$ when all the particles in ξ are at the origin.

The proof that the same equality holds also when the particles in ξ are not all at the origin is analogous and omitted. This proves that measures starting from $\mathcal{H}(c, n, \rho)$ are attracted by ν_ρ.

2.10.2 Corollary. *Let μ be a probability which is translationally invariant and ergodic with respect to space shifts. Assume also that $\eta(0) \in L^1(\mu)$. Then $\mu_t \to \nu_\rho$.*

CHAPTER III

HYDRODYNAMICS OF THE ZERO RANGE PROCESS

In this chapter we prove that the hydrodynamical behavior of the symmetric zero range process is described by a non linear diffusion equation. We use the Guo, Papanicolaou, Varadhan technique, [71], and the super-exponential estimates of Kipnis, Olla, Varadhan, [79].

§3.1 The finite volume zero range process.

The zero range process describes particles which jump from site to site with an intensity which depends only on the number of particles at the site from where they jump. We only consider finite volumes and nearest neighbor jumps. The scaling parameter is $\epsilon > 0$ and the hydrodynamic limit is obtained by letting $\epsilon \to 0$. For notational simplicity we assume that ϵ^{-1} is an integer, and we set $\mathbf{Z}_\epsilon = \mathbf{Z}$ modulo ϵ^{-1}. The state space is then $\mathbf{N}^{\mathbf{Z}_\epsilon}$. Since the number of particles is conserved, we can examine separately the subspaces with fixed number of particles. Each of these subspaces has a finite number of elements so we have a jump Markov process on a finite state space, the existence of dynamics is not a problem here. The configurations change because any particle may jump to the nearest neighbor sites, remember that $\mathbf{Z}_\epsilon$ is identified to the interval $[1, \epsilon^{-1}]$ with periodic boundary conditions. The intensity of any of the two possible jumps is $\frac{1}{2}c(\eta(x))$, if x is the site from where the particle jumps, and $\eta(x)$ is the number of particles at x. We shall denote by η the set of all $\eta(x)$. We assume that *$c(k)$ is a non decreasing function such that (1) $c(0) = 0$, $c(k) > 0$ for all $k > 0$, and (2) there is a constant $\beta \geq 0$ such that $c(k+1) \leq e^{\beta k} c(k)$, for all $k \geq 1$.*

The generator of this process is

$$(Lf)(\eta) = \sum_{x \in \mathbf{Z}_\epsilon} \frac{1}{2} c(\eta(x)) \left[(f(\eta^{x,x+1}) - f(\eta)) + (f(\eta^{x,x-1}) - f(\eta)) \right] \tag{3.1a}$$

f is any function on $\mathbf{N}^{\mathbf{Z}_\epsilon}$ and

$$\eta^{x,y} = \eta - \delta_x + \delta_y \tag{3.1b}$$

We use the following notation: $\eta \in \mathbf{N}^{\mathbf{Z}_\epsilon}$, $\delta_x \in \mathbf{N}^{\mathbf{Z}_\epsilon}$ is the configuration with just one particle at x, the sum and the difference of two configurations is the sum, respectively the difference, of their components.

3.1.1 Remarks.

If $c(k)$ is proportional to k the particles move independently of each other. If $k^{-1}c(k) \to 0$, for large k the interaction is attractive, since the particles jump less frequently than in the independent case. The extreme case is when $c(k) = 1$ for all $k \geq 1$, this model being isomorphic to the symmetric simple exclusion process, see Chapter VI, with a tagged particle. The opposite case is when there is a repulsive interaction among particles, which, for pair interactions, gives a jump intensity which grows exponentially. This last statement will become more clear after the following definition.

3.1.2 The Gibbs measures.

The Gibbs measure ν_z is the product measure on $\mathbb{N}^{\mathbb{Z}_\epsilon}$ such that

$$\nu_z(\eta(x) = k) = N\frac{z^k}{c(1)\cdots c(k)} \tag{3.2}$$

where *the fugacity* z in (3.2) is such that the sum over k is finite, N is then the normalization factor, namely such that the sum of (3.2) over k is normalized to 1. Notice that to have coherent notation we should have added a superscript ϵ to ν_z, since the space on which this measure is defined depends on ϵ.

Since $\nu_z(\{\eta\})c(\eta(x)) = \nu_z(\{\eta'\})c(\eta'(x\pm1))$ where $\eta' \equiv \eta - \delta_x + \delta_{x\pm1}$, these measures are invariant and reversible.

In the independent case, $c(k) = \text{const}.k$, the Gibbs measure is indeed a product of Poisson measures. All values of z are allowed. If $c(k) = 1$ for all $k \geq 1$ then ν_z is a product of geometric measures and only the values $z < 1$ are allowed. Finally if $c(k) = ke^{\beta k}$ all values of z are allowed and

$$\nu_z(\eta(x) = k) = \frac{1}{Z}z^k\frac{e^{-\beta k(k-1)/2}}{k!} \tag{3.3}$$

where *the partition function* Z is the normalizing factor in (3.3). The Gibbs factor in (3.3) corresponds to the energy of a non negative zero range pair potential. If its value at the origin is 1, we interpret β in (3.3) as $(KT)^{-1}$, where T is the temperature and K the Boltzmann constant.

3.1.3 The canonical Gibbs measures.

The canonical Gibbs measures are indexed by the positive integers N, for each N the corresponding measure has support on

$$\{\eta \in \mathbb{N}^{\mathbb{Z}_\epsilon} : |\eta| \equiv \sum_{x\in\mathbb{Z}_\epsilon} \eta(x) = N\}$$

i.e. the configurations with N particles. The canonical Gibbs measure is then obtained by conditioning on $|\eta| = N$ any Gibbs measures ν_z, in fact the resulting conditional probability does not depend on z. *The set of all the canonical Gibbs measures coincides with the set of all the extremal invariant measures.* In fact any two configurations η and η' with same number of particles are connected with positive probability, namely starting from η at time 0, the probability of η' at time 1 (or at any other positive time) is positive. By the Perron Frobenius theorem, see Proposition 3.1.4 below, this property ensures that there is a unique invariant measure on each subspace $\{|\eta| = N\}$. This shows that any invariant measure is a convex combination of canonical Gibbs measures, hence that these are all the extremal invariant measures.

3.1.4 Proposition. *Let $P_t(x \to y)$ be the transition probability of a Markov process with finite state space E. Assume that there are t_0 and δ positive such that for all x and y in E: $P_{t_0}(x \to y) \geq \delta$. Then there is a unique invariant measure μ on E and for any t*

$$\sum_{y\in E} |P_t(x \to y) - \mu(y)| \leq (1-\delta)^{(t/t_0-1)}$$

*Proof**. Let λ and ν be any two probabilities on E, let

$$\lambda_n(y) = \sum_x \lambda(x) P_{nt_0}(x \to y); \qquad \nu_n(y) = \sum_x \nu(x) P_{nt_0}(x \to y)$$

and

$$\begin{aligned} \|\lambda_n - \nu_n\| &= \sum_x |\lambda_n(x) - \nu_n(x)| \\ &= \sum_{x \in E_+(n)} [\lambda_n(x) - \nu_n(x)] + \sum_{x \in E_-(n)} [\nu_n(x) - \lambda_n(x)] \end{aligned}$$

where

$$E_\pm(n) = \{x \in E : \lambda_n(x) \gtrless \nu_n(x)$$

Then

$$\begin{aligned} \lambda_{n+1}(y) - \nu_{n+1}(y) &= \sum_{x \in E_+(n)} [\lambda_n(x) - \nu_n(x)] P_{t_0}(x \to y) + \sum_{x \in E_-(n)} [\lambda_n(x) - \nu_n(x)] P_{t_0}(x \to y) \\ &= \sum_{x \in E_+(n)} [\lambda_n(x) - \nu_n(x)][P_{t_0}(x \to y) - \delta] + \sum_{x \in E_-(n)} [\lambda_n(x) - \nu_n(x)][P_{t_0}(x \to y) - \delta] \end{aligned}$$

so that, summing over y,

$$\begin{aligned} \|\lambda_{n+1} - \nu_{n+1}\| &\le \sum_{x \in E_+(n)} [\lambda_n(x) - \nu_n(x)](1-\delta) + \sum_{x \in E_-(n)} [\nu_n(x) - \lambda_n(x)](1-\delta) \\ &\le (1-\delta)\|\lambda_n - \nu_n\| \end{aligned}$$

from this last inequality it is not difficult to conclude the proof of the Proposition. □

Since the function which multiplies N on the right hand side of (3.2) increases with z, then, for any fixed choice of the intensities $c(\cdot)$, there is a value $\bar{z}$ of the fugacity z (possibly equal to $+\infty$) such that ν_z^ϵ is well defined for all $z < \bar{z}$. In the sequel we shall tacitly assume that z always satisfies this relation. For such a z we define the density ρ as the following function of z:

$$\rho = \sum_{k=1}^{\infty} k\ \nu_z(\eta(x) = k) \tag{3.4}$$

(3.4) establishes a one to one correspondence between $z \in (-\infty, \bar{z})$ and $\rho \in (0, \rho_c)$, where ρ_c is the sup over all $z \in (-\infty, \bar{z})$ of the right side of (3.4). We shall denote by $z(\rho)$ the inverse of $\rho(z)$ as given by (3.4).

§3.2 The hydrodynamical limit.

We denote by $\rho(r)$, $r \in \mathcal{T}$, [$\mathcal{T}$ being the unit circle], a smooth positive function such that for all $r \in \mathcal{T}$, $z(\rho(r)) < \bar{z}$ and we set $z_{max} = \max_{r \in \mathcal{T}} z(\rho(r))$. For each $\epsilon > 0$ μ^ϵ is the product measure on $\mathbb{N}^{\mathbb{Z}_\epsilon}$ such that for all k

$$\mu^\epsilon(\eta(x) = k) = \nu_{z[\rho(\epsilon x)]}(\eta(1) = k) \tag{3.5}$$

We shall consider the process with initial distribution given by μ^ϵ and denote by $\mathbf{E}_{\mu^\epsilon}$ its expectation. *The density field* $X_t^\epsilon(\phi)$ is defined for all $t \geq 0$ and all $\phi \in \mathcal{S}$, $\mathcal{S}$ being the Schwartz space of all smooth functions on the unit circle $\mathcal{T}$, as

$$X_t^\epsilon(\phi) = \epsilon \sum_{x \in \mathbf{Z}_\epsilon} \phi(\epsilon x)\eta(x,t) \tag{3.6}$$

The same expression was introduced in the previous chapter for the independent particles in the symmetric case.

We think of the density field $X_t^\epsilon(\phi)$ as the canonical variable on $\mathcal{D}(\mathbb{R}_+, \mathcal{S}')$ and denote by $\mathbb{P}^\epsilon$ the law on $\mathcal{D}$ induced by the zero range process which starts from μ^ϵ. The main result in this chapter is

3.2.1 Theorem. *The law $\mathbb{P}^\epsilon$ on $\mathcal{D}(\mathbb{R}_+, \mathcal{S}')$ converges to $\mathbb{P}$, the law supported by the distribution valued trajectory $\rho_t(r)dr$, where*

$$\frac{\partial}{\partial t}\rho = \frac{1}{2}\frac{\partial}{\partial r}[D(\rho)\frac{\partial}{\partial r}\rho] \tag{3.7}$$

with initial condition $\rho(r)$. $D(\rho)$ is given by

$$D(\rho) = \frac{\partial}{\partial \rho}\mathbf{E}_{\nu_{z(\rho)}}(c(\eta(1))) \tag{3.8}$$

The remaining of this chapter is devoted to the proof of Theorem 3.2.1. We follow the strategy outlined in Chapter II. Tightness works as well as for the independent particles because of our assumption that the intensities $c(k)$ are non-decreasing. We have in fact

3.2.2 Proposition. *(A priori bounds). Let f be any increasing function on $\mathbb{N}^{\mathbf{Z}_\epsilon}$ [$\eta \geq \eta'$ meaning that $\eta(x) \geq \eta'(x)$, $\forall x \in \mathbf{Z}_\epsilon$] then for any ϵ and $t \geq 0$*

$$\mathbf{E}_{\mu^\epsilon}(f(\eta_t)) \leq \mathbf{E}_{\nu_{z_{max}}}(f(\eta)) \tag{3.9}$$

where $\mathbf{E}_\lambda$ is the expectation when the initial distribution is λ.

*Proof**. The proof is divided into three steps. In the first one we prove that μ^ϵ is "stochastically smaller" than ν_z, (in the course of this proof we shall write z for z_{max}). This means that there is a probability $Q(\eta,\eta')$ on $\mathbb{N}^{\mathbf{Z}_\epsilon} \times \mathbb{N}^{\mathbf{Z}_\epsilon}$ with marginals

$$\sum_{\eta'} Q(\eta,\eta') = \mu^\epsilon(\eta), \qquad \sum_{\eta} Q(\eta,\eta') = \nu_z(\eta')$$

(called a joint representation of μ^ϵ and ν_z) which is supported by the pairs $\eta \leq \eta'$. Notice that this implies (3.9) with $t = 0$, in fact

$$\mathbf{E}_{\nu_z}(f(\eta)) - \mathbf{E}_{\mu^\epsilon}(f(\eta)) = \sum_{\eta,\eta'} Q(\eta,\eta')[f(\eta') - f(\eta)] \geq 0$$

(It can be proven, in general, that the converse holds, namely two measures are stochastically ordered if the expectations of the increasing functions are ordered, see [90]).

Since both μ^ϵ and ν_z are product measures, it is enough to establish that, for any given x, $\lambda \le \nu$, stochastically, where λ and ν are the marginals of μ^ϵ and ν_z on $\{\eta(x)\}$, namely the probabilities on $\mathbb{N}$ defined by

$$\lambda(n) \equiv \mu^\epsilon(\eta(x) = n), \qquad \nu(n) \equiv \nu_z(\eta(x) = n)$$

To prove that $\lambda \le \nu$, we first observe that there is $k \ge 0$ such that $\lambda(n) > \nu(n)$ for $n < k$ and $\lambda(n) \le \nu(n)$ for $n \ge k$: we have in fact

$$\lambda(n) = \exp\{-\alpha(n) + hn - A\}$$

where $\alpha(n) = \log(c(1)\dots c(n))$, $h = \log z'$ (if $z' = z(\rho(\epsilon x))$ is the fugacity at x) and A the log of the normalization factor. For ν we have the same expression with new h and A larger or equal than the previous ones, hence the existence of k.

A geometrical representation will help to visualize the proof. We call Q the set of all the $\mathbb{N} \times \mathbb{N}$ matrices $q = (q(m,n))$ with non negative entries and such that the "total mass" in the n-th column is

$$\sum_m q(m,n) = \lambda(n)$$

for every n. To each $q \in Q$ we associate a probability μ_q by looking at the mass in each row:

$$\mu_q(m) = \sum_n q(m,n)$$

The desired joint representation is an element $q \in Q$ such that 1) $q \in Q^+$, this latter being the set of elements in Q with no mass below the diagonal ($q(m,n) = 0$ when $m < n$), and 2) $\mu_q = \nu$.

We shall construct such a q by an iterative procedure starting from $q_0 \in Q^+$ which has mass only on the diagonal:

$$q_0(m,n) = \delta_{m,n}\lambda(n)$$

To define the iteration we need a new notion: the m-th row of an element $q \in Q^+$ has a mass excess E if

$$E = \nu(m) - \sum_n q(m,n) \ge 0$$

The m-th row has a mass defect D if

$$D = \sum_n q(m,n) - \nu(m) \ge 0$$

Let $q^\star \in Q^+$ and suppose that there is $k^\star$ so that the rows $m < k^\star$ have a mass excess, and the others a mass defect. (Notice that q_0 fulfills such a property). We then define a map F on such $q^\star$'s as follows. Denote by $q = F(q^\star)$, then $q = q^\star$ if there is no positive mass excess and defect. Otherwise let m_0 be the first row with $E > 0$ and m_1 the first one with $D > 0$. Call n_0 the first value for which $q^\star(m_0, n_0) > 0$. Then letting $a \vee b = \min\{a, b\}$, we set

$$\gamma = (q^\star(m_0, n_0) \vee E) \vee D$$

q is then defined by moving the mass γ upward from (m_0, n_0) to (m_1, n_0), namely

$$q(m_0, n_0) = q^*(m_0, n_0) - \gamma, \quad q(m_1, n_0) = q^*(m_1, n_0) + \gamma$$

all the other entries are left unchanged.

Observe that $F(q^*)$ is still in the domain of F, so that we can apply again F and after a finite number of iterations we reach a value q where the mass excess and defect are all 0. Recalling that q_0 is in the domain of F, the orbit of F starting from q_0 is well defined and it ends on the desired joint representation, which completes the proof that ν is stochastically larger than λ.

The proof of the Proposition 3.2.2 proceeds by establishing that the evolution preserves the order, namely if $\eta \leq \eta'$ then for any t the law at time t starting from η is stochastically smaller than that starting from η'. To see this we define the evolution in

$$\{(\eta, \eta') \in \mathbb{N}^{\mathbf{Z}_\epsilon} \times \mathbb{N}^{\mathbf{Z}_\epsilon} : \eta \leq \eta'\}$$

as the jump process with generator

$$\tilde{L}f(\eta, \eta') = \sum_{x \in \mathbf{Z}_\epsilon} \sum_{y:|y-x|=1} \frac{1}{2}\Big(c(\eta(x))[f(\eta^{x,y}, \eta'^{x,y}) - f(\eta, \eta'))]$$
$$+ \{c(\eta'(x)) - c(\eta(x))\}[f(\eta, \eta'^{x,y}) - f(\eta, \eta'))]\Big)$$

where $\eta^{x,y}$ is defined in (3.1b); recall that $c(\eta'(x)) - c(\eta(x)) \geq 0$ because $c(\cdot)$ is non decreasing. In the process generated by $\tilde{L}$, whenever a particle of the configuration η jumps then the same occurs in η'; there are jumps in η' which do not occurr in η but the particle which jumps starts from a site x where $\eta'(x) > \eta(x)$, so that after the jump $\eta'(x) \geq \eta(x)$, and the same holds at the other sites. The action of $\tilde{L}$ on functions which depend only on η [respectively η'] is the same as that of L, see (3.1a), and from this it follows that the evolution preserves the order.

To prove (3.9) we simply notice that for what established so far $\mu^\epsilon \leq \nu_{z_{max}}$ so that we can construct a joint representation of these two measures supported on (η, η') with $\eta \leq \eta'$. We let it evolve with the process with generator $\tilde{L}$, so that the measure at any time t is still supported by ordered pairs and it is a joint representation of the evoluted at time t starting from μ^ϵ and $\nu_{z_{max}}$; (3.9) follows then from the invariance of $\nu_{z_{max}}$. □

3.2.3 Proposition. *For any $\delta > 0$ and $T > 0$ the following holds.*

$$\lim_{\epsilon \to 0} \mathbb{P}^\epsilon_{\mu^\epsilon}\Big(\sup_{t \leq T} \Big|X^\epsilon_t(\phi) - X^\epsilon_0(\phi) - \int_0^t ds \frac{1}{2}\epsilon \sum_x \phi''(\epsilon x) c(\eta(x, \epsilon^{-2}s))\Big| > \delta\Big) = 0 \tag{3.10}$$

Proof. From the definition (3.6) and (3.1) it follows that

$$\begin{aligned} \gamma^\epsilon_1(\phi, t) &\equiv \epsilon^{-2} L X^\epsilon_t(\phi) \\ &= \epsilon^{-2}\epsilon \sum_{x \in \mathbf{Z}_\epsilon} c(\eta(x,t))[\frac{1}{2}\phi(\epsilon x + \epsilon) + \frac{1}{2}\phi(\epsilon x - \epsilon) - \phi(\epsilon x)] \end{aligned} \tag{3.11}$$

Notice that by (3.9) the expectation of the square of $\gamma^\epsilon_1(\phi, t)$ is uniformly bounded.

For $\gamma_2^\epsilon(\phi,t)$ we get

$$\gamma_2^\epsilon(\phi,t) \equiv \epsilon^{-2}LX_t^\epsilon(\phi)^2 - 2X_t^\epsilon(\phi)\epsilon^{-2}LX_t^\epsilon(\phi)$$
$$= \epsilon\{\frac{1}{4}\epsilon\sum_x c(\eta(x,t))[(\frac{\phi(\epsilon(x+1))-\phi(\epsilon x)}{\epsilon})^2 + (\frac{\phi(\epsilon(x-1))-\phi(\epsilon x)}{\epsilon})^2]\} \tag{3.12}$$

Again, same argument as above, the expectation of $\gamma_2^\epsilon(\phi,t)^2$ is uniformly bounded. By the tightness criterion of Chapter II we can conclude that $\mathbb{P}^\epsilon$ converges by subsequences. As in Chapter II we also have that the expectation of $\gamma_2^\epsilon(\phi,t)$ [and of its square as well] vanishes as $\epsilon \to 0$ uniformly on t. From this the proposition follows. □

The above Proposition shows that the limiting laws of the density field are supported by continuous, distribution-valued trajectories. However, as a difference from Chapter II, (3.10) does not identify the limiting points of $\mathbb{P}^\epsilon$, at least directly, because the integrand in (3.12) is not a density field. The great simplification in the independent case is that the expected increments of the density fields are again density fields, so that the equations are automatically closed. This is no longer true in the interacting case. On the other hand if we try to write the martingale equation for the field $\{\epsilon\sum_x \psi(\epsilon x)c(\eta(x,t))\}$, $\psi\in\mathcal{S}$, we readily see that the expectation of the corresponding γ_i^ϵ diverges when $\epsilon\to 0$. The ϵ factors come out all right only for the density fields, because the particles number is conserved [this is not enough, though, and another condition, fulfilled by the zero range process, is needed. This is the *gradient condition* which will be discussed again in Chapter VIII].

If we assume that

$$\lim_{\epsilon\to 0}|X_t^\epsilon(\phi) - \int dr\rho(r,t)\phi(r)| = 0 \tag{3.13}$$

and

$$\lim_{\epsilon\to 0}|\frac{1}{2}\epsilon\sum_x \phi''(\epsilon x)c(\eta(x,\epsilon^{-2}s)) - \frac{1}{2}\int dr\phi''(r)\mathbb{E}_{\nu_{z(\rho(r,s))}}(c(\eta(0,0)))| = 0 \tag{3.14}$$

then, from (3.10), we derive, in weak form, the correct equation (3.7). The property (3.14) is a way to state "local equilibrium" since it says that the distribution at time $\epsilon^{-2}s$ around $\epsilon^{-1}r$ is approximately the Gibbs distribution with density $\rho(r,s)$. Of course such an assumption presumes already the knowledge of $\rho(r,s)$ and it is a much stronger property than the same statement we want to prove. We may weaken (3.14) by replacing $\rho(r,s)$ by the empirical particle density, namely the density which is obtained by counting the actual number of particles at time $\epsilon^{-2}s$ in an interval I centered at $\epsilon^{-1}r$. We have to choose the size of I: it has to diverge when $\epsilon\to 0$ to avoid statistical errors. If we choose it infinitesimally smaller than ϵ^{-1}, then we will not be able to relate the density in I to a density field, so we choose the size of I equal to $\epsilon^{-1}\ell$ and then let $\ell\to 0$, but after $\epsilon\to 0$.

To proceed further, it is clearly convenient to rewrite (3.10) in terms of bounded variables and true space averages, this is the content of the next Proposition.

3.2.4 Proposition. *For any $\delta > 0$ we have that*

$$\limsup_{R\to\infty}\limsup_{\ell\to 0}\limsup_{\epsilon\to 0}\mathbb{P}_{\mu^\epsilon}\Big(\int_0^T dt|\epsilon\sum_x \psi(\epsilon x)c(\eta(x,\epsilon^{-2}t))$$
$$-\epsilon\sum_x\psi(\epsilon x)\frac{1}{\epsilon^{-1}\ell}\sum_{|y-x|\le\frac{1}{2}\epsilon^{-1}\ell} c_R(\eta(y,\epsilon^{-2}t))| > \delta\Big) = 0 \tag{3.15}$$

where $c_R(k) = \min(c(k), R)$.

Proof. We have that

$$c(\eta(x)) - c_R(\eta(x)) = 1(\eta(x) \geq R)c(\eta(x))$$

which is a non decreasing function of $\eta(x)$. Therefore for any t

$$\mathbb{E}_{\mu^\epsilon}\Big(c(\eta(x,t)) - c_R(\eta(x,t))\Big) = \mathbb{E}_{\mu^\epsilon}\Big(1(\eta(x,t) \geq R)c(\eta(x,t))\Big)$$
$$\leq \mathbb{E}_{\nu_{z_{max}}}\Big(1(\eta(0,0) \geq R)c(\eta(0,0))\Big)$$

which goes to zero as $R \to \infty$.

We can then replace c by c_R with an error which vanishes uniformly on ϵ as $R \to \infty$. On the other hand, for each R

$$\lim_{\ell \to 0} \lim_{\epsilon \to 0} \sup_t \mathbb{E}_{\mu^\epsilon}\Big(\epsilon \sum_x c_R(\eta(x,t)|\psi(\epsilon x) - \frac{1}{\epsilon^{-1}\ell} \sum_{|y-x| \leq \epsilon^{-1}\ell/2} \psi(\epsilon y)|\Big) = 0$$

so that we can also replace ψ by its average. Integrating by parts the sum over y, we then complete the proof of (3.15). □

Our purpose, next, is to estimate the difference

$$\frac{1}{\epsilon^{-1}\ell} \sum_{|y-x| \leq \frac{1}{2}\epsilon^{-1}\ell} c_R(\eta(y,t)) - \nu_{z(\rho)}(c_R(\eta(0))) \tag{3.16}$$

with ρ chosen as the empirical density, namely

$$\rho = \frac{1}{\epsilon^{-1}\ell} \sum_{|y-x| \leq \frac{1}{2}\epsilon^{-1}\ell} \eta(y,t) \tag{3.17}$$

We shall now make some heuristic considerations in order to estimate the difference in (3.16). Let us first assume that the initial measure is not μ^ϵ but a true Gibbs measure with density ρ, same ρ as in (3.16): surprisingly such a case will cover also the case we are interested in. Since $\nu_{z(\rho)}$ is invariant, we can set $t = 0$ in (3.16). Since $\nu_{z(\rho)}$ is a product measure we have the sum of $\approx \epsilon^{-1}$ independent (bounded) variables, so that the probability that (3.16) is in absolute value larger than any given $\delta > 0$ is bounded by $\exp(-a\epsilon^{-1})$, $a > 0$, times some constant. One might naively think that the space-time average

$$\frac{1}{\epsilon^{-2}T} \int_0^{\epsilon^{-2}T} dt \frac{1}{\epsilon^{-1}\ell} \sum_{|y-x| \leq \frac{1}{2}\epsilon^{-1}\ell} c_R(\eta(y,t)) - \mathbb{E}_{\nu_{z(\rho)}}(c_R(\eta(0,0)))$$

has much smaller probability (of being larger in absolute value than $\delta > 0$), because it involves roughly ϵ^{-3} variables. But this is only a hope, just think of a fluctuation for which the inital density is $\rho' \neq \rho$, : since the dynamics preserves the particles number, such a deviation will persist so that (3.16) remains different from 0. The probability of the above inital fluctuation is bounded by $\exp(-c\epsilon^{-1})$, so we have again an exponential estimate.

However if instead of fixing ρ in (3.16) we choose it as in (3.17), then we can avoid the above "pathology". However we are not yet in business. Let us in fact consider the independent case that we have already studied in the previous chapter. Consider the case when initially there is a density fluctuation such that the density to the right of x is ρ' and to the left ρ'', both different from ρ. Again such a fluctuation has an exponential bound $\exp\{-c\epsilon^{-1}\}$, for some $c > 0$. On the other hand we know that the hydrodynamical equation associated to the system is the heat equation, so that we know that the difference (3.16) will be significantly different from 0 at all times (if the average of ρ' and ρ'' is different from ρ).

If we consider also space averages over x, then only those x close to the discontinuity of the initial profile in the above example will contribute, so that if we choose ℓ small their contribution will be made small too. The analysis of this particular case indicates therefore a bound for the space time average of the absolute value of (3.16) of the form $\exp\{-c(\ell)\epsilon^{-1}\}$, with $c(\ell) \to \infty$ as $\ell \to 0$. If such a super-exponential bound is verified in our case, we would be through: in fact, as we shall see, the Radon-Nikodym derivative of μ^ϵ with respect to $\nu_{z(\rho)}$ is bounded by $\exp\{a\epsilon^{-1}\}$, for some finite a, so that the super-exponential estimate holds as well for μ^ϵ, we shall be back on this below.

Let us now be more detailed. We first need some new notation.

3.2.5 Definition.

For any space interval I in $\mathbb{Z}_\epsilon$ let

$$\mathcal{E}(c_R(\cdot), I, t) \equiv |I|^{-1} \sum_{x\in I} c_R(\eta(x,t)) \tag{3.18a}$$

$$\tilde{c}_R(\mathcal{E}(I,t)) \equiv \mathbb{E}_{\nu_z}(c_R(\eta(x,t))) \qquad z = \rho(\mathcal{E}(I,t)) \tag{3.19a}$$

where

$$\mathcal{E}(I,t) \equiv |I|^{-1} \sum_{x\in I} \eta(x,t) \tag{3.19b}$$

the symbol $\mathcal{E}$ stands for *empirical average.*

3.2.6 Theorem. *(The super-exponential estimate). For any positive t and δ and any allowed value of z [i.e. $z < \bar{z}$]*

$$\limsup_{\ell\to 0}\ \limsup_{\epsilon\to 0}\ \epsilon \log\Big(\mathbb{P}^\epsilon_{\nu_z}\Big[\frac{1}{t}\int_0^t ds W^\epsilon_s > \delta\Big]\Big) = -\infty \tag{3.20a}$$

where (see Definition 3.2.5)

$$W^\epsilon_s = \epsilon \sum_{x\in\mathbb{Z}_\epsilon} |\mathcal{E}(c_R(\cdot), \tau_x I, s) - \tilde{c}_R(\mathcal{E}(\tau_x I, s))| \tag{3.20b}$$

and $I = \big[1, [\epsilon^{-1}\ell]\big]$, $\tau_x I$ is the translate of I by x in $\mathbb{Z}_\epsilon$.

Remarks.

(i) The same super-exponential estimate holds for any bounded cylindrical function replacing c_R, as it will be evident from the proof of Theorem 3.2.2.

(ii) The super-exponential estimate does not hold at a single time, namely the probability that $\{W_0^\epsilon > \delta\}$ (we can take the time equal to 0 because ν_z is invariant) is not super-exponential. Let in fact $\gamma_{z(\rho)}$ be the the Gibbs measure when the intensities are $c'(\cdot)$, where $c'(k) = c(k)$ for all k except $k = n$, for some n. Then $\tilde{c}_R(\rho) \neq \mathbb{E}_{\gamma_{z(\rho)}}(c_R(\eta(x))$ and, from the law of large numbers for the measure γ_z, for δ small enough but positive

$$\lim_{\epsilon\to 0} \mathbb{P}^\epsilon_{\gamma_z}(W_0^\epsilon > \delta) = 1$$

On the other hand we can prove that

$$\left|\epsilon \log(\frac{d\gamma_z}{d\nu_z})\right| \leq c$$

for some constant c. Therefore it *only* costs an exponential price to be in $\{W_s^\epsilon > \delta\}$ [i.e. its probability is not less than $e^{-c\epsilon^{-1}}$], but super-exponentially more to stay there for longer times.

(iii) In §3.8 the reader may find an alternative way to identify the limiting density field, which uses entropy inequalities.

Before proving Theorem 3.2.6 we show how to conclude the proof of Theorem 3.2.1. For this purpose we need the following:

3.2.7 Corollary. *(to Theorem 3.2.6) The same estimate (3.20) holds with ν_z replaced by μ^ϵ, μ^ϵ being as in Theorem 3.2.1.*

Proof. Set for notational simplicity z in Theorem 3.2.6 equal to

$$z_{max} = \max_{r\in\mathcal{T}} z(\rho(r)) < \bar{z}$$

by assumption. By (3.2)

$$\frac{d\mu^\epsilon}{d\nu_z}(\eta) = N_\epsilon \prod_{x\in\mathbf{Z}_\epsilon} [\frac{z(\rho(\epsilon x))}{z}]^{\eta(x)}$$

where N_ϵ is the normalization factor. Hence the Radon-Nikodym derivative is bounded by N_ϵ and

$$N_\epsilon = (\int d\nu_z \prod_{x\in\mathbf{Z}_\epsilon} (\frac{z(\rho(\epsilon x))}{z})^{\eta(x)})^{-1} \leq [\nu_z(\{\eta(x) = 0 \ \forall x\})]^{-1} \leq e^{a\epsilon^{-1}}$$

with some finite a, because $z < \bar{z}$. Call $A_\epsilon = \{t^{-1}\int_0^t ds W_s^\epsilon > \delta\}$ then

$$\mathbb{P}_{\mu^\epsilon}(A_\epsilon) \leq \mathbb{P}_{\nu_z}(A_\epsilon) e^{a\epsilon^{-1}}$$

hence the Corollary. □

3.2.8 Proof of Theorem 3.2.1.

From (3.12), (3.15), the Corollary to Theorem 3.2.6 and changing the sum over x into an integral (with error vanishing as $\epsilon \to 0$ because of the a-priori bound), it follows that for any $\delta > 0$ and $T > 0$

$$\limsup_{R\to\infty} \limsup_{\ell\to 0} \limsup_{\epsilon\to 0} \mathbb{P}^\epsilon_{\mu^\epsilon}(\sup_{t\leq T} |X_t^\epsilon(\phi) - X_0^\epsilon(\phi) \\ - \int_0^t ds \int_{\mathbb{R}} dr \frac{1}{2}\phi''(r)\tilde{c}_R(X_s^\epsilon(\chi_{\ell,r})| > \delta) = 0 \qquad (3.21)$$

where $\chi_{\ell,0}$ is a suitable test function which approximates the characteristic function of the interval of size ℓ centered around the origin, and $\chi_{\ell,r}$ is its translated by r. Since we have already proven tightness we can take the limit as $\epsilon \to 0$ along a converging subsequence and replace $\mathbb{P}^\epsilon_{\mu^\epsilon}$ by its limit $\mathbb{P}$ in (3.21), hence dropping the limit over ϵ.

On the other hand we can show that $\mathbb{P}$ has support on distributions which are given by L_2 densities with respect to the Lebesgue measure. To see this we use the a-priori bounds to state that, [$\|\cdot\|_2$ being the $L_2(dr)$ norm]

$$\mathbb{E}_{\mu^\epsilon}\big(\sup_{\|\phi\|_2=1} X^\epsilon_t(\phi)^2\big) \leq c\epsilon \sum_x \phi(\epsilon x)^2$$

Let T be any positive number, let ρ_t be any trajectory in $C([0,\infty], \mathcal{S}')$ and denote by $\|\rho_t\|_2$ its L_2 norm, defined as

$$\|\rho_t\|_2 = \sup_{\phi\in\mathcal{S}} \frac{<\rho_t,\phi>}{\|\phi\|_2}$$

which might be infinite. We then have

$$\mathbb{E}\Big(\int_0^T dt \|\rho_t\|_2^2\Big) = \mathbb{E}\Big(\int_0^T dt \sup_{\|\phi\|_2=1} |<\rho_t,\phi>|^2\Big) = \mathbb{E}\Big(\int_0^T dt \sup_{\|\phi\|_2=1} |X_t(\phi)|^2\Big) \leq c \qquad (3.22)$$

where $\mathbb{E}$ denotes the expectation with respect to $\mathbb{P}$.

Hence $\mathbb{P}$ is supported by L_2 densities for almost all t. Calling $\rho(r,t)$ such densities we have from (3.21) that

$$\mathbb{E}\Big(\int_0^T dt \Big| \int dr \phi(r)[\rho(r,t)-\rho(r,0)] - \int_0^T ds \int dr' \phi''(r')\tilde{c}(\rho(r',s))\Big|\Big) = 0 \qquad (3.23)$$

where

$$\tilde{c}(\rho) = \mathbb{E}_{\nu_{z(\rho)}}(c(\eta(1))$$

This is (3.7) in weak form and if we had a uniqueness theorem for this problem we would have concluded the theorem (modulo the proof of Theorem 3.2.6). We have not been able to find this statement in the PDE literature, but on the other hand we did not try too hard. In fact it is possible to prove that $\mathbb{P}$ is supported by trajectories in H_2 and in this class uniqueness is known to hold. However such a-priori estimates require some extra work and we prefer to focus on Theorem 3.2.6. So hoping that uniqueness for the problem in (3.23) holds, we give Theorem 3.2.1 as proven. □

3.2.9 Proof of Theorem 3.2.6.

In the Paragraphs §3.3, 3.4, 3.5 we shall prove the following Proposition.

3.2.10 Proposition. *For any a,*

$$\limsup_{\ell\to 0}\ \limsup_{\epsilon\to 0} \epsilon \log \mathbb{E}_{\nu_z}\big(e^{2t^{-1}\int_0^t ds[a\epsilon^{-1}W^\epsilon_s - \lambda N_s]}\big) \leq 0 \qquad (3.24)$$

From Proposition 3.2.10 Theorem 3.2.6 easily follows. In fact, by using the exponential Chebyshev inequality we have for any positive a and for any λ such that $e^{2\lambda\eta(x)}$ is integrable with respect to the measure ν_z,

$$\begin{aligned}\mathbb{P}_{\nu_z}(t^{-1}\int_0^t dsW_s^\epsilon > \delta) &\le e^{-\epsilon^{-1}a\delta}\mathbb{E}_{\nu_z}(e^{t^{-1}\int_0^t dsa\epsilon^{-1}W_s^\epsilon}) \\ &\le e^{-\epsilon^{-1}a\delta}\left[\mathbb{E}_{\nu_z}(e^{2\lambda N})\right]^{\frac{1}{2}}\left[\mathbb{E}_{\nu_z}(e^{2t^{-1}\int_0^t ds[a\epsilon^{-1}W_s^\epsilon-\lambda N_s]})\right]^{1/2}\end{aligned} \tag{3.25}$$

where $N_s = N_0 \equiv N \equiv \sum_x \eta(x)$, denotes the total number of particles.

Since

$$\lim_{\epsilon\to 0}\epsilon \log \mathbb{E}_{\nu_z}(e^{2\lambda N})$$

is finite, Theorem 3.2.6 follows from the arbitrarity of a. □

§3.3 The reduction to an eigenvalue problem.

In this paragraph we us the Feynman-Kac formula to reduce the proof of (3.24) to an eigenvalue problem.

3.3.1 Proposition. *For any fixed a and t, we have*

$$\mathbb{E}_{\nu_z}(e^{2t^{-1}\int_0^t ds[a\epsilon^{-1}W_s^\epsilon-\lambda N_s]}) \le e^{\epsilon^{-1}M(t\epsilon^{-1}L+V_a)} \tag{3.26}$$

where $M(\cdot)$ is the maximal eigenvalue of $(\cdot)$ and V_a is the operator on $L_2(\mathbb{N}^{\mathbb{Z}_\epsilon}, \nu_z)$ which multiplies by the function (also denoted by) V_a:

$$V_a = 2(aW_0^\epsilon - \lambda\epsilon N) \tag{3.27}$$

Proof. We call ψ_s the integrand on the left of (3.26) and denote by $< \cdot, \cdot >$ the scalar product in $L_2(\mathbb{N}^{\mathbb{Z}_\epsilon}, \nu_z)$. Eq.(3.26) then follows from the Feynman-Kac identity

$$\mathbb{E}_{\nu_z}(e^{\int_0^t ds\psi_s}) = < 1, T(t)1 > \tag{3.28}$$

$$T(t) = e^{(L^\epsilon+\psi_0)t} \qquad\qquad L^\epsilon = \epsilon^{-2}L$$

Notice that $L^\epsilon + \psi_0$ is self-adjoint, so that we can use the spectral decomposition to define $T(t)$.

For the sake of completeness we sketch the proof of (3.28). Set $T^0(t) = e^{L^\epsilon t}$. Then

$$T(t) = T^0(t) + \int_0^t dsT(s)\psi_0 T^0(t-s)$$

Since the number of particles is conserved, it is enough to verify (3.28) on each subspace with a fixed number of particles. On these subsets all the operators are bounded hence the expansions below are trivially convergent. Iterating the above integration by parts formula we get for the right hand side of (3.28)

$$1 + \sum_{n=1}^{\infty}\int_0^t ds_1 \cdots \int_0^{s_{n-1}} ds_n \mathbb{E}_{\nu_z}(\psi_{s_n}\cdots\psi_{s_1})$$

which is the same series obtained when expanding the exponential on the left of (3.28). Therefore (3.26) is proven. □

3.3.2 A variational principle.

From (3.24) and (3.26) it follows that Theorem 3.2.6 is a consequence of the following Proposition.

3.3.3 Proposition. *For any fixed t and a,*

$$\limsup_{\ell\to 0}\ \limsup_{\epsilon\to 0} M(t\epsilon^{-1}L+V_a)\leq 0 \tag{3.29}$$

Before proving Proposition 3.3.3, we make a few remarks. First of all we notice that

$$M(t\epsilon^{-1}L+V_a)=\sup_{<\phi,\phi>=1} <\phi,(t\epsilon^{-1}L+V_a)\phi> \tag{3.30}$$

We denote by $D(\phi)$ the *Dirichlet form* [to simplify notation we write ν for ν_z in the sequel]

$$D(\phi)=-\frac{2}{z}<\phi,L\phi>=\int d\nu(\eta)\sum_{x\in\mathbf{Z}_\epsilon}\left[\phi(\eta+\delta_x)-\phi(\eta+\delta_{x+1}\right]^2 \tag{3.31}$$

(3.31) is obtained by using (3.1), collecting the terms relative to the jumps from x to $x+1$ and viceversa and then using the explicit expression for ν as given in (3.2).

3.3.4 Strategy of the proof of Proposition 3.3.3.

If $D(\phi)$ is larger than $C\epsilon$, $C>4(tz)^{-1}a\|W_0^\epsilon\|_\infty$, see (3.27), then $\epsilon^{-1}D(\phi)$ is the leading term in (3.30), hence its right hand side is negative for such ϕ's. On the other hand when $D(\phi)=0$, $<\phi,V_a\phi>$ is the expectation of V_a with respect to the measure that has density ϕ^2 (with respect to ν). Since $D(\phi)=0$ implies that ϕ depends only on the total number of particles, then, by the local central limit theorem, $W_0^\epsilon\to 0$ in probability with respect to this new measure as we shall see. The intermediate cases are where the difficulty is, since we do not know how to characterize those ϕ such that $D(\phi)\leq C\epsilon$. This difficulty is circumvented by using a clever trick introduced in [71] which reduces the problem to a finite state space, the *reduction to finite blocks.*

For what said so far we can restrict ϕ to $D(\phi)\leq C\epsilon$. Observe that

$$D(\phi)\geq D(|\phi|) \tag{3.32}$$

and that the term with V_a in (3.30) depend on ϕ via $|\phi|$. Then, denoting hereafter by ρ a non negative function such that $\int d\nu\rho=1$ and defining

$$\bar{M}=\sup_{D(\sqrt{\rho})\leq C\epsilon}\int d\nu V_a\rho \tag{3.33}$$

by (3.30) and (3.32) it suffices to show that $\bar{M}\to 0$ when $\epsilon\to 0$ and then $\ell\to 0$.

3.3.5 Convexity of the Dirichlet form.

Let a_i b_i, $i=1,\cdots n$, be non negative reals, then

$$\left[(\sum_{i=1}^n a_i)^{\frac{1}{2}}-(\sum_{i=1}^n b_i)^{\frac{1}{2}}\right]^2\leq\sum_{i=1}^n\left[\sqrt{a_i}-\sqrt{b_i}\right]^2 \tag{3.34}$$

which is proven by expliciting the square on the left hand side and using Cauchy-Schwartz for the products.

The inequality (3.34) proves that $D(\sqrt{\rho})$ is a convex function of ρ, i.e. if

$$\rho = \sum_{i=1}^{n} \lambda_i \rho_i$$

$\lambda_i \geq 0$, then

$$D(\sqrt{\rho}) \leq \sum_{i=1}^{n} \lambda_i D(\sqrt{\rho_i}) \tag{3.35}$$

3.3.6 Reduction to translationally invariant test functions.

A first consequence of (3.35) is the (3.36) below. Let τ_x denote the translation by x, then

$$D(\sqrt{\epsilon \sum_{x \in \mathbf{Z}_\epsilon} \tau_x \rho}) \leq \epsilon \sum_{x \in \mathbf{Z}_\epsilon} D(\sqrt{\tau_x \rho}) = D(\sqrt{\rho}) \tag{3.36}$$

hence we can restrict the sup in (3.33) to translationally invariant ρ's, i.e.

$$\bar{M} = \sup_{\tau_1 \rho = \rho, D(\sqrt{\rho}) \leq C\epsilon} \int d\nu \rho V_a \tag{3.37}$$

because V_a is translationally invariant.

Since the *sup* in (3.37) is over translationally invariant densities, we can replace V_a in (3.37), cf. (3.27), by

$$aW - \lambda \mathcal{E}(I)$$

where

$$W = |\mathcal{E}(c_R, I) - \tilde{c}_R(\mathcal{E}(I))| \tag{3.38}$$

[we omit writing the time in the argument of $\mathcal{E}$, since it is identically 0 hereafter].

We further notice that the the integral in (3.37) is non positive if

$$\int d\nu \, \rho \, \mathcal{E}(I) \geq C'$$

where $C' = \lambda^{-1} C$, cf. 3.3.4.

We denote by sup^+ the *sup* over all ρ such that (1) ρ is translationally invariant, (2) $D(\sqrt{\rho}) \leq C\epsilon$ and (3) $\int d\nu \rho(\eta)\eta(1) \leq C'$. For what said it is enough to show that

$$M' = sup^+ \int d\nu \rho W \tag{3.39}$$

vanishes in the limit $\epsilon \to 0$ and $L \to 0$.

3.3.7 Reduction to finite blocks.

We write

$$I = \bigcup_{i=1}^{N} \mathcal{B}_i \tag{3.40}$$

where the $\mathcal{B}_i$ are disjoint intervals of length k, except possibly for $\mathcal{B}_N$ whose length is $\leq k$. However for notational simplicity we assume that also $\mathcal{B}_N$ has length k. We have

$$\begin{aligned} W &\leq \left|\frac{1}{N}\sum_{i=1}^{N} \mathcal{E}(c_R, \mathcal{B}_i) - \tilde{c}_R(\mathcal{E}(I))\right| \\ &\leq \frac{1}{N}\sum_{i=1}^{N}\{|\mathcal{E}(c_R, \mathcal{B}_i) - \tilde{c}_R(\mathcal{E}(\mathcal{B}_i))| + |\tilde{c}_R(\mathcal{E}(\mathcal{B}_i)) - \tilde{c}_R(\mathcal{E}(I))|\} \end{aligned} \tag{3.41}$$

and Taylor-expanding to first order

$$\begin{aligned} |\tilde{c}_R(\mathcal{E}(\mathcal{B}_i)) - \tilde{c}_R(\mathcal{E}(I))| &\leq c|\mathcal{E}(\mathcal{B}_i) - \mathcal{E}(I)| \\ &\leq c\frac{1}{N}\sum_{j=1}^{N} |\mathcal{E}(\mathcal{B}_i) - \mathcal{E}(\mathcal{B}_j)| \end{aligned} \tag{3.42}$$

where

$$c = \left\|\frac{d\tilde{c}_R}{d\rho}\right\|_\infty$$

From (3.41) and (3.42), using the translationally invariance of ρ we are reduced to prove the following two inequalities:

$$\liminf_{k\to\infty}\ \limsup_{\epsilon\to 0}\ sup^+ \int d\nu\rho|\mathcal{E}(c_R, \mathcal{B}_0) - \tilde{c}_R(\mathcal{E}(\mathcal{B}_o))| = 0 \tag{3.43}$$

$$\liminf_{k\to\infty}\ \limsup_{\ell\to 0}\ \limsup_{\epsilon\to 0}\ sup^+ \sup_{ki\leq\epsilon^{-1}\ell} \int d\nu\rho|\mathcal{E}(\mathcal{B}_0) - \mathcal{E}(\mathcal{B}_i)| = 0 \tag{3.44}$$

where $\mathcal{B}_i = \{ki+1, \cdots, k(i+1)\}$.

§3.4 The 1-block estimate.

We start by (3.43). Notice that the integrand in (3.43) depends only on $\eta(x)$ with $x \in \mathcal{B}_0$. We then set

$$\rho_{\mathcal{B}_0}(\eta_{\mathcal{B}_0}) = \int d\nu(\eta_{\mathcal{B}_0^c})\rho(\eta_{\mathcal{B}_0}, \eta_{\mathcal{B}_0^c}) \tag{3.45}$$

where $\mathcal{B}_0^c$ is the complement of $\mathcal{B}_0$, $\eta_F = \{\eta(x), x \in F\}$ and $d\nu(\eta_{\mathcal{B}_0^c})$ is the relativization of $d\nu$ to $\{\eta_{\mathcal{B}_0^c}\}$. Setting

$$D_{\mathcal{B}_0}(\sqrt{\rho_{\mathcal{B}_0}}) = \int d\nu(\eta_{\mathcal{B}_0})\sum_{x=1}^{k-1}[(\rho_{\mathcal{B}_0}(\eta_{\mathcal{B}_0} + \delta_x))^{\frac{1}{2}} - (\rho_{\mathcal{B}_0}(\eta_{\mathcal{B}_0} + \delta_{x+1}))^{\frac{1}{2}}]^2 \tag{3.46}$$

We consider $\rho(\eta_{\mathcal{B}_0}, \eta_{\mathcal{B}_0^c})$ as a function of $\eta_{\mathcal{B}_0}$ for each fixed value of the *parameter* $\eta_{\mathcal{B}_0^c}$, then $D_{\mathcal{B}_0}(\sqrt{\rho})$ is a function of $\eta_{\mathcal{B}_0^c}$, and, from (3.35), we get

$$D_{\mathcal{B}_0}(\sqrt{\rho_{\mathcal{B}_0}}) \leq \int d\nu(\eta_{\mathcal{B}_0^c})D_{\mathcal{B}_0}(\sqrt{\rho})$$

and expliciting $D_{\mathcal{B}_0}$ as in (3.46)

$$\leq \sum_{x=1}^{k-1} \int d\nu(\eta)\left[\sqrt{\rho(\eta+\delta_x)} - \sqrt{\rho(\eta+\delta_{x+1})}\right]^2$$

Since ρ is translationally invariant

$$D(\sqrt{\rho}) = \epsilon^{-1} \int d\nu(\eta)\left[\sqrt{\rho(\eta+\delta_x)} - \sqrt{\rho(\eta+\delta_{x+1})}\right]^2 \tag{3.47}$$

hence

$$D_{\mathcal{B}_0}(\sqrt{\rho_{\mathcal{B}_0}}) \leq k\epsilon D(\sqrt{\rho}) \leq kC\epsilon^2 \tag{3.48}$$

The left hand side of (3.43) is therefore bounded by

$$\liminf_{k\to\infty} \sup_{D(\sqrt{\rho})=0} \sup_{\int d\nu\rho\mathcal{E}(\mathcal{B}_0)\leq C'} \int d\nu\ \rho\ |\mathcal{E}(c_R,\mathcal{B}_0) - \tilde{c}_R(\mathcal{E}(\mathcal{B}_0))| \tag{3.49}$$

where C' is a suitable constant. We sketch below the steps necessary to obtain (3.49). We fix k below and consequently $\mathcal{B}_0$. All densities, in the 4 steps below, are only functions of $\eta_{\mathcal{B}_0}$ but for notational simplicity we shall omit writing the subscript $\mathcal{B}_0$.

(1). Fix k and ϵ in (3.43). Then the *sup* is achieved at some $\rho \equiv \rho^{(\epsilon)}$, since it is the *sup* of a continuous function on a compact set ($\int d\nu\rho\mathcal{E}(\mathcal{B}_0) \leq C'$).

(2). For fixed k the density $\rho^{(\epsilon)}$ in (1) is a function of ϵ. The set of measures $d\nu\rho^{(\epsilon)}$ obtained by varying the parameter ϵ is compact, therefore the *limsup* is obtained by taking the limit over a sequence $\epsilon_n \to 0$ such that the corresponding sequence of measures $d\nu\rho^{(n)}$ converges. It is also clear that the limit has a density ρ with respect to ν, and that $\rho^{(n)}$ converges pointwise to ρ. In fact the inverse images of $\mathcal{E}(\mathcal{B}_0)$ have finitely many elements all having positive probability with respect to the measure ν.

(3). Since the integrand in (3.43) is a continuous function of $d\nu\rho^{(n)}$, cf. (2) above, the *limsup* is $\int d\nu\rho|\mathcal{E}(c_R,\mathcal{B}_0) - \tilde{c}_R(\mathcal{E}(\mathcal{B}_0))|$

(4). We also have

$$D_{\mathcal{B}_0}(\sqrt{\rho}) \leq \lim_{n\to\infty} D_{\mathcal{B}_0}(\sqrt{\rho^{(n)}}) = 0$$

3.4.1 Conclusion of the proof of the 1-block estimate.

Fix k in (3.49), denote by u the possible values of $\mathcal{E}(\mathcal{B}_0)$ and let

$$\Gamma(u) = \int d\nu_{z(\rho)} 1(\mathcal{E}(\mathcal{B}_0) = u)$$

Finally, let $\hat{\nu}_u$ be the canonical Gibbs measure with support on the configurations in $\mathcal{B}_0$ such that $\mathcal{E}(\mathcal{B}_0) = u$.

Then, using Cauchy-Schwartz, we have to control

$$|\sum_u \Gamma(u) \int d\hat{\nu}_u \left[c_R(\eta(1)) - \tilde{c}_R(u)\right]\left[c_R(\eta(2)) - \tilde{c}_R(u)\right]|$$

since the diagonal terms when expanding $\left[\mathcal{E}(c_R,\mathcal{B}_0) - \tilde{c}_R(\mathcal{E}(\mathcal{B}_0))\right]^2$ give vanishing contribution when $k \to \infty$, while the non diagonal ones are all equal to the term written above. For each fixed u, $\hat{\nu}_u \to \nu_{z(u)}$ as $k \to \infty$ (weakly), such a property, called the equivalence of the ensembles, is proved using the local central limit theorem. By the tightness of $\Gamma(u)$ we then prove that the limit of the above expression vanishes for $k \to \infty$.

§3.5 The two blocks estimate.

The condition $D(\sqrt{\rho}) \le C\epsilon$ implies, like in (3.48), that

$$D_{\mathcal{B}_0,\mathcal{B}_i}(\sqrt{\rho_{0,i}}) \le C2k\epsilon^2$$

$\rho_{0,i}$ being as in (3.45) with $\mathcal{B}_0$ replaced by $\mathcal{B}_0 \cup \mathcal{B}_i$, while the sum in (3.46) is replaced by a sum over $1 \le x \le k-1$ and $ki+1 \le x \le k(i+1)-1$. This condition implies that the limiting $\rho_{0,i}$ will depend on $\mathcal{E}(\mathcal{B}_0)$ and $\mathcal{E}(\mathcal{B}_i)$ but does not say anything about $|\mathcal{E}(\mathcal{B}_0) - \mathcal{E}(\mathcal{B}_i)|$. To establish a relation on such a difference we notice that

$$\begin{aligned} D_{1,ki+1} &\equiv \int d\nu \big[\sqrt{\rho(\eta+\delta_{ki+1})} - \sqrt{\rho(\eta+\delta_1)}\big]^2 \\ &\le ki \sum_{x=1}^{ki} \int d\nu \big[\sqrt{\rho(\eta+\delta_{x+1})} - \sqrt{\rho(\eta+\delta_x)}\big]^2 \\ &\le (ki)^2 C\epsilon^2 \le (\epsilon^{-1}\ell)^2 C\epsilon^2 \end{aligned} \tag{3.50}$$

Therefore the sup^+ in (3.44) can be replaced by a sup over all $\rho_{0,i}$ such that (1) $\int d\nu\rho_{0,i}[\mathcal{E}(\mathcal{B}_0) + \mathcal{E}(\mathcal{B}_i)] \le C'$, (2)$D_{\mathcal{B}_0,\mathcal{B}_i}(\sqrt{\rho_{0,i}}) \le C2k\epsilon^2$ and (3) $D_{1,ki+1}(\sqrt{\rho_{0,i}}) \le C^2\ell^2$.

For each fixed k, ϵ and ℓ the sup of the integral in (3.44) over densities $\rho_{0,i}$ which satisfy the three conditions stated above, is the same for all i. We can therefore fix $i = 1$ and consider only this case. Notice that conditions (2) and (3) above for $i = 1$, imply that the Dirichlet form $D_{\mathcal{B}_0\cup\mathcal{B}_1}(\sqrt{\rho_{0,1}})$ vanishes when $\epsilon \to 0$ and $\ell \to 0$. The above being the Dirichlet form relative to the block $1, \cdots, 2k$, namely it contains also the term

$$\big[(\rho_{0,1}(\eta+\delta_k))^{\frac{1}{2}} - (\rho_{0,1}(\eta+\delta_{k+1}))^{\frac{1}{2}}\big]^2$$

which was not present in $D_{\mathcal{B}_0,\mathcal{B}_1}(\sqrt{\rho_{0,i}})$.

We can now proceed just like in the proof of (3.43) and in this way we complete the proof of Proposition 3.3.3. □

§3.6 An entropy inequality.

Recalling (3.22) and the subsequent discussion leading to the statement on the super exponential estimate, we see that the main point, for identifying the limiting density field, is to prove that

$$\int_0^t ds\, \epsilon \sum_{x\in\mathbf{Z}_\epsilon} \mathbf{E}^\epsilon_{\mu^\epsilon} |\mathcal{E}(c_R(\cdot), I, s) - \tilde{c}_R(I,s)| \tag{3.51}$$

vanishes when $\epsilon \to 0$ and then $\ell \to 0$. Call r_s the Radon-Nikodym derivative of the law of the process at time $\epsilon^{-2}s$ with respect to the invariant measure ν^ϵ_z, the same appearing in the super-exponential estimate. Set

$$\rho = t^{-1} \int_0^t ds r_s \tag{3.52}$$

We can then rewrite (3.51) as

$$t \int d\nu\rho |\mathcal{E}(c_R(\cdot), I) - \tilde{c}_R(I)| \tag{3.53}$$

We shall now prove that

$$D(\sqrt{\rho}) \le c\epsilon$$

and this will reduce the problem to that solved in the second part of the proof of Theorem 3.2.

We introduce the *entropy*

$$H(t) = \int d\nu_z (r_t \log r_t - r_t + 1) \tag{3.54}$$

Recalling that the densities evolve according to the adjoints of the generators, but that in our case the generator is self-adjoint, we get

$$\frac{dr_t}{dt} = \epsilon^{-2} L^\epsilon r_t$$

After a change of variables, as when deriving (3.31), we obtain

$$\begin{aligned}\frac{dH(t)}{dt} &= -\epsilon^{-2} e(t) \\ &= -\epsilon^{-2} \int d\nu \sum_{x\in \mathbb{Z}_\epsilon} \left[r_t(\eta+\delta_x) - r_t(\eta+\delta_{x+1})\right] \log \left[\frac{r_t(\eta+\delta_x)}{r_t(\eta+\delta_{x+1})}\right]\end{aligned} \tag{3.55}$$

where the first equality defines $e(t)$. From (3.57) below, H does not increase and, being non negative by definition, cf.(3.54), we deduce that

$$\int_0^t ds e(s) = \epsilon^2[H(0) - H(t)] \le \epsilon^2 H(0) \le c\epsilon \tag{3.56}$$

[c being a suitable constant] since μ^ϵ has entropy bounded by $c\epsilon^{-1}$. We use the inequality

$$(a-b)\log\frac{a}{b} \ge 2(\sqrt{a} - \sqrt{b})^2 \tag{3.57}$$

[To prove it, take $a \ge b > 0$, call $x = a/b$, then the inequality becomes:

$$(x-1)\log x \ge 2(\sqrt{x} - 1)^2 \iff (\sqrt{x}+1)\log\sqrt{x} - (\sqrt{x}-1) \ge 0$$

which holds because the left hand side in the last inequality is an increasing function of $\sqrt{x}$, for $x \ge 1$, which vanishes at $x = 1$]. Then, from (3.57) and (3.56) we get

$$\int_0^t ds\, D(\sqrt{r_s}) \le \frac{1}{2}\int_0^t ds\, e(r_s) \le \frac{1}{2}c\epsilon$$

$$t^{-1}\int_0^t ds\, D(\sqrt{r_s}) \le \frac{1}{2}ct^{-1}\epsilon, \quad \text{and, by (3.35) and (3.52),} \quad D(\sqrt{\rho}) \le \frac{1}{2}ct^{-1}\epsilon$$

We have therefore reduced the problem to that studied when proving Theorem 3.2. In fact we know that

$$\int d\nu\rho\epsilon \sum_{x\in\mathbb{Z}_\epsilon} \eta(x) \le \text{ const.}$$

because this is so for the initial measure μ^ϵ, and the total particles number is conserved. In this way we are reduced to the sup^+, as in the proof of the super-exponential estimate, cf. (3.39).

§3.7 Bibliographical notes.

In this chapter we have simply translated into the language of the zero range process what done in [71] for the Ginzburg-Landau models, see Chapter VIII, and in [79] about the super-exponential estimates on the simple exclusion process. The analysis of the zero range processes with entropy methods can also be found in [132]. The first results on the hydrodynamics of the Ginzburg-Landau models have been obtained by Fritz, [66]. Rost and Fritz have also derived hydrodynamics for zero range processes, but we are not aware of any published paper. The special case when $c(k) = 1$ for $k \geq 1$, which is isomorphic to the simple symmetric exclusion process with a tagged particle has been solved in [62a] using coupling techniques; in [62b] the convergence of the fluctuations fileds is proven.

The entropy methods for deriving hydrodynamics have been applied to several interesting cases, like (1) interacting brownians, [130], (2) interacting Ornstein-Uhlenbeck particles, [105], (3) analysis of the non equilibrium steady states of some lattice gases in the presence of a temperature gradient, (there are two preprints on this subject by Eyink, Lebowitz and Spohn, (1990)), (4) interacting Ginzburg-Landau processes, [117], to which we shall refer also in Chapter VIII. See also [122] for a discussion in the context of deterministic evolutions. More recently there have been results also for "non gradient systems", [133] , [113], and by Varadhan, private communication. Other results have been obtained by Yau with a variant of the method, which uses inequalities on the relative entropy, [134]. In this way Olla, Varadhan and Yau, in a paper in preparation, have proven hydrodynamics also for hyperbolic systems, deriving Euler-like equations.

CHAPTER IV

PARTICLE MODELS FOR REACTION–DIFFUSION EQUATIONS

In this Chapter we consider a system of symmetric random walks which undergo a birth-death process. We prove that the behavior of the system in the macroscopic limit is described by a reaction-diffusion equation. This result is proven by studying the BBGKY hierarchy of equations for the correlation functions.

The reaction-diffusion equations that we consider have the form

$$\frac{\partial \rho}{\partial t} = \frac{1}{2}\frac{\partial^2 \rho}{\partial r^2} - V'(\rho) \tag{4.1a}$$

$$\rho(r,0) = \rho_0(r) \tag{4.1b}$$

where $r \in \mathbb{R}$, $t \geq 0$,

$$-V'(\rho) = F_+(\rho) - F_-(\rho) \tag{4.2}$$

$F_\pm$ are polynomials with non negative coefficients whose degrees, $\deg(F_\pm)$, are such that

$$deg(F_+) < deg(F_-), \quad F_\pm(0) = 0 \tag{4.3}$$

$\rho_0(r)$ is a non negative C^2 function uniformly bounded with its derivatives. In this setup (4.1) has a unique non negative C^2 solution, uniformly bounded with its derivatives.

Phenomenologically (4.1) is used to describe a system of molecules which diffuse independently and undergo chemical reactions with the particles of a reservoir. Each reaction requires that a given number of molecules and of particles are suitably close to each other and, as a result of the reaction, molecules may disappear or new molecules appear. We simplify the model by assuming that there are so many particles in the reservoir that they can always provide the necessary elements for the reactions to take place. We then obtain an equation like (4.1) where $\rho(r,t)$ denotes the *local, macroscopic density* of molecules at r, t. The first and second terms on the right hand side of (4.1) describe respectively the diffusion of the molecules and the effect of the chemical reactions with the reservoir. Each monomial in $F_\pm$ is related to a different chemical reaction, its degree indicates the number of molecules needed for the reaction to take place and its numerical coefficient the intensity of the reaction; the sign attached to $F_\pm$ in (4.2) corresponds to the creation or disappearence of a molecule.

A large variety of systems are described by (4.1), for instance by interpreting the reactive term as due to true births and deaths one has a model for population dynamics. (4.1) is also used as a model for propagating waves, as when

$$-V'(\rho) = \rho - \rho^2 \tag{4.4}$$

The equation, in this case, is the KPP (Kolmogorov, Petrovskii, Piscounov) equation, it has traveling wave solutions, namely solutions which depend on r and t via $r - vt$, v being a constant which represents the velocity of the wave. The KPP equation has also been derived from other particle systems, see Chapter VI and [18], to which we refer for a discussion on the physical motivations and on the many open questions concerning the microscopic structure of the traveling waves.

An important remark, when constructing models for (4.1), is that (4.1) does not have a space-time scaling symmetry: if $\rho(r,t)$ solves a purely diffusive equation, then

$$\rho_\epsilon(r,t) = \rho(\epsilon^{-1}r, \epsilon^{-2}t) \tag{4.5}$$

is still a solution of the same equation, but if ρ solves (4.1a), then ρ_ϵ satisfies (4.1a) with a potential $V_\epsilon = \epsilon^{-2}V$. This means that the same ϵ used to scale time and space also relates the ratio between the intensities of the diffusive and the reaction processes, this will be exactly reflected in the generator of the particle model for (4.1).

§4.1 The particle model.

The configuration space of the system is $\mathbb{N}^{\mathbb{Z}}$, its elements are the particle configurations denoted by $\eta \in \mathbb{N}^{\mathbb{Z}}$. ($\delta_x \in \mathbb{N}^{\mathbb{Z}}$ below is the configuration with just one particle at x).

For each $0 < \epsilon \leq 1$, we introduce a Markov process whose generator L^ϵ is

$$L^\epsilon f(\eta) = \epsilon^{-2} L_0 f(\eta) + L_c f(\eta) \tag{4.6}$$

(f above is a cylinder function) where L_0 is the generator of the symmetric independent process defined in (2.1), namely

$$L_0 f(\eta) = \frac{1}{2}\sum_x \eta(x)[f(\eta+\delta_{x+1}-\delta_x) + f(\eta+\delta_{x-1}-\delta_x) - 2f(\eta)] \tag{4.7}$$

and L_c is the generator of the birth-death process

$$L_c f(\eta) = \sum_x \Big(q_+(\eta(x))[f(\eta+\delta_x) - f(\eta)] + q_-(\eta(x))[f(\eta-\delta_x) - f(\eta)]\Big) \tag{4.8a}$$

where $q_\pm$ are polynomials satisfying the same condition (4.3) as $F_\pm$, namely they have positive coefficients and

$$deg(q_+) < deg(q_-), \quad q_\pm(0) = 0 \tag{4.8b}$$

We prove in the next paragraph existence and uniqueness for the process when it starts from configurations with finitely many particles and convergence of the expectations of cylinder functions when a cutoff on the initial number of particles is removed.

We shall then study the macroscopic limit, by scaling the space as ϵ^{-1}, the prefactor ϵ^{-2} in front of L_0 corresponds to scaling times like the square of the space, as required by the diffusive limit. By virtue of this scaling, the independent updatings occur, in the average, ϵ^{-2}-times more frequently than the birth-death events. It is therefore reasonable to expect that the Poisson measures (invariant for the independent process) play an important role also for the full process, (with generator L^ϵ), at least if ϵ is sufficiently small. With this in mind we choose the initial measure μ^ϵ on $\mathbb{N}^{\mathbb{Z}}$ as a product of Poisson measures in such a way that

$$\mathbb{E}_{\mu^\epsilon}(\eta(x)) = \rho_0(\epsilon x) \tag{4.9}$$

We shall see, in Theorem 4.3.1 below, that μ_t^ϵ, the distribution of the process at time t, is still, but only approximately, a product of Poisson distributions, whose averages, rescaled as in (4.9), in the limit when $\epsilon \to 0$ satisfy the reaction-diffusion equation (4.1). However, $F_\pm$ and $q_\pm$ are not equal, but only related, as expressed by (4.29).

To prove convergence in the macroscopic limit, we do not follow the methods of Chapter III. In fact, to apply the GPV technique, we need a-priori bounds on the moments of the occupation numbers that we do not know how to prove without controlling the hierarchical equation for the correlation functions. But once we have this, the result already follows, even in a stronger sense.

§4.2 Construction of the process.

For $a \in \mathbb{N}$, let $q_-^a(n) = q_-(n)$ and

$$q_+^a(n) = \begin{cases} q_+(n) & \text{if } n \leq a \\ 0 & \text{otherwise} \end{cases} \tag{4.10}$$

Let then $L^{\epsilon,a} = \epsilon^{-2}L_0 + L_c^a$, where L_c^a is defined by (4.8a) with $q_\pm^a$ replacing $q_\pm$. It is easy to see that $L^{\epsilon,a}$ is the generator of a jump Markov process on Ω, where

$$\Omega = \{\eta \in \mathbb{N}^{\mathbb{Z}} : \|\eta\| \equiv \sum_{x\in\mathbb{Z}} \eta(x) < \infty\}$$

4.2.1 Proposition. *Let $\eta \in \Omega$ and $\mathbb{P}_\eta^{\epsilon,a}$ be the law of the process with generator $L^{\epsilon,a}$ starting from η. Then $\mathbb{P}_\eta^{\epsilon,a}$ converges weakly when $a \to \infty$ to $\mathbb{P}_\eta^\epsilon$, which is the law of a jump process in Ω with generator L^ϵ.*

*Proof**. Clearly the proof of the Proposition requires a control on the total number of particles at any time t. Therefore we need to estimate the distribution of

$$\sup_{0\leq t\leq T} |\eta_t|$$

for any fixed $T > 0$, uniformly in the cutoff a. We use the martingale inequalities (2.38) and (2.39), so that

$$\mathbb{E}_\eta^{\epsilon,a}\left(\sup_{0\leq t\leq T} |\eta_t|\right) \leq 8|\eta|^2 + 8T \sup_{0\leq t\leq T} \mathbb{E}_\eta^{\epsilon,a}\left(\gamma_1^{\epsilon,a}(t)^2 + |\gamma_2^{\epsilon,a}(t)|\right) \tag{4.11a}$$

where by (2.21) and (2.22)

$$\gamma_1^{\epsilon,a} = L_c^a|\eta|, \qquad \gamma_2^{\epsilon,a} = L_c^a(|\eta|^2) - 2|\eta|L_c^a|\eta| \tag{4.11b}$$

because the process with generator L_0 does not change the total number of particles.

We have

$$L_c|\eta| = \sum_{x\in\mathbf{Z}} \Big(q_+(\eta(x)) - q_-(\eta(x)) \Big)$$

(analogous expression holds for L_c^a). The expression for $L_c|\eta|^2$ can also be written explicitly, so that what we need to bound are polynomials in the $\eta(x)$'s. It is convenient to rewrite them in terms of the Poisson polynomials, thus we introduce the correlation functions

$$u^{\epsilon,a}(\xi, t|\eta) = \mathbf{E}_\eta^{\epsilon,a}\big(D(\xi, \eta_t)\big) \tag{4.12}$$

where $\xi \in \Omega$ and $D(\xi, \eta)$ is the Poisson polynomial defined in (2.53) and (2.55).

Notice that we can express $|\eta|$ in term of D functions:

$$|\eta| = \sum_{x\in\mathbf{Z}} D(\delta_x, \eta)$$

Same holds for γ_i^ϵ, $i = 1, 2$, defined as in (4.11) but with L_c, and not L_c^a. The corresponding expectations in (4.11a) are finite sums of expression of the form

$$A_t(h,k) = \sum_{x\neq y} D(\xi^{x,y}, \eta_t), \qquad \xi^{x,y} = h\delta_x + k\delta_y \tag{4.13a}$$

and

$$B_t(h) = \sum_{x} D(\xi^x, \eta_t), \qquad \xi^x = h\delta_x \tag{4.13b}$$

for h and k varying in some finite set. If we consider L_c^a instead of L^c, then the outcomes are the same polynomials in η times a characteristic function which, according to (4.10), distinguishes the cases $\eta(x) \gtrless a$: by bounding it by 1 we obtain again the expressions in (4.13).

We shall derive in the sequel an equation for the correlation functions, the so called BBGKY hierarchy, and from its analysis we shall prove that given $\eta \in \Omega$, ϵ, T, h and k there is a constant c such that

$$\sup_{t\leq T} \mathbf{E}_\eta^{\epsilon,a}\big(A_t(h,k) + B_t(h)\big) \leq c, \qquad \text{for all } a \tag{4.14}$$

which implies that, for a suitable constant c,

$$\mathbf{E}_\eta^{\epsilon,a}\Big(\sup_{0\leq t\leq T} |\eta_t|\Big) \leq c, \qquad \text{for all } a$$

We next construct a process in $[0,T]$ with values in Ω: given any a we define a measure in $\{|\eta_t| \leq a, \forall t \leq T\}$ which coincides with $\mathbf{P}_\eta^{\epsilon,a}$ restricted to the same set. Then, by the Caratheodory theorem, see for instance Ch. III.5.4 of [54], this defines a probability on Ω, which is easily verified to define a jump process with generator (4.6). Therefore the Proposition is proved, modulo (4.14). □

The proof of (4.14) is based on the analysis of the following equation for the correlation functions:

4.2.2 Lemma. *For any $\eta \in \Omega$, $\xi \in \Omega$ and $t > 0$,*

$$u^{\epsilon,a}(\xi,t|\eta) = \sum_{\xi'} P_t^\epsilon(\xi \to \xi')u^{\epsilon,a}(\xi',0|\eta) + \int_0^t ds \sum_{\xi'} P_{t-s}^\epsilon(\xi \to \xi')\mathbf{E}_\eta^{\epsilon,a}\big(L_c^a D(\xi',\eta_s)\big) \tag{4.15}$$

where P_t^ϵ is the transition probability for the independent process with generator $\epsilon^{-2}L_0$, and L_c^a in (4.15) acts on the η-variables.

*Proof**. As in Proposition 2.9.4, we call ξ-process the process with generator L_0 starting from ξ as in (4.15), and η-process the process with generator $L^{\epsilon,a}$ starting from η. We also denote by $\mathcal{E}$ the expectation with respect to the product of the ξ and the η-process. Then

$$\frac{d}{ds}\mathcal{E}\big(D(\xi_{t-s},\eta_s)\big) = \mathcal{E}\big(L_c^a D(\xi_{t-s},\eta_s)\big), \quad (L_c^a \text{ acts on } \eta) \tag{4.16}$$

because, by (2.57), L_0 does not contribute to the derivative. Since

$$u^{\epsilon,a}(\xi,t|\eta) = \mathcal{E}\big(D(\xi_0,\eta_t)\big)$$

by integrating (4.16) over s from 0 to t we get (4.15). □

The analysis of (4.15) stands on some simple inequalities for the functions $L_c D$.

4.2.3 Lemma. *Given $\xi \in \Omega$, denote by $\xi_{(x)} = \xi(x)\delta_x$ the element in Ω which has only $\xi(x)$ particles in x, and none elsewhere. Then*

$$L_c D(\xi,\eta) = \sum_x D(\xi - \xi_{(x)},\eta)L_c D(\xi_{(x)},\eta) \tag{4.17}$$

and same equality holds with L_c replaced by L_c^a.

Furthermore, there are positive, non decreasing, coefficients $d(n)$, $n \geq 1$, such that

$$L_c D(\xi_{(x)},\eta) \leq d(\xi(x)) \begin{cases} 1 & \text{if } \xi(x) > 0 \\ D(\xi_{(x)},\eta) & \text{if } \xi(x) = 1 \\ D(\xi_{(x)} - 1,\eta) & \text{if } \xi(x) > 1 \end{cases} \tag{4.18}$$

Same inequalities hold with L_c replaced by L_c^a.

Proof. (4.17) follows directly from the definition of L_c (same for L_c^a). To prove (4.18), we call $k = \xi(x)$ and $n = \eta(x)$, then the contribution of the births to $L_c D$ is

$$q_+(n)\big[(n+1)\cdots(n-k+2) \;-\; n\cdots(n-k+1)\big]$$

that of the deaths

$$q_-(n)\big[(n-1)\cdots(n-k) \;-\; n\cdots(n-k+1)\big]$$

so that

$$L_c D(\xi_{(x)},\eta) = kq_+(n)\, n\cdots(n-k+2) \;-\; kq_-(n)\,(n-1)\cdots(n-k+1) \tag{4.19}$$

The right hand side, by (4.8b), goes to $-\infty$ when $n\to\infty$, hence the first case in (4.18). If $k=1$, L_cD equals $q_+(n)-q_-(n)\le cn$, for some c: in fact the left hand side in this last inequality goes to $-\infty$ as $n\to\infty$ and it equals 0 for $n=0$. If $k>1$, we write the last term in (4.19) as

$$\begin{cases} 2\big[n\,q_+(n)-q_-(n)(n-1)\big] & \text{if } k=2 \\ k\,(n-1)\cdots(n-k+2)\big[n\,q_+(n)-q_-(n)(n-k+1)\big] & \text{if } k>2 \end{cases}$$

and the square bracket term in both cases is again bounded by some constant (depending on k) times n. This proves the third case in (4.18) and completes the proof of the Lemma, because the analysis of L_c^a is similar. □

We shall use Lemma 4.2.3 to prove the following Proposition.

4.2.4 Proposition. *Let $d(n)$ be as in Lemma 4.2.3. Let $\underline{x}\in\mathbb{Z}^n$ and let $U(\underline{x})=\xi$, see (2.2b) for notation, then*

$$u^{\epsilon,a}(\xi,t|\eta)\le e^{2d(n)t}\big[g^\epsilon(x_1,t)g^\epsilon(x_2,t)+n\tilde g^\epsilon(x_1,x_2,t)\big]\prod_{i=3}^{n}\Big(g^\epsilon(x_i,t)+d(n)\Big) \tag{4.20a}$$

where

$$g^\epsilon(x,t)=\sum_y P_t^\epsilon(x\to y)\eta(y) \tag{4.20b}$$

$$\tilde g^\epsilon(x_1,x_2,t)=\int_0^t ds\sum_{z,y}P_{t-s}^\epsilon(x_1\to z)P_{t-s}^\epsilon(x_2\to z)P_s^\epsilon(z\to y)\eta(y) \tag{4.20c}$$

where $P_t^\epsilon(x\to y)$ is the transition probability for a single symmetric random walk moving with intensity ϵ^{-2}.

Proof. We write $\underline{x}$ instead of $U(\underline{x})$ in the argument of the correlation functions below. We are going to see that (4.15) yields

$$u^{\epsilon,a}(\underline{x},t|\eta)\le\prod_{i=1}^{n}g^\epsilon(x_i,t)+\int_0^t ds\, d(n)\sum_{\underline{y}}P_{t-s}^\epsilon(\underline{x}\to\underline{y})\Big(2u^{\epsilon,a}(\underline{y},s) + 1(y_1=y_2)u^{\epsilon,a}(\underline{y}^{(2)},s)+\sum_{j=3}^{n}u^{\epsilon,a}(\underline{y}^{(j)},s)\Big) \tag{4.21a}$$

where

$$P_{t-s}^\epsilon(\underline{x}\to\underline{y})=\prod_{i=1}^{n}P_{t-s}^\epsilon(x_i\to y_i) \tag{4.21b}$$

and the configuration $\underline{y}^{(i)}$ is obtained from $\underline{y}$ by erasing the particle with label i.

We use (4.15) and (4.17) to write $L_c^{\epsilon,a}D(\xi,\eta)$, with $\xi=U(\underline{y})$. The first term in the integral comes from

$$1(y_1\ne y_2; y_j\ne y_1,y_2,\ j>2)\sum_{i=1}^{2}D(\xi-\xi_{(y_i)},\eta)L_c^{\epsilon,a}D(\xi_{(y_i)},\eta) \tag{4.21c}$$

We then use the second inequality in (4.18) and bound the characteristic function in (4.21c) by 1. The second term on the right hand side of (4.21c) comes from the case when $y_1 = y_2$ and $y_j \neq y_1$, for all $j > 2$: we use the third inequality in (4.18) and drop the condition that $y_j \neq y_1$. The third term takes into account the action of the generator on the sites y_j, $j > 3$, for which we use again the third inequality in (4.18). If there is multiple occupancy, we are overcounting the terms, but since they are non negative, we certainly do not decrease the bound.

From (4.21a), recalling (4.20b), we then get

$$u^{\epsilon,a}(\underline{x},t|\eta) \leq e^{2d(n)t}\prod_{i=1}^{n} g^{\epsilon}(x_i,t) + \int_0^t ds\, e^{2d(n)(t-s)} \sum_{\underline{y}} P^{\epsilon}_{t-s}(\underline{x} \to \underline{y})$$
$$\times\, d(n)\Big(1(y_1 = y_2)u^{\epsilon,a}(\underline{y}^{(2)},s) + \sum_{j=3}^{n} u^{\epsilon,a}(\underline{y}^{(j)},s)\Big)$$

and, by iteration, we obtain (4.20a), we omit the details. □

Proof of (4.14). We need to bound the expectation of

$$D(\xi^{x,y},\eta_t) \qquad \xi^{x,y} = h\delta_x + k\delta_y$$

in such a way that the sum over x and y is finite. We define $\underline{x}$ so that $\xi^{x,y} = U(\underline{x})$ and $x_1 = x$, $x_2 = y$. Since η has finitely many particles, we have

$$\sum_{x,z} P^{\epsilon}_t(x \to z)\eta(z) < \infty, \qquad \sum_{x,y}\sum_{z,z'} P^{\epsilon}_{t-s}(x \to z)P^{\epsilon}_{t-s}(y \to z)P^{\epsilon}_s(z \to z')\eta(z') < \infty$$

From this and (4.20a), we then prove (4.14). □

We have thus completed the proof of Proposition 4.2.1, hence the process is well defined in Ω. Next we study the process starting from configurations with infinitely many particles. Given any configuration η we denote by $\eta^{(N)}$ the configuration

$$\eta^{(N)}(z) = \begin{cases} \eta(z) & \text{if } z \leq N \\ 0 & \text{otherwise} \end{cases}$$

We have

4.2.5 Proposition. *Given any positive c and n, let $\eta \in X(c,n)$, see (2.6). Then for any $\xi \in \Omega$, $u^{\epsilon}(\xi,t|\eta)$ exists and it is finite where*

$$u^{\epsilon}(\xi,t|\eta) = \lim_{N\to\infty} u^{\epsilon}(\xi,t|\eta^{(N)}) \tag{4.22}$$

Proof. The process starting from $\eta^{(M)}$ is stochastically larger [see for instance the proof of Proposition 3.2.2.] than the one starting from $\eta^{(N)}$, if $M \geq N$. This holds individually for the processes with generators L_0 and L_c, hence also for their sum, as in (4.6). Therefore we need only to prove that $\mathbb{E}^{\epsilon}_{\eta^{(N)}}(D(\xi,t))$ is uniformly bounded in N. This follows from (4.20), because

$$\sum_{y} P^{\epsilon}_t(x \to y)\eta(y) < \infty$$

In fact $P_t^\epsilon(x \to y)$ decays at least exponentially in $|x-y|$ because the probability that in the interval $[0, \epsilon^{-2}t]$ there are n events out of an exponential distribution of intensity one, decays exponentially as $n \to \infty$. □

By arguments similar to those used in Chapter II, it is possible to prove that, for all ξ, $u^\epsilon(\xi, t|\eta)$ is the expectation of $D(\xi, \eta_t)$ with respect to a probability, which is then interpreted as the transition probability of the process with generator (4.6). It can also be shown that this measure is supported by the union over all the sets $X(c, n)$ defined in (2.6). Since we do not need these facts for studying the macroscopic limit, we omit their proofs.

We now consider the process with random initial configuration, distributed according to μ^ϵ, so that it is supported by the union of the sets $X(c, n)$. We define the correlation functions for such an initial measure as

$$\begin{aligned} u^\epsilon(\xi, t|\mu^\epsilon) =& \mathbf{E}_{\mu^\epsilon}(D(\xi, \eta_t)) \\ \equiv& \int \mu^\epsilon(d\eta) \lim_{N\to\infty} \mathbf{E}^\epsilon_{\eta^{(N)}}(D(\xi, \eta_t)) = \lim_{N\to\infty} \int \mu^\epsilon(d\eta)\mathbf{E}^\epsilon_{\eta^{(N)}}(D(\xi, \eta_t)) \end{aligned} \tag{4.23}$$

Since $\mathbf{E}^\epsilon_{\eta^{(N)}}(D(\xi, \eta_t))$ is a non decreasing function of N, the interchange of limits is justified. It might in principle be that the limit is infinite, but this is not so, in fact we have:

4.2.6 Corollary. *(of Proposition 4.2.4). Let $d(n)$ be as in Lemma 4.2.3, then*

$$u^\epsilon(\xi, t|\mu^\epsilon) \le e^{2d(n)t}\left[d(n) + \|\rho_0\|\right]^{n-2} \left(\|\rho_0\|^2 + nt\|\rho_0\|\right) \tag{4.24}$$

and it satisfies the equation

$$u^\epsilon(\xi, t|\mu^\epsilon) = \sum_{\xi'} P_t^\epsilon(\xi \to \xi')u^\epsilon(\xi', 0|\mu^\epsilon) + \int_0^t ds \sum_{\xi'} P_{t-s}^\epsilon(\xi \to \xi')\mathbf{E}^\epsilon_{\mu^\epsilon}(L_c D(\xi', \eta_s)) \tag{4.25}$$

Proof. By Proposition 4.2.4 we can take the limit as $a \to \infty$ of both sides of (4.15) with $\eta = \eta^{(N)}$, so that we get

$$\begin{aligned} \int \mu^\epsilon(d\eta)\Big(u^\epsilon(\xi, t|\eta^{(N)}) - \sum_{\xi'} 1(\xi'(x) = 0, \ \forall\ |x| > N) P_t^\epsilon(\xi \to \xi')D(\xi, \eta) \\ - \int_0^t ds \sum_{\xi'} P_{t-s}^\epsilon(\xi \to \xi')\mathbf{E}^\epsilon_{\eta^{(N)}}(L_c D(\xi', \eta_s))\Big) = 0 \end{aligned} \tag{4.26}$$

Since μ^ϵ is a product of Poisson measures

$$\int \mu^\epsilon(d\eta)D(\xi, \eta_0) = u^\epsilon(\xi, 0|\mu^\epsilon) = \prod_x \rho_0(\epsilon x)^{\xi(x)} \tag{4.27}$$

Using this in (4.26), we are reduced to the same equation (4.15) but with $\rho(\epsilon x)$ in the place of $\eta(x)$: from (4.23) and (4.20) we then obtain (4.24). For what said after (4.23), the correlation functions converge when $N \to \infty$, and by (4.24) they are bounded, hence (4.25) follows by taking the limit of (4.26), as $N \to \infty$. □

§4.3 The macroscopic limit.

The main theorem in this Chapter is

4.3.1 Theorem. *Let μ^ϵ be a product of Poisson measures with averages as in (4.9), ρ_0 being a C^2-function bounded with its first and second derivatives. Denote by*

$$\Omega_{n,L} = \{\xi \in \Omega : |\xi| = n, \xi(x) = 0 \text{ whenever } |x| > \epsilon^{-1}L\}$$

Then for all L, T and n, ($D(\xi,\eta)$ is defined in (2.53) and (2.55))

$$\lim_{\epsilon\to 0} \sup_{\xi\in\Omega_{n,L}} \sup_{0\le t\le T} | \mathbb{E}^\epsilon_{\mu^\epsilon}(D(\xi,\eta_t)) - \prod_x \rho_t(\epsilon x)^{\xi(x)}| = 0 \tag{4.28}$$

where $\rho_t(r)$ solves (4.1) and

$$F_\pm(\rho) = \mathbb{E}_{\nu_\rho}(q_\pm(n)) \tag{4.29}$$

In (4.29) ν_ρ denotes the Poisson distribution on $\mathbb{N}$ with average ρ.

4.3.2 Remarks.

The limit in (4.28) is uniform also in L, for such a statement we need the assumption that the first and the second derivatives of ρ_0 are bounded, which is otherwise unnecessary. We will not prove such a statement. Some remarks about the relation between $q_\pm$ and $F_\pm$: it is somewhat unexpected that $q_-(n) = n^2$ does not reproduce $F_-(\rho) = \rho^2$, but rather $F_-(\rho) = \rho^2 + \rho$. This is a consequence of the local structure of the interaction, in other models where a mean-field like limit is considered, see for instance [6], and [7], $F_\pm = q_\pm$. In our case the monomials in $F_\pm$ are produced by the Poisson polynomials in $q_\pm$: ρ^n in $F_\pm$ comes from $D_n(\eta)$ in $q_\pm$.

The proof of Theorem 4.3.1 is obtained by studying the BBGKY equation (4.25). We already know from (4.24) that the correlation functions are bounded uniformly in ϵ. We shall prove equicontinuity which implies convergence by subsequences. Next we recognize that any limiting point solves a "limiting BBGKY hierarchy", for which we prove uniqueness. The proof will be completed by checking that the correlation functions defined by taking products of ρ's solution of (4.1) solve the limiting hierarchy. In the next paragraph we prove equicontinuity.

§4.4 Equicontinuity of the correlation functions.

For notational simplicity we write $u^\epsilon(\underline{x}, t|\mu^\epsilon)$ for $u^\epsilon(U(\underline{x}), t|\mu^\epsilon)$. For any $n \ge 1$, we then define,

$$\gamma^\epsilon(\underline{r}, t) = u^\epsilon(\underline{x}, 0|\mu^\epsilon), \qquad \underline{r} = \epsilon\underline{x},\ \underline{x} \in \mathbb{Z}^n \tag{4.30}$$

By linear interpolation $\gamma^\epsilon(\underline{r}, t)$ is defined for all $\underline{r} \in \mathbb{R}^n$. From (4.24), $\gamma^\epsilon(\underline{r}, t)$ is bounded, once n and $T > 0$ are fixed and $t \le T$.

4.4.1 Theorem. *Given any $n \ge 1$, L and T positive, the family $\gamma^\epsilon(\underline{r}, t)$, $\underline{r} \in \mathbb{R}^n$: $|r| \le L$, $0 \le t \le T$, is equibounded and equicontinous.*

In the remaining of this paragraph we shall prove the above Theorem. The boundedness has been already proven, the proof of the equicontinuity is elementary, so the more experienced reader may skip it and go directly to the next paragraph. We report below some detail.

We want to prove that given any $T > 0$, for any $\zeta > 0$ there exists $\delta > 0$ so that

$$\sup_{\substack{|t-t'|\leq\delta \\ t,t'\leq T}} \sup_{|\underline{r}-\underline{r}'|\leq\delta} |\gamma^\epsilon(\underline{r},t) - \gamma^\epsilon(\underline{r}',t)| \leq \zeta \tag{4.31}$$

We shall use the following well known properties of a symmetric random walk with transition probability $P_t^\epsilon(x \to x')$:

4.4.2 Lemma. *For any ζ^* and ℓ^* positive there is t^* so that for all $t \leq t^*$, ϵ, x:*

$$\sum_{x'} P_t^\epsilon(x \to x') 1(|x - x'| > \epsilon^{-1}\ell^*) < \zeta^* \tag{4.32}$$

Furthermore there is c so that for all x, y, t and ϵ

$$|P_t^\epsilon(x \to z) - P_t^\epsilon(y \to z)| \leq \frac{c}{\sqrt{\epsilon^{-2}t}} |x - y| \tag{4.33}$$

*Proof**. By the central limit theorem:

$$\lim_{s\to\infty} \sum_{|y|>L\sqrt{s}} P_s(0 \to y) = \int_{|r|\geq L} dr \frac{1}{\sqrt{2\pi}} e^{-r^2/2}$$

where P is the transition probability of a symmetric random walk with jump intensity one. Hence there are L^* and s^* so that for all $s \geq s^*$

$$\sum_{|y|>L^*\sqrt{s}} P_s(0 \to y) < \zeta^*$$

We choose t^* so small that $\ell^*/\sqrt{t^*} > L^*$. Since the left hand side of (4.32) can be written as

$$\sum_{x'} P_s(x \to x') 1(|x - x'| > \ell^* \sqrt{s/t})$$

if $s \geq s^*$, the left hand side of (4.32) is bounded by ζ^*.

It remains therefore to consider the values of ϵ and t such that $\epsilon^{-2}t < s^*$. Given s^* there is R so that

$$\sup_{s\leq s^*} \sum_{|y|\geq R} P_s(0 \to y) < \zeta^*$$

(4.32) is thus verified for $\epsilon^{-2}t < s^*$ and $\epsilon^{-1}\ell^* \geq R$. We are therefore left only with the case $\epsilon^{-2}t < s^*$ and $\epsilon^{-1}\ell^* < R$, hence $s = \epsilon^{-2}t < (R/\ell^*)^2 t^*$. We choose $t^* < \tau$ where τ is so small that

$$\sup_{s\leq(R/\ell^*)^2\tau} \sum_{y} P_s(0 \to y) 1(|y| > \ell^*) < \zeta^*$$

which a fortiori holds when summing over $|y| \geq \epsilon^{-1}\ell^*$ with $\epsilon < 1$. The proof of (4.32) is therefore concluded.

By the local central limit theorem, see [58], there is a constant c such that

$$\sum_{x} |P_t(0 \to x) - G_t(x)| \leq \frac{c}{\sqrt{t}} \tag{4.34}$$

where

$$G_t(r) = \frac{e^{-r^2/2t}}{\sqrt{2\pi t}} \tag{4.35}$$

From this (4.33) easily follows. □

4.4.3 Proof of Theorem 4.4.1.

We prove equicontinuity by showing separately that both terms on the right hand side of (4.25) are equicontinuous. We start from the first one, and since $\|\rho_0\| = \sup \rho_0(r) < \infty$, it is enough to prove the equicontinuity of

$$g^\epsilon(x,t) = \sum_z P_t^\epsilon(x \to z)\rho_0(\epsilon z) \tag{4.36}$$

Notice that $g^\epsilon(x,t) \le \|\rho_0\|$. We set $\zeta'\|\rho_0\| \le 10^{-2}\zeta$ and introduce $\ell > 0$ so that

$$\sup_{|z-z'|\le\epsilon^{-1}\ell} |\rho_0(\epsilon z) - \rho_0(\epsilon z')| \le \zeta'\|\rho_0\| \tag{4.37}$$

for all $\epsilon > 0$, (this can be done because of the assumption that $\rho_0(r)$ is bounded with its first derivative). We choose τ_0 so that (4.32) holds with $\ell^* = \ell$, $\zeta^* = \zeta'$ and all $t \le \tau_0$. We then fix $\delta > 0$ so that $\delta < \tau_0/2$, $\delta < \ell$, $\delta < c^{-1}\zeta'\sqrt{\tau_0/2}$ (c is defined in (4.33)). Furthermore we set $\ell' = c^{-1}\zeta'\sqrt{\tau_0/2}$ and we require that for all $t \le \delta$, (4.32) holds with $\ell^* = \ell'$ and $\zeta^* = \zeta'$.

Since $\delta < \tau_0/2$ and $|t - t'| \le \delta$, it follows that either both $t, t' \le \tau_0$ or both $t, t' \ge \tau_0/2$. We start by considering the case $t, t' \le \tau_0$. We have

$$\begin{aligned}
|g^\epsilon(x,t) - g^\epsilon(x,0)| &= \Big|\sum_z P_t^\epsilon(x \to z)\big(\rho_0(\epsilon z,0) - \rho_0(\epsilon x,0)\big)\Big| \\
&\le \zeta'\|\rho_0\| + \sum_z P_t^\epsilon(x \to z)1(|z - x| \le \epsilon^{-1}\ell)|\rho_0(\epsilon z,0) - \rho_0(\epsilon x,0)| \\
&\le 2\zeta'\|\rho_0\|
\end{aligned}$$

hence

$$|g^\epsilon(x,t) - g^\epsilon(y,t')| \le 4\zeta'\|\rho_0\| + |g^\epsilon(x,0) - g^\epsilon(y,0)| \le 5\zeta'\|\rho_0\|$$

which, by the choice of ζ', proves (4.31).

Case $t, t' > \tau_0/2$. To be definite let $t' > t$. Then, proceeding as before and exploiting the choice of δ,

$$|g^\epsilon(y,t') - g^\epsilon(y,t)| \le \sup_{|z-y|\le\epsilon^{-1}\ell'} |g^\epsilon(z,t) - g^\epsilon(y,t)| + \zeta'\|\rho_0\|$$

Hence, using (4.33), for any $\tau_0/2 \le t \le t' \le t + \delta$:

$$\begin{aligned}
\sup_{|x-y|\le\epsilon^{-1}\delta} |g^\epsilon(y,t') - g^\epsilon(x,t)| &\le \zeta'\|\rho_0\| + \sup_{|x-y|\le\epsilon^{-1}(\delta+\ell')} |g^\epsilon(y,t) - g^\epsilon(x,t)| \\
&\le \zeta'\|\rho_0\| + \frac{c}{\sqrt{\epsilon^{-2}\tau_0/2}}\epsilon^{-1}(\delta + \ell')\|\rho_0\| \le 3\zeta'\|\rho_0\|
\end{aligned}$$

This concludes the analysis of g^ϵ, the first term in (4.25).

For the second term on the right hand side of (4.25), we notice that from the first inequality in (4.18) it follows that there is $\bar{c}$ such that

$$\sup_{0\le s\le T} \mathbf{E}_{\mu^\epsilon}(L_c D(\xi,\eta_s)) \le \bar{c}$$

so that it is bounded as follows: (as usual we write $\underline{x}$ insted of $U(\underline{x})$ in the argument of D)

$$\left|\int_0^t ds \sum_{\underline{z}} [P^\epsilon_{t-s}(\underline{x} \to \underline{z})\mathbf{E}_{\mu^\epsilon}(L_c D(\underline{z},\eta_s)) - P^\epsilon_{t'-s}(\underline{y} \to \underline{z})\mathbf{E}_{\mu^\epsilon}(L_c D(\underline{z},\eta_s))\right|$$
$$\leq \bar{c}\sum_{i=1}^n \int_0^t ds \sum_z |P^\epsilon_{t-s}(x_i \to z) - P^\epsilon_{t'-s}(y_i \to z)|$$

It is therefore sufficient to prove the equicontinuity for each term of the sum. The proof is similar to that for g^ϵ and we omit the details. We have thus completed the proof of Theorem 4.4.1. □

§4.5 The limiting hierarchy.

From Theorem 4.4.1 it follows that in any sequence $\epsilon_k \to 0$ there is a subsequence along which $\gamma^\epsilon(\underline{r},t)$, $\underline{r} \in \mathbf{R}^n$, converges for all $n \geq 1$, $r_1,\dots,r_n$ and $t \geq 0$. The convergence is uniform on the compacts. Let $\gamma_t(\underline{r})$ be its limit, then Theorem 4.3.1 will be proven once we show that

$$\gamma_t(\underline{r}) = \prod_{i=1}^n \rho_t(r_i) \tag{4.38}$$

ρ_t being the solution to (4.1) with $F_\pm(\rho)$ given by (4.29).

To prove (4.38) we first derive a hierarchy of equations for the limiting correlation functions $\gamma_t(r_1,\dots,r_n)$ and then prove that such a hierarchy has a unique solution given by the right hand side of (4.38). We go back to (4.25) and we write it as

$$\begin{aligned} u^\epsilon(\xi,t|\mu^\epsilon) =& \sum_{\xi'} P^\epsilon_t(\xi \to \xi')u^\epsilon(\xi',0|\mu^\epsilon) + \int_0^t ds \sum_{\xi'} P^\epsilon_{t-s}(\xi \to \xi')1(\xi'(x) \leq 1\,\forall x) \\ &\times \sum_{x:\xi'(x)=1} \mathbf{E}^\epsilon_{\mu^\epsilon}(D(\xi' - \xi'_{(x)},\eta_s)[q_+(\eta(x,s)) - q_-(\eta(x,s))]) \\ &+ \int_0^t ds \sum_{\xi'} P^\epsilon_{t-s}(\xi \to \xi')1(\exists x\ :\ \xi'(x) > 1\,)\mathbf{E}^\epsilon_{\mu^\epsilon}(L_c D(\xi',\eta_s)) \end{aligned} \tag{4.39}$$

We observe that there is $c > 0$ such that

$$\sum_{\xi'} P^\epsilon_{t-s}(\xi \to \xi')1(\exists x\ :\ \xi'(x) > 1\,) \leq c|\xi|(|\xi|-1)\frac{1}{\sqrt{\epsilon^{-2}(t-s)}} \tag{4.40}$$

as it follows from the local central limit theorem, see (4.34). By (4.24) $\mathbf{E}^\epsilon_{\mu^\epsilon}(L_c D(\xi',\eta_s))$ is bounded, hence, by (4.40), the third term on the right hand side of (4.39) vanishes in the limit $\epsilon \to 0$.

We can express the polynomials $q_\pm$ in terms of the Poisson polynomials, namely, there exist coefficients a^+_ℓ, $\ell \leq \deg(q_+)$, a^-_ℓ, $\ell \leq \deg(q_-)$, such that

$$q_\pm(\cdot) = \sum_\ell a^\pm_\ell D_\ell(\cdot) \tag{4.41}$$

Using (4.41) the integrand in the second term on the right hand side of (4.39) becomes a linear combination of correlation functions. We then take the limit as $\epsilon \to 0$ along the subsequence which produces $\gamma_t(\underline{r})$ and we obtain

$$\gamma_t(\underline{r}) = \prod_{i=1}^{n}[\int G_t(r_i - r)\rho_0(r)dr] + \int_0^t ds \int \prod_{i=1}^{n}[G_{t-s}(r_i - r_i')dr_i'] \times \sum_{j=1}^{n}\sum_{\ell}\{(A_{j,\ell}^{+}\gamma_s)(\underline{r}') - (A_{j,\ell}^{-}\gamma_s)(\underline{r}')\}] \tag{4.42}$$

where $G_t(r)$ is defined in (4.35), (we have used once again the local central limit theorem, (4.34)), and

$$(A_{i,\ell}^{\pm}\gamma_s)(\underline{r}) = a_{\ell}^{\pm}\gamma_s(\underline{r}_{i,\ell}), \qquad r_{i,\ell} = (r_1, \cdots, r_{i-1}, r_i, \ldots, r_i, r_{i+1}, \ldots r_n) \tag{4.43}$$

where r_i in $r_{i,\ell}$ is repeated ℓ times.

4.5.1 A solution to the hierarchy.

We shall first see that

$$\bar{\gamma}_t(\underline{r}) = \prod_{i=1}^{n}\rho_t(r_i) \tag{4.44}$$

solves (4.42). We have

$$\frac{d}{dt}\bar{\gamma}_t(\underline{r}) = \sum_{j=1}^{n}\frac{1}{2}\frac{\partial^2}{\partial r_j^2}\bar{\gamma}_t(\underline{r}) + \sum_{j=1}^{n}\Big(\prod_{i\neq j}\rho_t(r_i)\Big)[F_+(\rho_t(r_j)) - F_-(\rho_t(r_j))] \tag{4.45}$$

By (4.29) and (4.41)

$$F_{\pm}(\rho_t(r)) = \mathbb{E}_{\nu_{\rho_t(r)}}(q_{\pm}(\eta(0)) = \sum_{\ell} a_{\ell}^{\pm}\rho_t(r)^{\ell} \tag{4.46}$$

Inserting (4.46) into (4.45) and then writing an integral version of (4.45) with respect to the semigroup with generator

$$\sum_{j=1}^{n}\frac{1}{2}\frac{\partial^2}{\partial r_j^2}$$

we just find (4.42). Hence we have proven that (4.44) is a solution of the hierarchy (4.42).

To complete the proof of Theorem 4.3.1 we need to prove uniqueness for (4.42). We do not know this in the class of correlation functions which satisfy the bound (4.24).

We shall see in 4.5.3 below that there is uniqueness for (4.42) in the smaller class of $\gamma_t(\underline{r})$ which satisfy the following bound: for any $T > 0$, there exists c so that

$$\sup_{\underline{r}\in\mathbb{R}^n}\sup_{t\leq T}\gamma_s(\underline{r}) \leq c^n \tag{4.47}$$

We next prove that the bound (4.47) is satisfied by any limiting set of correlation functions $\{\gamma_t(\underline{r})\}$.

4.5.2 Improved bounds on the limiting correlation functions.

To prove (4.47) we go back to (4.39) and using the first inequality in (4.18) we get

$$u^{\epsilon}(\xi, t|\mu^{\epsilon}) \leq \sup_{\xi'} u^{\epsilon}(\xi', 0|\mu^{\epsilon}) + \int_0^t ds\, d(1)\, |\xi| \sup_{\xi':|\xi'|=|\xi|-1} u_s^{\epsilon}(\xi')$$
$$+ \int_0^t ds \Big|\sum_{\xi'} P_{t-s}^{\epsilon}(\xi \to \xi')1(\exists x\ :\ \xi'(x) > 1\ )\mathbb{E}_{\mu^{\epsilon}}^{\epsilon}(L_c D(\xi', \eta_s))\Big|$$

As we have already seen the last term vanishes when $\epsilon \to 0$, so that taking the limit $\epsilon \to 0$ along the converging subsequence we get

$$\gamma_t(\underline{r}) \le \|\rho_0\|^n + \int_0^t ds\, d(1)\, n \sup_{\underline{r}\in\mathbb{R}^{n-1}} \gamma_t(\underline{r})$$

from which the bound (4.47) easily follows.

4.5.3 Uniqueness of solutions of the hierarchy.

We shall now prove uniqueness for (4.42) in the time interval $[0,T]$ for correlation functions $\{\gamma_t(\underline{r})\}$ verifying (4.47) for any given c. We first rewrite (4.42) in compact notation.

Let $\mathcal{V}$ be the space of sequences $v = (v^{(n)}(\underline{r}))_{n\ge 1}$ where for each n and $\underline{r} \in \mathbb{R}^n$ $v^{(n)}(\underline{r})$ is uniformly bounded. On $\mathcal{V}$ we consider the operators $\mathcal{G}_t$, $t \ge 0$, and $\mathcal{L}$ defined as

$$\underline{r}\in\mathbb{R}^n,\quad (\mathcal{G}_t v)^{(n)}(\underline{r}) = \int dr'_1 \dots dr'_n [\prod_{i=1}^n G_t(r_i - r'_i)]\, v^{(n)}(\underline{r}')$$

$$\underline{r}\in\mathbb{R}^n,\quad (\mathcal{L}v)^{(n)}(\underline{r}) = \sum_{j=1}^n \sum_{\ell} \{(A^+_{j,\ell} v^{(n+\ell-1)})(\underline{r}_{j,\ell}) - (A^-_{j,\ell} v^{(n+\ell-1)})(\underline{r}_{j,\ell})\}$$

see (4.43). Therefore if γ_t and λ_t, $t \le T$, are two solutions of (4.42) with the same initial datum, both satisfying the bound (4.47), we have that for all N

$$\gamma_t - \lambda_t = \int_0^T ds_1 \cdots \int_0^{s_{N-1}} ds_N\, \mathcal{G}_{t-s_1}\mathcal{L}\mathcal{G}_{s_1-s_2}\cdots\mathcal{G}_{s_{N-1}-s_N}\mathcal{L}(\gamma_{s_N} - \lambda_{s_N}) \tag{4.49}$$

We introduce the seminorms

$$\|v\|_n = \sup_{1\le i\le n} \sup_{\underline{r}\in\mathbb{R}^i} |v^{(i)}(\underline{r})| \tag{4.50}$$

and notice that for all $n \ge 1$

$$\|\mathcal{L}v\|_n \le c_1 n \|v\|_{n+k}$$

where $k = \deg(q_-)$ and c_1 is a suitable constant. Since

$$\|\mathcal{G}_t v\|_n = \|v\|_n \qquad n\ge 1\ t\ge 0$$

then, for any N,

$$\begin{aligned}\|\gamma_t - \lambda_t\|_n \le \int_0^t ds_1 \cdots \int_0^{s_{N-1}} ds_N\, c_1^N\, n(n+k)\cdots \\ \cdots (n+k(N-1))\|\gamma_{s_N} - \lambda_{s_N}\|_{n+kN} \\ \le \frac{(c_1 t)^N}{N!} k^N \frac{(n+N)!}{n!} c^{n+kN}\end{aligned} \tag{4.51}$$

c being the coefficient in (4.47). From (4.51), letting $N \to \infty$, we deduce that $\gamma_t = \lambda_t$ for small values of t, namely when $t \le \tau$ and τ is such that $\tau c_1 c^k k < 1$.

Observe that (4.42) has the semigroup property, namely if γ_t is a solution to (4.40) and if $t > \tau > 0$ then

$$\gamma_t = \mathcal{G}_{t-\tau}\gamma_\tau + \int_\tau^t ds\, \mathcal{G}_{t-s}\mathcal{L}\gamma_s$$

We can then repeat the previous argument since $\gamma_\tau = \lambda_\tau$ proving that $\gamma_{2\tau} = \lambda_{2\tau}$. The argument can be iterated till times for which the a priori bound (4.47) holds. This completes the proof of the uniqueness of solutions to (4.42) and consequentely the proof of Theorem 4.3.1. □

§4.6 Bibliographical notes.

The model we have considered in this chapter has been studied in [17], and for particular choices of the reacting potential, in [47] and [48]. We have followed closely the proof in [17], correcting an error, present in [17], on some infinite volume estimates.

The present model is a local version of a mean field-like process considered in [6], [7] and [104].

A particle model for the KPP equation has been introduced and studied in [18], see the Remarks below (4.4). Spin models for reaction diffusion equations are also considered in Chapter VI, they were first introduced in [36]. Reaction-diffusion equations in more than one dimension, with only deaths, have been obtained in the "low density limit": the death operator is not depressed as in our model, but the initial densities should be suitably small, vanishing as $\epsilon \to 0$. A continuous version is discussed in [84], in [100] several other scalings are considered. Similar problems are discussed in [128] where a general survey on the correlation functions technique can be found.

The fluctuations from the limiting behavior of the density field have been studied in [95] for local mean field models, and in [16] for the model that we have considered in this chapter.

In the last years several models have been proposed for the simulation of reaction diffusion equations on a computer. Since in the experiments there is no limit $\epsilon \to 0$, the scaling parameter does not appear explicitly. If the ϵ's are inserted, the convergence can be proven using the arguments presented here and in Chapter VI. The analysis when ϵ is small but fixed, raises several interesting questions, but it looks quite hard.

We conclude these notes by mentioning two general references on reaction-diffusion equations, [123] and [8].

CHAPTER V

PARTICLE MODELS FOR THE CARLEMAN EQUATION

The Carleman equation is

$$\frac{\partial}{\partial t}\rho(r,\sigma,t)+\sigma\frac{\partial}{\partial r}\rho(r,\sigma,t)=\rho(r,-\sigma,t)^2-\rho(r,\sigma,t)^2 \tag{5.1a}$$

$$\rho(r,\sigma,0)=\rho_0(r,\sigma) \tag{5.1b}$$

where $r\in\mathbb{R}$, $\sigma=\pm1$ and $\rho_0(r,\sigma)\geq 0$; (5.1) is a particular case of a discrete velocity Boltzmann equation

$$\frac{\partial}{\partial t}f+v\cdot\nabla f=\lambda Q(f) \tag{5.2}$$

where $f=f(r,v,t)$, $r\in\mathbb{R}^d$, $v\in V$, V a finite set of vectors in $\mathbb{R}^d$ that we call velocities, and $Q(f)$ a non linear operator which satisfies suitable conditions. (5.2) is a discretized version of the continuous Boltzmann equation, where the velocity $v\in\mathbb{R}^d$. The operator Q is called the collision operator, it describes the outcome of the collisions between the particles. We refer to Chapter IV, Part I, of [125] for a general discussion on the Boltzmann equation and just restrict our attention to (5.1).

This equation is very special, in fact a global existence theorem holds for (5.1), [76], which is not available, in general, for (5.2). Indeed, a global existence theorem has been recently proved by R. Di Perna and P.L. Lions for the true Boltzmann equation, but existence holds in a weak sense, while in the Carleman equation the solution has the same regularity as at time 0. Global existence for (5.1) is a consequence of monotonicity: let ρ and ρ' be two solutions of (5.1). If $\rho(r,\sigma,0)\leq\rho'(r,\sigma,0)$ for all r and σ, then $\rho(r,\sigma,t)\leq\rho'(r,\sigma,t)$ for all r, σ and t. This property may be easily checked, its origin comes from the fact that in the collision term, i.e. in the right hand side of (5.1), there are only products of ρ's computed at the same particle state (r,σ). This is not generally true for the collision operator Q in (5.2). To conclude the proof of global existence, assume first that the initial datum is in L^∞ (and also that it is in C^1). Since the functions of (r,σ,t) which are identically equal to a constant are solutions of (5.1), we can bound the initial datum $\rho(r,\sigma,0)$ by a constant solution. Therefore $\rho(r,\sigma,t)$, if existing, is bounded by the same constant. By this a-priori estimate we can find a global solution: local solutions in fact can be constructed for time intervals whose length only depends on the L^∞ norm of the initial values, but due to the a-priori bound we can continue them to arbitrary times. Observe finally that information propagates with speed 1 in the Carleman equation, so that it is not even necessary to assume that the initial datum is in L^∞. We will not insist on this point, since for simplicity we are going to assume that $\rho_0(r,\sigma)$ is periodic with period 2, or, equivalently, that (5.1) is restricted to $[-1,1]$ with periodic boundary conditions.

The Carleman equation has several unpleasant features. The most evident one is that collisions involve particles with the same velocity, (after the collision the velocities are reversed). This can be read from (5.1) by looking at its right hand side. The first term is the gain term: it gives the rate of increase of particles with velocity σ at (r,t) which is proportional to the density of the pair of particles at (r,t) with velocity $-\sigma$, (in fact, assuming factorization, this is equal to the square of the density of particles with velocity $-\sigma$), hence the interpretation that these particles collide and change their velocities. The loss term, the second one on the right hand side of (5.1), has analogous meaning. The above collision rules are quite unrealistic, but there is little to do in one dimension and in any case the discrete velocity Boltzmann equations are often nothing more than caricatures of real systems.

Also the one dimensionality of the equation is unpleasant, physically the Boltzmann equation describes rarefied fluids, where the free path of a particle is comparable to the macroscopic lengths, technically this means that the Knudsen number is of the order of unity, but in one dimension this cannot be achieved because the particles do not have room to avoid each other. If however the evolution is stochastic, the particles have a possibility to pass through each other undisturbed, and if this probability is suitably chosen the Knudsen number may stay finite. There are other, deeper reasons for having random evolutions: after the Uchyiama's analysis, [129], of the Broadwell gas, it is clear that in general discrete velocity Boltzmann equations cannot be derived from deterministic dynamics, we shall discuss with more details these arguments in §7.1.

As in the reaction diffusion equation of Chapter IV, we find in (5.1) that the non linearity affects only terms with no derivatives, while, in common with the non linear diffusion equation of Chapter III, there is the fact that the L^1 norm is conserved by the streaming part. Both facts play an important role in our analysis. The first one makes it possible, as in Chapter IV, to write perturbative expansions. The particles conservation, hence the absence of dissipation, prevents us however from having the a priori bounds of Chapter IV. Without them we cannot repeat the proof of global convergence. We face the typical difficulties when deriving the Boltzmann equation where the perturbative series are proved to converge only for short times. A real difficulty, which has not been overcome for the true Boltzmann equation, except for the result of Illner and Pulvirenti, [77], on a gas expanding in the vacuum. However for stochastic particle systems the situation has been improved. There are convergence results at all times for which the solution of the limiting equation stays bounded, [26], [35], see §7.1 for further details. Our purpose here is to present these methods in the simpler frame of the Carleman equation, where the main ideas remain the same.

There are two difficulties when trying to apply the entropy methods of Chapter III. One is the hyperbolicity of (5.1): no hope for the two blocks estimate. However this point may be bypassed with a variant, which uses relative entropy with respect to local equilibirum distributions, as recently proposed by Yau, [134] and Olla, Varadhan and Yau, (in preparation). The other difficulty is to prove a-priori L^2 bounds for the occupation numbers, which, at the moment, do not seem to have been obtained. But even if all these problems may be circumvented, still we think it is useful to have a solution in terms of correlation functions, since they give a much more detailed and precise description of the evolution. Of course, we cannot use the method of the correlation functions to go beyond perturbations of independent-like evolutions; full interactions as in Chapter

III are not within the range of the method, but when the system is simple enough, as in the present case, then the method really works well.

§5.1 The microscopic model.

For each ϵ, $0 < \epsilon \leq 1$, the state space of the system is $\Omega_\epsilon = \mathbb{N}^{\mathbb{Z}_\epsilon \times \{-1,1\}}$ where $\mathbb{Z}_\epsilon$ is $\mathbb{Z}$ modulo $2[\epsilon^{-1}]+1$. The elements of Ω_ϵ are particles configurations denoted by $\eta = \{\eta(x),\, x = (q,\sigma),\, |q| \leq 2[\epsilon^{-1}]+1\}$, equivalently $q \in \mathbb{Z}$, and $\eta((q,\sigma)) = \eta((q + 2[\epsilon^{-1}]+1, \sigma))$. By analogy with (5.1) we shall refer to (q,σ) as position and velocity variables. $x = (q,\sigma)$ is called a single particle state. *Caution: here x denotes the pair (q,σ) and it is not a site as in the previous chapters.*

The evolution is defined by the jump Markov process on Ω_ϵ with generator

$$L^\epsilon = \epsilon^{-1} L_0 + L_c \tag{5.3}$$

where, the sum of q's below is modulo $2[\epsilon^{-1}]+1$,

$$L_0 f(\eta) = \sum_{x=(q,\sigma)} \eta(x)[f(\eta - \delta_{(q,\sigma)} + \delta_{(q+\sigma,\sigma)}) - f(\eta)] \tag{5.4}$$

$$L_c f(\eta) = \frac{1}{2} \sum_{x=(q,\sigma)} \eta(x)(\eta(x)-1)[f(\eta - 2\delta_{(q,\sigma)} + 2\delta_{(q,-\sigma)}) - f(\eta)] \tag{5.5}$$

L_0 describes a process which is a direct product of two independent evolutions of the type considered in Chapter II. The process of the particles with $\sigma = 1$ has $p = 1$ and $q = 0$, while for $\sigma = -1$ the opposite holds. Therefore particles with velocity σ after exponential times of intensity 1 jump by σ, these times being all mutually independent.

L_c describes the collisions: all the pairs of particles at same site with same velocity, independently flip their velocities after an exponential time of intensity 1. This is not the only possible choice, it is natural, as well, to consider the case when only one of the two particles in the pair flips its velocity. As far as the macroscopic behavior is concerned the two models are equivalent, as proven in a previous version of this paper, [42]. We shall only discuss here the generator (5.5).

As in Chapter IV we assume that the initial measure μ^ϵ is a product of Poisson distributions and that

$$\mathbb{E}_{\mu^\epsilon}(\eta(q,\sigma)) = \rho_0(\epsilon q, \sigma)$$

where $\rho_0 \geq 0$ is a periodic function on $[-1,1]$ with continuous derivative. Such a condition may only be imposed when ϵ^{-1} is an integer, we do not miss much in generality if we restrict ϵ to satisfy such a condition, as we shall tacitly do from now on. We denote by $\mathbb{P}^\epsilon$ the law of the process on Ω_ϵ with generator L^ϵ and by $\mathbb{E}^\epsilon$ its expectation. We also write $\mathbb{P}^\epsilon_\mu$ and $\mathbb{E}^\epsilon_\mu$ to underline, if nedeed, that the initial measure is μ.

§5.2 The correlation functions.

We just adapt to the present case the definitions given in 2.9.2 and 2.9.3 for the independent particles. So we first define, for any $(\xi,\eta) \in \Omega_\epsilon \times \Omega_\epsilon$, $D(\xi,\eta)$ as in (2.53)-(2.55) with $x = (q,\sigma)$.

We define the correlation functions as

$$u^\epsilon(\xi, t|\mu^\epsilon) = \mathbb{E}^\epsilon_{\mu^\epsilon}(D(\xi, \eta_t)) \tag{5.6}$$

η_t denoting the state of the process at time t.

We write $u^\epsilon(\xi,t|\lambda)$ if the initial measure is λ and $u^\epsilon(\xi,t|\eta)$ if the initial measure is supported by η, same notation as in Chapter II. We use also the same terminology, by saying that the n-body correlation functions correspond to $|\xi| = n$, or that they have degree n.

A final remark: if the initial measure is supported by a single configuration, then the expectations in (5.6) are well defined, because the total number of particles is finite. Also when the initial measure is μ^ϵ, (5.6) is well defined, in fact the total number of particles, given ϵ, has also a Poisson distribution. In general if the initial probability to have N particles decays faster than any power of N, then the correlation functions of any degree are well defined.

The main result in this chapter is

5.2.1 Theorem. *Let μ^ϵ be the probability on Ω_ϵ defined in §5.1 and let $\mathbb{E}^\epsilon_{\mu^\epsilon}$ be the expectation with respect to the process which starts from μ^ϵ and has generator given by (5.3). Then for all $t > 0$ all $n \geq 1$*

$$\lim_{\epsilon\to 0}\sup_{|\xi|=n} |u^\epsilon(\xi,t|\mu^\epsilon) - \prod_{x=(q,\sigma)} \rho(\epsilon q,\sigma,t)^{\xi(x)}| = 0 \tag{5.7}$$

where $\rho(r,\sigma,t)$ solves (5.1).

The remaining of the chapter is devoted to the proof of Theorem 5.2.1.

§5.3 Convergence at short times. (The Lanford technique).

Copying Lanford's derivation of the Boltzmann equation for a hard spheres gas, [83], we can prove the validity of the theorem for short times. As we shall see the same method cannot be used to extend the result to longer times, even in the case we are considering.

5.3.1 Duality for interacting particles.

We define the ξ-process as the process with generator $\epsilon^{-1}L_0^*$ where

$$(L_0^* f)(\eta) = \sum_{(q,\sigma)} \eta(q,\sigma)[f(\eta - \delta_{(q,\sigma)} + \delta_{(q-\sigma,\sigma)}) - f(\eta)] \tag{5.8}$$

The η-process has generator L^ϵ. We suppose that the ξ process starts from ξ and the η process from μ^ϵ, (we could take as well any other measure whose correlation functions exist, as for instance measures supported by single configurations). We denote by $\mathbb{E}^{\epsilon,*}_\xi$ the expectation with respect to the process with generator L_0^* starting from the configuration ξ.

The first dual process that we consider is the direct product of the ξ and η processes. Denoting by $\mathcal{E}^\epsilon$ its expectation we have

$$\frac{d}{ds}\mathcal{E}^\epsilon(D(\xi_{t-s},\eta_s)) = \mathcal{E}^\epsilon(L_c D(\xi_{t-s},\eta_s)) \tag{5.9}$$

where L_c acts on $D(\xi_{t-s},\eta_s)$ as a function of η_s.

From (5.9) we get

$$u^\epsilon(\xi,t|\mu^\epsilon) = \mathbb{E}^{\epsilon,*}_\xi(u^\epsilon(\xi_t,0|\mu^\epsilon)) + \int_0^t ds\,\mathcal{E}^\epsilon(L_c D(\xi_{t-s},\eta_s)) \tag{5.10}$$

It is clear that $L_c D(\xi_{t-s}, \eta_s)$ is a linear combination of products of η's and that we can always rewrite it as a sum of D functions. In this way we obtain a closed set of equations for the family of all the correlation functions with different degrees. In the physical literature this is known as *the BBGKY hierarchy of equations*. While clear in principle, writing the explicit form of this equation requires some work. We have

$$L_c D(\xi, \eta) = \sum_q D(\xi - \xi_q, \eta - \eta_q) L_c D(\xi_q, \eta_q) \tag{5.11}$$

where ξ_q is the configuration consisting of $\xi(q,1)$ and $\xi(q,-1)$ particles at q with velocity 1, respectively -1; η_q is defined analogously.

To have more compact formulas we set $\xi(q,1) = k$, $\xi(q,-1) = h$, $\eta(q,1) = n$, $\eta(q,-1) = m$. We have

$$\begin{aligned} L_c D(\xi_q, \eta_q) \equiv L_c D_k(n) D_h(m) &= \frac{1}{2} n(n-1)[D_k(n-2) D_h(m+2) - D_k(n) D_h(m)] \\ &+ \frac{1}{2} m(m-1)[D_k(n+2) D_h(m-2) - D_k(n) D_h(m)] \end{aligned}$$

To reconstruct D functions we need to re-express the right hand side in such a way that the arguments of the Poisson polynomials are n and m. We have

$$\begin{aligned} &n(n-1) D_k(n-2) = D_{k+2}(n) \\ &\frac{1}{2} n(n-1) D_k(n) = \frac{1}{2} D_{k+2}(n) + k D_{k+1}(n) + \frac{1}{2} k(k-1) D_k(n) \\ &\frac{1}{2} D_h(m+2) = \frac{1}{2} D_h(m) + \frac{1}{2} h(h-1) D_{h-2}(m) + h D_{h-1}(m) \end{aligned}$$

We then get

$$\begin{aligned} L_c D_k(n) D_h(m) = \ & k[D_{k-1}(n) D_{h+2}(m) - D_{k+1}(n) D_h(m)] \\ &+ h[D_{k+2}(n) D_{h-1}(m) - D_k(n) D_{h+1}(m)] \\ &+ \frac{1}{2} k(k-1)[D_{k-2}(n) D_{h+2}(m) - D_k(n) D_h(m)] \\ &+ \frac{1}{2} h(h-1)[D_{k+2}(n) D_{h-2}(m) - D_k(n) D_h(m)] \end{aligned} \tag{5.12}$$

Using (5.12) we can rewrite the right hand side of (5.10) in terms of correlation functions at time s and solve the equation by iteration: since we can study the series starting from single configurations (and only after that integrating over μ^ϵ), we do not have to worry about convergence, because the correlation functions with degree larger than the total number of particles vanish. But the problem comes up again when taking the limit as $\epsilon \to 0$, can we take the limit term by term in the above series? First hope is that we do not need to take into account the signs in (5.12) as well as the positions of the particles in ξ_{t-s} taking sup's over the ξ's. But this hope does not last long. If we just consider the contribution of the terms in (5.12) which contain the factors D_{h+2} and D_{k+2}, we readily see that the iterative series diverges because of the factors $k(k-1)$ and $h(h-1)$ in (5.12).

We therefore need to take into account the signs in (5.12), but this is not so awful as it may look: in fact, by the way they are written, the incriminated terms can be interpreted as due to the action of L_c on the ξ variables, h and k. We can therefore aggregate them to the ξ process which remains a Markov semigroup, so that the L^∞ norm does not grow. In this new duality the last two lines in (5.12) will be absent and the above procedure, with sup norms and all signs changed into pluses, does indeed work. Let us see in details how all this can be achieved.

5.3.2 A second duality.

In this case the ξ-process is defined by the generator $\epsilon^{-1}L_0^\star + L_c$, its expectation is denoted by $\bar{\mathbb{E}}^{\epsilon,\star}$. The η process, as before, has generator $\epsilon^{-1}L_0 + L_c$. The new dual process is the direct product of these two and we denote by $\bar{\mathcal{E}}^\epsilon$ its expectation. We have from (5.12)

$$\frac{d}{ds}\bar{\mathcal{E}}^\epsilon(D(\xi_{t-s},\eta_s)) = \sum_{x=(q,\sigma)} \bar{\mathcal{E}}^\epsilon\Big(\xi(x,t-s)[D(\xi_{t-s}-\delta_{(q,\sigma)}+2\delta_{(q,-\sigma)},\eta_s) - D(\xi_{t-s}+\delta_{(q,\sigma)},\eta_s)]\Big) \tag{5.13}$$

hence

$$u^\epsilon(\xi,t|\mu^\epsilon) = \bar{\mathbb{E}}_\xi^{\epsilon,\star}(u^\epsilon(\xi_t,0|\mu^\epsilon)) + \int_0^t ds \sum_{(q,\sigma)} \bar{\mathbb{E}}_\xi^{\epsilon,\star}\Big(\xi(q,\sigma,t-s)$$
$$\times\,[u^\epsilon(\xi_{t-s}-\delta_{(q,\sigma)}+2\delta_{(q,-\sigma)},s|\mu^\epsilon) - u^\epsilon(\xi_{t-s}+\delta_{(q,\sigma)},s|\mu^\epsilon)]\Big) \tag{5.14}$$

As claimed at the end of the last subparagraph, we have now an expression linear in ξ and this allows to iterate without danger (5.14). Before doing this let us see how and why this miracle has happened. Let us go back to the second proof of 2.9.4 and to the interpretation of the equation for the correlation functions for independent particles. This was based on the relation between the n-body correlation functions computed at (ξ,t) and the probability of having n given labelled particles at the sites specified by ξ at time t. Due to the independence of the evolution one had only to trace back these n particles to the random positions they occupied at the initial time and compute the probability of finding them there. In our case, due to the interaction, this will no longer be true. There are two sources of errors, one comes from the "external collisions" of our particles with the others in the system, the second one is due to the "internal collisions" of the n particles among themselves. The external collision are proportional to n, the internal ones to n^2. But these latter can be inserted in the dual process, by imposing the collision rule among the n particles. In this way we have a much more complicated dual process: when n is equal to the total number of particles, it is just the original process (with velocities reversed). This is the price for not having anymore the dangerous factors n^2. If one looks back at the proofs we have given, one can see that they just reflect the above heuristic picture.

5.3.3 Lemma. *Given $c^\star > 0$ and $0 < \epsilon \le 1$, suppose that there is a probability λ on Ω_ϵ such that for all $k \ge 1$*

$$\sup_{|\xi|=k} u^\epsilon(\xi,0|\lambda) \le (c^\star)^k \tag{5.15}$$

Then, setting $t_0 = (4c^\star)^{-1}$, for all $k \geq 1$

$$\sup_{0\leq t\leq t_0}\sup_{|\xi|=k} u^\epsilon(\xi,t|\lambda) \leq 2(2c^\star)^k \tag{5.16}$$

Proof. Given $N > 0$, let λ_N be the restriction of λ to the subset $\{\eta\ :\ |\eta| \equiv \sum_x \eta(x) \leq N\}$. We denote below by $u^{\epsilon,N}(\xi,t)$ the correlation functions corresponding to λ_N. Since $u^{\epsilon,N}(\xi,t)$ is a non decreasing function of N, we have for all t and ξ,

$$\lim_{N\to\infty} u^{\epsilon,N}(\xi,t) = u^\epsilon(\xi,t|\lambda)$$

the limit in principle could be infinite.

From (5.14) it follows that

$$u^{\epsilon,N}(\xi,t) = \bar{\mathbb{E}}_\xi^{\epsilon,\star}(u^{\epsilon,N}(\xi_t,0))$$
$$+\int_0^t ds \sum_{x=(q,\sigma)} \bar{\mathbb{E}}_\xi^{\epsilon,\star}\big(\xi(x,t-s)[u^{\epsilon,N}(\xi_{t-s}+2\delta_{(q,-\sigma)}-\delta_{(q,\sigma)},s)-u^{\epsilon,N}(\xi_{t-s}+\delta_{(q,\sigma)},s)]\big) \tag{5.17}$$

If $|\xi| = k$ by (5.15) we have

$$u^{\epsilon,N}(\xi,t) \leq (c^\star)^k + \int_0^t ds\ k \sup_{|\xi|=k+1} u^{\epsilon,N}(\xi,s)$$
$$\leq (c^\star)^k \sum_{\ell=0}^{N}(tc^\star)^\ell(k+\ell)![k!\ell!]^{-1} \leq (c^\star)^k\sum_{\ell=0}^{\infty}(4)^{-\ell}2^{k+\ell} \tag{5.18}$$

The second inequality uses the fact that $u_t^{\epsilon,N}(\xi) = 0$ if $|\xi| > N$. For the third inequality we have used the bound

$$\frac{(k+\ell)!}{k!\ell!} \leq 2^{k+\ell}$$

Notice that the last bound in (5.18) is uniform on ξ and N, (5.16) is therefore proven. □

5.3.4 Conclusion of the proof of Theorem 5.2.1 at short times.

Let $c^\star = \sup \rho_0(r)$, then, for any ϵ, μ^ϵ satisfies the condition (5.15). $u^\epsilon(\xi,t|\mu^\epsilon)$ is therefore expressed as a series, the n-th term collecting those with n time integrals. Such a series is uniformly convergent for $t \leq t_0$, by Lemma 5.3.3. We can therefore take the limit as $\epsilon \to 0$ term by term. Notice that the expectations $\bar{\mathbb{E}}_{\xi'}^{\epsilon,\star}$ appearing when iterating (5.17) have configurations ξ' with at most $|\xi| + n$ particles (because we are considering terms with n integrals). Call $\bar{P}_t^{\epsilon,\star}(\xi \to \xi')$ the transition probability associated to the ξ process defined in the second duality, 5.3.2, and $P_t^{\epsilon,\star}(\xi \to \xi')$ the one for the ξ process defined in 5.3.1. By classical estimates on random walks, for any t there is c so that for all n and ϵ:

$$\sup_{t\leq t_0}\sup_{|\xi|=n}\sum_{\xi'}\left|\bar{P}_t^{\epsilon,\star}(\xi\to\xi') - P_t^{\epsilon,\star}(\xi\to\xi')\right| \leq cn^2\sqrt{\epsilon} \tag{5.19}$$

A proof of (5.19) is given below.

5.3.5 Proof of (5.19).*

We consider the labelled versions of the processes. Assume ξ has n particles, we then consider a configuration $\underline{x} = (x_1, \ldots, x_n)$ such that $\xi = U(\underline{x})$, see (2.2). We call, by an abuse of notation, $\bar{\mathbb{P}}^{\epsilon,*}_{\underline{x}}$ the law of the labelled particles corresponding to the ξ process defined in 5.3.2 and call $\mathbb{P}^{\epsilon,*}_{\underline{x}}$ the law of the ξ process defined in 5.3.1. Both process start from the same $\underline{x}$. To make it short, we shall call interacting, respectively independent, the particles in the two processes.

The proof of (5.19) is based on a coupling, that is on a joint representation of the interacting and independent processes whose marginals reproduce these two processes. We introduce a third one, $(\mathcal{X}, \mathcal{P}^{\epsilon}_{\underline{x}})$, which is the direct product of a realization of the independent process starting from $\underline{x}$ and $n(n-1)/2$ exponential laws all with the same intensity $1/2$. These last processes are labelled by all the unordered pairs of distinct labels in $\{1, \ldots, n\}$ and we denote the canonical variables in this space by $T_{i,j}$. We next define the variables $t_{i,j}(t)$ as the total time till t spent by the particles i and j in the same state. We then introduce the stopping time τ as the first time t when there are (i,j) such that $t_{i,j}(t) = T_{i,j}$.

We define the coupling so that the independent process is the same as the particle process in $\mathcal{X}$, while the interacting process is equal to the particle process in $\mathcal{X}$ only up to the random time τ. If $\underline{x}'$ is the state of the particles in $\mathcal{X}$ at time τ and $t_{i,j}(\tau) = T_{i,j}$, we then define $\underline{x}''$ from $\underline{x}'$ by reversing the velocities of both particles i and j. The interacting process after τ is then defined independently of what happens in $\mathcal{X}$ as a copy of the interacting process which starts at τ from $\underline{x}''$. We leave it to the reader to verify that this is a coupling.

From the definition we have

$$\sum_{\underline{y}} \left|\bar{P}^{\epsilon}(\underline{x} \to \underline{y}) - P^{\epsilon}(\underline{x} \to \underline{y})\right| \le \mathcal{P}^{\epsilon}_{\underline{x}}(\tau \le t) \le \sum_{(i,j)} \mathcal{P}^{\epsilon}_{\underline{x}}(T_{i,j} \le t_{i,j}(t))$$

$$\le \frac{n(n-1)}{2} \sup_{(x_1,x_2)} \mathcal{P}^{\epsilon}_{(x_1,x_2)}(T_{1,2} \le t_{1,2}(t))$$

where the last probability refers to the case when there are only two particles. Of course the sup is achieved when the velocities of the two particles are the same. Then these are just two independent particles with intensity $1/2$, jumping, say, to the right. Calling $q_1(s)$ and $q_2(s)$ their positions, we have:

$$t_{1,2}(t) = \int_0^t ds \mathbb{1}(\{q_1(s) = q_2(s)\})$$

Since

$$\mathcal{P}^{\epsilon}_{(x_1,x_2)}(T_{1,2} \le a) \le a/2$$

we have

$$\mathcal{P}^{\epsilon}_{(x_1,x_2)}(T_{1,2} \le t_{1,2}(t)) \le 1/2\ \mathcal{E}^{\epsilon}_{(x_1,x_2)}\Big(\int_0^t ds \mathbb{1}(\{q_1(s) = q_2(s)\})\Big)$$

By interchanging the expectation with the time integral and using the local central limit theorem we prove (5.19). □

§5.4 The v and the ρ functions.

The a priori-bounds in Lemma 5.3.3 on the correlation functions $u^\epsilon(\xi,t|\mu^\epsilon)$, $0<\epsilon\le 1$, are much worse for $t\le t_0$ than their limiting values when $\epsilon\to 0$. In fact from Theorem 5.2.1 proven for $t\le t_0$, it follows that for any $t\le t_0$,

$$\lim_{\epsilon\to 0}\sup_{|\xi|=n} u^\epsilon(\xi,t|\mu^\epsilon)=\|\rho_t\|_\infty^n\le\|\rho_0\|_\infty^n$$

If such a bound would hold for finite ϵ's too, and not only in the limit, we would have right away Theorem 5.2.1 at all times, just by iteration. But for having better bounds we need to go back to the equation for the correlation functions in order to derive more explicit estimates. In (5.19) we have achieved this by expressing the expectations with respect to the interacting ξ-process in terms of the independent one. But the error terms are too large, they grow as $\sqrt{\epsilon}|\xi|^2$, hence, before the limit $\epsilon\to 0$, their sum is divergent. This is the same divergence we had found earlier in 5.3.1 when the duality relation involved the independent ξ-process.

This same problem is present also in the derivation of the true Boltzmann equation, see Lanford, [83], where convergence is proven only for short times, the proof being essentially the same as the above one. The extension to longer times remains, in general, an open problem, maybe the open problem in this field. But our model is infinitely simpler than the hard sphere gas from which the Boltzmann equation is derived, and it does not seem the best strategy to stick to a technique which applies to much more complex cases, without exploiting the peculiar features of our model.

Let us now briefly describe the approach used for proving Theorem 5.2.1. Our goal is to find an iterative scheme where the convergence till time T is proven in a closed way, namely in such a way that the conditions at the initial time are also satisfied at time T: we can then start again and reach arbitrarily long times.

Exploiting the Markov property of the evolution, our iterative procedure will consist of consecutive conditionings on the particles configuration at the iteration times T, $2T$, and so forth: we therefore need to prove convergence starting from singletons. In general single configurations behave badly with respect to sup norms. We have in fact, even for our nice initial measure μ^ϵ:

$$\lim_{\epsilon\to 0}\mathbb{P}_{\mu^\epsilon}\big(\{\eta(x)\le(\log\epsilon^{-1})^a\}\big)=\begin{cases}0 & \text{if } a<1\\ 1 & \text{if } a>1\end{cases}$$

Hence, for typical configurations, $\sup\eta(x)>|\log\epsilon|^a$, $0<a<1$, and the corresponding convergence time t_0 in Lemma 5.3.3 vanishes as $\epsilon\to 0$. But the sup norm is maybe too pessimistic, not for too many x, $\eta(x)>\log\epsilon^{-1}$, and since the evolution is random, single states may not be really relevant.

This is the second key point of our approach, namely to derive estimates which show that single configurations η behave like smoothened configurations η', obtained by convoluting η with some nice kernel whose support diverges as $\epsilon\to 0$. Things should be done in such a way that the sup norm of η' is bounded uniformly with probability which goes to 1 as $\epsilon\to 0$. If we can also prove that this holds not only at time 0 but with large probability with respect to the measure at time T starting from any η whose η' is nice, we have in our hands the desired iterative scheme.

5.4.1 Definition: the ρ-functions.

Given $\epsilon > 0$ and a measure λ on Ω_ϵ we define $\rho^\epsilon(q,\sigma,t|\lambda)$ as the solution of

$$\frac{d}{dt}\rho^\epsilon(q,\sigma,t|\lambda) = \rho^\epsilon(q,-\sigma,t|\lambda)^2 - \rho^\epsilon(q,\sigma,t|\lambda)^2 + \epsilon^{-1}[\rho^\epsilon(q-\sigma,\sigma,t|\lambda) - \rho^\epsilon(q,\sigma,t)] \tag{5.20a}$$

$$\rho^\epsilon(x,0|\lambda) = \mathbb{E}_\lambda(\eta(x)) \tag{5.20b}$$

assuming that the expected total number of particles with respect to λ is finite. If λ is supported by η we simply write $\rho^\epsilon(q,\sigma,t|\eta)$.

The right hand side of (5.20) is a particular discretization of (5.1), obtained by averaging $L^\epsilon\eta(q,\sigma)$ with respect to the product measures which has averages $\rho^\epsilon(x,t|\eta)$. Products of Poisson measures are not left invariant by the evolution, because of the presence of L_c in the generator. But we may hope that this is approximately true because the other term in the generator is $\epsilon^{-1}L_0$, (under which product of Poisson measures are transformed into themselves, see Theorem 2.9.5). It is also natural then to expect that the average occupation numbers are not too different from the solutions of (5.20). Therefore one of the ingredients in the proof of Theorem 5.2.1 will be to show that for most of the initial configurations η, $\rho^\epsilon(q,\sigma,t|\eta)$ converges to the solution of (5.1). The other crucial point, as mentioned above, is to prove that the measure at time t is close to the measure which is the product of Poisson measures with averages $\rho^\epsilon(q,\sigma,t|\eta)$. For this we need to define a distance between measures with the property that if the above two measures are close, then their typical configurations, considered as initial conditions in (5.2), should give rise to solutions which are close till macroscopic times. This is achieved in the following definition.

5.4.2 Definition: the v-functions.

Let $u^\epsilon(\xi|\mu)$ and $u^\epsilon(\xi|\nu)$, $\xi \in \Omega_\epsilon$, be the correlation functions of two probabilities, μ and ν, on Ω_ϵ. Let $\underline{x}$ be any labelling of ξ, namely such that $U(\underline{x}) = \xi$, see (2.2), and write $u^\epsilon(\underline{x}|\mu)$ for $u^\epsilon(U(\underline{x})|\mu)$. We also introduce the following notation: let $\underline{x} \in \mathbb{Z}^n$, J a subset of $\{1,\ldots,n\}$ and J' its complement, then we write $\underline{x}_J$ for the configuration in $\mathbb{Z}^{|J|}$ consisting of $\{x_j, j \in J\}$ ordered by the increasing values of j.

We set

$$d_{\mu,\nu}(\xi) = \sum_{J\subset\{1,\ldots,n\}} (-1)^{|J|} u^\epsilon(\underline{x}_{J'}|\mu) u^\epsilon(\underline{x}_J|\nu) \tag{5.21}$$

(5.21) recalls the definition of the cumulant functions and of the truncated correlation functions in statistical mechanics.

We use special notation when $\mu = \lambda_t^\epsilon$, (the measure at time t of the process with generator L^ϵ and initial measure λ), and ν the product measure with averages $\rho^\epsilon(q,\sigma,t|\lambda)$:

$$\begin{aligned} v_t^\epsilon(\xi|\lambda) &= d_{\lambda_t^\epsilon,\nu}(\xi) \\ &= \sum_{J\subset\{1,\ldots,n\}} (-1)^{|J|} u^\epsilon(\underline{x}_{J'}|\lambda) \prod_{i\in J} \rho^\epsilon(x_i,t|\lambda) \end{aligned} \tag{5.22}$$

If λ is supported by a single configuration η we also write $v_t^\epsilon(\xi|\eta)$. Sometimes we drop λ or η in the argument of the ρ and of the v-functions. Finally if $|\xi| = n$ we say that $v_t^\epsilon(\xi|\eta)$ has degree n or that it has n bodies.

Remarks. Notice that if $\xi = U(\underline{x})$ and $\underline{x} = (x_1, \ldots, x_n)$ with the x_i's distinct, then

$$v_t^\epsilon(\xi|\lambda) = \mathbf{E}_\lambda^\epsilon\Big(\prod_{i=1}^n [\eta(x_i,t) - \rho^\epsilon(x_i,t;\lambda)]\Big)$$

Good factorization properties of the evolution make the v functions small, even smaller when their degree increases. If the x_i's are not all distinct, then the smallness of v^ϵ is implied by the measure being close to a product of Poisson measures. This is what happens in our case, as shown by the following theorem, whose proof is postponed to the end of the chapter.

5.4.3 Theorem. *For each $\epsilon > 0$ let $\eta \in \Omega_\epsilon$ be such that*

$$\sup_{x \in \mathbf{Z}_\epsilon \times \{-1,1\}} \eta(x) \leq \epsilon^{-\zeta} \tag{5.23}$$

Then for any $\beta > 0$ and any $n \geq 1$ there is c so that for all $t \leq 2\epsilon^\beta$ and for all η satisfying (5.23)

$$\sup_{|\xi|=n} |v_t^\epsilon(\xi|\eta)| \ \leq \ c\,(\frac{\epsilon}{t})^{n/4}\,\epsilon^{-\zeta n} \tag{5.24}$$

Theorem 5.4.3 shows that the process follows the deterministic equation (5.20) in a very strong sense. This however cannot be used directly to prove Theorem 5.2.1 because the result holds only for the infinitesimally short times $t \leq \epsilon^\beta$, $\beta > 0$. On the other hand (5.24) allows to characterize the typical configuration at time ϵ^β and this is the beginning of an iterative procedure which will make it possible to reach finite times.

5.4.4 Definition: the seminorm $\|\cdot\|$.

For any function f on Ω_ϵ we introduce the seminorm

$$\|f\| = \sup_{x \in \mathbf{Z}_\epsilon \times \{-1,1\}} \Big| \sum_{x' \in \mathbf{Z}_\epsilon \times \{-1,1\}} P^{\epsilon,*}_{\epsilon^{1/4}}(x \to x') f(\eta(x'))\Big| \tag{5.25}$$

where $P_t^{\epsilon,*}(x \to x')$ is the probability that a particle starting from $x = (q,\sigma)$ and jumping by $-\sigma$ after exponential times of intensity 1, is at $x' = (q', \sigma)$ at time $\epsilon^{-1}t$.

The choice $t = \epsilon^{1/4}$ in the above definition is not mandatory, other choices would have worked as well. Notice that $P^{\epsilon,*}_{\epsilon^{1/4}}(x \to x')$ is essentially supported by the sites q' which are in an interval centered at $q - \sigma\epsilon^{-1}\epsilon^{1/4}$ with length $\backsim (\epsilon^{-1}\epsilon^{1/4})^{1/2} = \epsilon^{-3/8}$, so that the above is, roughly speaking, the sup-norm of the average of f in all the intervals of size $\epsilon^{-3/8}$.

5.4.5 Lemma. *Let $\gamma = \frac{1}{16}$, $\beta = \frac{1}{100}$ and $\zeta = \frac{1}{10^6}$, (the Lemma holds for more general choices of γ, β and ζ: γ has to be fixed small enough, then β should be accordingly smaller and ζ, in turns, much smaller). Assume that $\eta(x) \leq \epsilon^{-\zeta}$and let $\rho^\epsilon(x,t|\eta)$ solve (5.20) with λ supported by η. Then for any n there is c so that for all $\epsilon^\beta \leq t \leq 2\epsilon^\beta$*

$$\mathbf{P}_\eta^\epsilon\big(\|\rho^\epsilon(\cdot,t|\eta) - \eta(\cdot,t)\| > \epsilon^\gamma\big) \leq c\epsilon^n \tag{5.26}$$

The proof uses the Chebyshev inequality, the estimate (5.24) on the v-functions, an a-priori bound on the ρ-functions established in Lemma 5.6.1 below, and the local central limit theorem for random walks. For the sake of completeness we give the details below.

5.4.6 Proof.*

To avoid confusion we denote by η_0 the initial configuration η in Lemma 5.4.5. We use the Chebyshev inequality to bound the left hand side of (5.26) by

$$\sum_{x\in\mathbb{Z}_\epsilon\times\{-1,1\}} \epsilon^{-2k\gamma} \sum_{x_1\cdots x_{2k}} \prod_{i=1}^{2k} P^{\epsilon,\star}_{\epsilon^{1/4}}(x\to x_i)\mathbf{E}^\epsilon_{\eta_0}(\prod_{i=1}^{2k}[\eta(x_i,t)-\rho^\epsilon(x_i,t|\eta_0)]) \tag{5.27}$$

Before deriving a bound for (5.27), we bound some of the terms which appear in the expression (5.27): this will give an idea of the proof in the general case.

(i) *Contribution to (5.27) coming from the sum when all the $x_1,\dots,x_{2k}$ are distinct.* By (5.24) we have

$$\begin{aligned}\sum_{x\in\mathbb{Z}_\epsilon\times\{-1,1\}} \epsilon^{-2k\gamma} \sum_{x_1\neq\cdots\neq x_{2k}} \prod_{i=1}^{2k} P^{\epsilon,\star}_{\epsilon^{1/4}}(x\to x_i)\mathbf{E}^\epsilon_{\eta_0}(\prod_{i=1}^{2k}[\eta(x_i,t)-\rho^\epsilon(x_i,t|\eta_0)])\\ \le \epsilon^{-1}\epsilon^{-2k\gamma}\sup_{|\xi|=n}|v^\epsilon_t(\xi)|\le c\epsilon^{-1}\epsilon^{-2k\gamma}\epsilon^{(1-1/4)2k/4-\varsigma 2k}\end{aligned} \tag{5.28}$$

which is smaller than $c\epsilon^n$ for k large enough.

(ii) *Contribution to (5.27) coming from the sum over pairwise equal $x_1,\dots,x_{2k}$.* Using the fact that for all $\epsilon^\beta\ge t>0$, and all $x\in\mathbb{Z}_\epsilon\times\{-1,1\}$,

$$P^{\epsilon,\star}_t(x\to x_i)\le\frac{c}{\sqrt{\epsilon^{-1}t}} \tag{5.29}$$

we have that

$$\begin{aligned}\sum_x\epsilon^{-2k\gamma}\sum_{x_1\neq\cdots\neq x_k}\prod_{i=1}^{k}P^{\epsilon,\star}_{\epsilon^{1/4}}(x\to x_i)^2\mathbf{E}^\epsilon_{\eta_0}(\prod_{i=1}^{k}[\eta(x_i,t)-\rho^\epsilon(x_i,t|\eta_0)]^2)\\ \le\epsilon^{-1}\epsilon^{-2k\gamma}\left(\frac{1}{\sqrt{\epsilon^{-1}\epsilon^{1/4}}}\right)^k\sup_{x_1\neq\cdots\neq x_k}\mathbf{E}^\epsilon_{\eta_0}(\prod_{i=1}^{k}[\eta(x_i,t)-\rho^\epsilon(x_i,t|\eta_0)]^2)\end{aligned} \tag{5.30}$$

Recalling (5.22) we define

$$w^\epsilon_t(2,\eta(x)|\eta_0)=[\eta(x,t)^2-\eta(x,t)]-2\eta(x,t)\rho^\epsilon(x,t|\eta_0)+\rho^\epsilon(x,t|\eta_0)^2$$

so that

$$\mathbf{E}^\epsilon_{\eta_0}(w^\epsilon_t(2,\eta(x)|\eta_0))=v^\epsilon(2\delta_x,t|\eta_0)$$

where $2\delta_x$ is the configuration with only two particles both in x.

We have

$$[\eta(x,t)-\rho^\epsilon(x,t|\eta_0)]^2=w^\epsilon_t(2,\eta(x)|\eta_0)+w^\epsilon_t(1,\eta(x)|\eta_0)+\rho^\epsilon(x,t|\eta_0)$$

where

$$w_t^\epsilon(1,\eta(x)|\eta_0) = \eta(x,t) - \rho^\epsilon(x,t|\eta_0)$$

Using (5.29), (5.24) and that $\rho^\epsilon(x,t|\eta_0) \le \epsilon^{-\zeta}$ by Lemma 5.6.1 below, we bound the right hand side of (5.30) by

$$c\epsilon^{-1}\epsilon^{-2k\gamma}\left(\frac{1}{\sqrt{\epsilon^{-1+1/4}}}\right)\epsilon^{-\zeta k}$$

where c is a suitable constant. By choosing k large enough this is bounded, for any given n, by $c\epsilon^n$, for a suitable c.

We now turn to the estimate of the full expression (5.27). Recalling (5.22), (2.53) and (2.55) we define

$$w_t^\epsilon(n,\eta(x)|\eta_0) = \sum_{J\subset\{1,\dots,2k\}} (-1)^{|J|} D_{|J'|}(\eta(x,t))\rho^\epsilon(x,t|\eta_0)^{|J|}, \quad J' = \text{complement of } J \tag{5.31}$$

setting $w_t^\epsilon(0,\eta(x)|\eta_0) \equiv 1$. Then

$$v_t^\epsilon(\xi|\eta_0) = \mathbb{E}_{\eta_0}^\epsilon\left(\prod_x w_t^\epsilon(\xi(x),\eta(x)|\eta_0)\right) \tag{5.32}$$

We have an inversion formula:

$$\eta(x,t)^n = \sum_{h,k} a_{h,k}(n) w_t^\epsilon(h,\eta(x)|\eta_0)\rho^\epsilon(x,t|\eta_0)^k \tag{5.33a}$$

where $a_{h,k}(n) = 0$ when $h > n$ and $k > n-h$, $a_{n,0}(n) = 0$. Analogously

$$[\eta(x,t) - \rho^\epsilon(x,t|\eta_0)]^n = \sum_{h,k} b_{h,k}(n) w_t^\epsilon(h,\eta(x)|\eta_0)\rho^\epsilon(x,t|\eta_0)^k \tag{5.33b}$$

where $b_{h,k}(n) = 0$ when $h > n$, $k > n-h$, $b_{n,0}(n) = 0$ as before, and $b_{h,k}(1) = \delta_{h,1}\delta_{k,0}$, because

$$\eta(x,t) - \rho^\epsilon(x,t|\eta_0) = w_t^\epsilon(1,\eta(x)|\eta_0) \tag{5.33c}$$

Let $\xi = U(\underline{x})$, $\underline{x} = (x_1,\dots,x_{2k})$, see (2.2), then

$$\prod_{i=1}^{2k}[\eta(x_i,t) - \rho^\epsilon(x_i,t|\eta_0)] = \prod_{x:\xi(x)>0}\sum_{h,k} b_{h,k}(|\xi(x)|) w_t^\epsilon(h,\eta(x)|\eta_0)\rho^\epsilon(x,t|\eta_0)^k$$

Recalling that $\rho^\epsilon(x,t|\eta_0) \le \epsilon^{-\zeta}$, see Lemma 5.6.1, we have, using (5.24) and the properties of the coefficients $b_{h,k}(n)$ stated above,

$$\mathbb{E}_{\eta_0}^\epsilon\left(\prod_{i=1}^{2k}[\eta(x_i,t) - \rho^\epsilon(x_i,t|\eta_0)]\right) \le c\left(\frac{\epsilon}{t}\right)^{n(\underline{x})}\epsilon^{-\zeta(2k-n(\underline{x}))}$$

where c is a suitable constant and $n(\underline{x})$ is the number of indices i such that $x_j \ne x_i$ for all $j \ne i$.

We then have that (5.27) is bounded by

$$c\epsilon^{-1}\epsilon^{-2k\gamma}\sum_{n=0}^{2k}\left(\frac{\epsilon}{t}\right)^n\epsilon^{-\zeta(2k-n)}\left(\frac{1}{\sqrt{\epsilon^{-1+1/4}}}\right)^{(2k-n)/2}$$

In (i) we had considered the case $n = 2k$, in (ii) the case $n = 0$, proving a better bound than the above one, $\epsilon^{-\zeta k}$ instead of $\epsilon^{-2k\zeta}$. Since $t \ge \epsilon^\beta$, recalling the values of β, γ and ζ we deduce (5.26) after choosing k large enough. □

§5.5 Convergence at all times.

Lemma 5.4.5 cannot be iterated in a direct way because $\|\rho^\epsilon(\cdot,t|\eta)-\eta(\cdot,t)\|\le\epsilon^\gamma$ does not imply that $\eta(x,t)\le\epsilon^{-\zeta}$, condition necessary for re-applying Lemma 5.4.5. We shall prove that $\rho^\epsilon(\cdot,t|\eta)$ is bounded by a constant, for typical η's, and then, by using (5.26), that with large probability $\eta(\cdot,t)$ cannot be much larger than $\rho^\epsilon(\cdot,t|\eta)$, in particular that it is smaller than $\epsilon^{-\zeta}$. We first need a definition

5.5.1 Definition: The good set H^ϵ.

For each fixed $T>0$ and $1\le r\le 2$ we define

$$H^\epsilon=\{\eta^{(k)}\equiv\eta_{kr\epsilon^\beta},\ \ kr\epsilon^\beta\le T,\ \text{so that 1), 2), 3) below hold }\}\tag{5.34}$$

where

1) $\|\eta_0-\rho_0^\epsilon\|\le\epsilon^\gamma\quad \rho_0^\epsilon(q,\sigma)=\rho_0(\epsilon q,\sigma)$.
2) $\eta^{(k)}(x)\le\epsilon^{-\zeta}$ for all x and for all $k\ge1$ and such that $kr\epsilon^\beta\le T$.
3) $\|\eta^{(k)}-\rho^\epsilon(\cdot,r\epsilon^\beta|\eta^{(k-1)})\|\le\epsilon^\gamma$ for all $k\ge1$ and such that $kr\epsilon^\beta\le T$.

5.5.2 Proposition. *For any n there is a c so that*

$$\mathbf{P}^\epsilon_{\mu^\epsilon}(H^\epsilon)\ge 1-c\epsilon^n\tag{5.35}$$

The proof of the proposition is postponed to the end of this paragraph. Next we establish a regularity property of the trajectories in H^ϵ inherited from the good properties of the solutions to (5.20), which indeed has much better smoothening properties than (5.1) itself.

5.5.3 Proposition. *For any $T>0$ there is a constant $\bar c$ such that the following holds. Given any $r\in[1,2]$ and any $0<\epsilon\le1$ let $\eta^{(k)}$, $kr\epsilon^\beta\le T$, be any sequence in H^ϵ.*

Then for all k as above and $x=(q,\sigma)$

$$|\rho^\epsilon(x,r\epsilon^\beta|\eta^{(k)})-\rho(\epsilon q,\sigma,(k+1)r\epsilon^\beta)|\le\bar c\epsilon^\gamma k\le\bar c\epsilon^{\gamma-\beta}T\tag{5.36}$$

where $\rho(r,\sigma,t)$ solves (5.1) with initial datum ρ_0.

Of course for (5.36) to hold with a given k it is enough to assume 1), 2) and 3) only for $h\le k$.

Since $\gamma=1/16$ and $\beta=10^{-2}$, the last expression in (5.36) vanishes as $\epsilon\to0$, for any given T.

We also postpone the proof of Proposition 5.5.3 to the end of the paragraph and conclude that of Theorem 5.2.1 Fix $t>0$ and, given $\epsilon>0$, let $r\in[1,2]$ be such that $t=kr\epsilon^\beta$, then

$$\mathbf{E}^\epsilon_{\mu^\epsilon}(D(\xi,\eta_t))=\mathbf{E}^\epsilon_{\mu^\epsilon}\Big([1-\chi]D(\xi,\eta_t)\Big)+\mathbf{E}^\epsilon_{\mu^\epsilon}\Big(\chi\mathbf{E}^\epsilon_{\eta_{(k-1)r\epsilon^\beta}}(D(\xi,\eta_{r\epsilon^\beta}))\Big)\tag{5.37}$$

where χ is the characteristic function of H^ϵ_{k-1}, H^ϵ_k being the subset of H^ϵ where conditions 1), 2) and 3) are imposed only for $h\le k$. Using the Cauchy-Schwartz inequality and (5.35), for any given n we bound the first term on the right hand side of (5.37) by

$$\left[\mathbf{E}^\epsilon_{\mu^\epsilon}([1-\chi]D(\xi,\eta_t))\right]^2\le c\epsilon^n\mathbf{E}^\epsilon_{\mu^\epsilon}(D(\xi,\eta_t)^2)\le c\epsilon^n\mathbf{E}^\epsilon_{\mu^\epsilon}\Big([\sum_x\eta(x,t)]^{2|\xi|}\Big)$$
$$\le c\epsilon^n\mathbf{E}^\epsilon_{\mu^\epsilon}\Big([\sum_x\eta(x,0)]^{2|\xi|}\Big)\le c\epsilon^{n-2|\xi|}$$

the value of the constant c changing from line to line. In the third inequality we have used that the total number of particles is conserved and in the fourth one that the distribution of $\sum \eta(x,0)$ is Poisson with mean $\sum \rho_0(\epsilon q,\sigma)$. By choosing n large enough, the last term vanishes when $\epsilon \to 0$. For the second term on the right hand side of (5.37) we write

$$\mathbb{E}^\epsilon_{\mu^\epsilon}\left(\chi \mathbb{E}^\epsilon_{\eta_{(k-1)r\epsilon^\beta}}(D(\xi,\eta_{r\epsilon^\beta}))\right) \le \sup_{\eta\in H^\epsilon_{k-1}} \left|\mathbb{E}^\epsilon_\eta\left([D(\xi,\eta_{r\epsilon^\beta}) - \prod_x \rho^\epsilon(x,r\epsilon^\beta|\eta)^{\xi(x)}]\right)\right| + \mathbb{E}^\epsilon_{\mu^\epsilon}\left(\chi[\prod_x \rho^\epsilon(x,r\epsilon^\beta|\eta)^{\xi(x)}]\right)$$

By (5.33) $D(\xi,\eta_{r\epsilon^\beta}) - \prod_x \rho^\epsilon(x,r\epsilon^\beta|\eta)^{\xi(x)}$ is equal to a linear combination of products of w^ϵ-functions, see (5.31), and ρ^ϵ-functions, each term containing at least one w^ϵ. By Proposition 5.5.3 $\rho^\epsilon(x,r\epsilon^\beta|\eta) \le c$ because $\sup_{r,\sigma}\rho(r,\sigma,t) \le \|\rho_0\|_\infty \le c$, therefore by Theorem 5.4.3

$$\sup_{\eta\in H^\epsilon_{k-1}} \sup_{|\xi|=n} \mathbb{E}^\epsilon_\eta\left([D(\xi,\eta_{r\epsilon^\beta}) - \prod_x \rho^\epsilon(x,r\epsilon^\beta|\eta)^{\xi(x)}]\right) \le c\epsilon^{(1-\beta)(1/4-\zeta)}$$

By using again (5.36) we get

$$\sup_{|\xi|=n} \chi\left|\prod_x \rho^\epsilon(x,r\epsilon^\beta|\eta)^{\xi(x)} - \prod_x \rho(\epsilon q,\sigma,rk\epsilon^\beta)^{\xi(x)}\right| \le c\epsilon^{\gamma-\beta}$$

Collecting all the above estimates we obtain the proof of Theorem 5.2.1, once Theorem 5.4.3 and Propositions 5.5.2-5.5.3 are also proven. □

5.5.4 Proof of Proposition 5.5.3.

We need the following estimate: there is c such that [$\pm\epsilon x$ below shorthands $(\epsilon q,\pm\sigma)$, $x=(q,\sigma)$]

$$|\rho(\epsilon x,t) - \sum_y P^{\epsilon,\star}_{t-kr\epsilon^\beta}(x\to y)\rho(\epsilon y,kr\epsilon^\beta) - \int_{kr\epsilon^\beta}^t ds \sum_y P^{\epsilon,\star}_{t-s}(x\to y)[\rho(-\epsilon y,s)^2 - \rho(\epsilon y,s)^2]| \le c_1\sqrt{\epsilon} \tag{5.38}$$

where $P^{\epsilon,\star}_t$ was introduced in 5.4.4.

The inequality (5.38) is a direct consequence of (1) the partial derivative $\rho_r(r,\sigma,t)$ with respect to r of the solution of (5.1) is uniformly bounded when t is in the compacts, and (2) the following inequality holds for any $s>0$ and (q,σ),

$$\sum_{q'} P^{\epsilon,\star}_s(q\to q')|q' - (q-\sigma\epsilon^{-1}s)| \le c\epsilon^{-\frac{1}{2}}s$$

The statement (2) is a consequence of the local central limit theorem, cf. also (4.34). To prove (1) we take the derivative with respect to r of both sides of (5.1a) and observe that a) $\rho(r,\sigma,t)$ is uniformly bounded as it is at time 0, and b) $|\rho_r(r,\sigma,0)|$ is also uniformly bounded, by assumption. One then finds a linear equation on $\rho_r(r,\sigma,t)$ with bounded coefficients, hence the solution is uniformly bounded on the compacts.

Denoting as usual, by $\|\rho_0\|_\infty$ the sup norm of ρ_0 we define

$$c^\star = e^{6\|\rho_0\|_\infty T}[3 + c_1 + \|\rho_0\|_\infty] \tag{5.39a}$$

and consider ϵ so small that

$$c^{\star}\epsilon^{-\beta}T\epsilon^{\gamma}+\|\rho_0\|_{\infty}+2\epsilon^{1/4-2\zeta}\leq 2\|\rho_0\|_{\infty} \tag{5.39b}$$

We are going to prove that (5.36) holds for ϵ as above and $\bar{c}=c^{\star}$, from this Propostion 5.5.3 easily follows. We shall proceed by induction on k. We assume that (5.36) holds with $k=k^{\star}-1$, $(k^{\star}+1)r\epsilon^{\beta}\leq T$ and we shall prove that it holds with $k=k^{\star}$.

We define

$$h^{\epsilon}(x,s)=|\rho^{\epsilon}(x,s-kr\epsilon^{\beta},\eta^{(k)})-\rho(\epsilon x,s)|,\qquad kr\epsilon^{\beta}<s\leq(k+1)r\epsilon^{\beta} \tag{5.39c}$$

$$h^{\epsilon}_{t}=\sup_{x\in\mathbb{Z}_{\epsilon}\times\{-1,1\}}h^{\epsilon}(x,t) \tag{5.39d}$$

Let $kr\epsilon^{\beta}+\epsilon^{1/4}\leq t\leq(k+1)r\epsilon^{\beta}$, $k\leq k^{\star}$, then

$$h^{\epsilon}(x,t)\leq[h^{\epsilon}_{kr\epsilon^{\beta}}+\epsilon^{\gamma}]+c_1\sqrt{\epsilon}+2\epsilon^{1/4-2\zeta}+\|\rho_0\|_{\infty}\epsilon^{1/4}+\int_{kr\epsilon^{\beta}+\epsilon^{1/4}}^{t}ds6\|\rho_0\|_{\infty}\,h^{\epsilon}_{s} \tag{5.40}$$

The first term on the right hand side of (5.40) is obtained using condition 3) on the sequence $\eta^{(k)}$; for the second one we use (5.38). The third term arises from the integral equation for $\rho^{\epsilon}(x,t|\eta^{(k-1)})$: the integral extended to $[(kr\epsilon^{\beta},kr\epsilon^{\beta}+\epsilon^{1/4}]$ is bounded using Lemma 5.6.1 below. The analogous integral in the equation for $\rho(\epsilon q,\sigma,t)$ gives rise to the fourth term on the right hand side of (5.40). For the integral term in (5.40) we have used the following bounds:

$$\begin{aligned}|\rho^{\epsilon}(-\epsilon y,s|\eta^{(k)})^2-\rho^{\epsilon}(\epsilon y,s|\eta^{(k)})^2-&\rho(-\epsilon y,s+kr\epsilon^{\beta})^2+\rho(\epsilon y,s+kr\epsilon^{\beta})^2|\\&\leq 2h^{\epsilon}_{s}(\|\rho_0\|_{\infty}+\sup_{x}\rho^{\epsilon}(\epsilon x,s|\eta^{(k)}))\end{aligned}$$

We then obtain (5.40) once we prove that

$$\sup_{x}\rho^{\epsilon}(\epsilon x,s|\eta^{(k)})\leq 2\|\rho_0\|_{\infty},\quad\text{for all }\epsilon^{1/4}\leq s\leq r\epsilon^{\beta} \tag{5.41}$$

To prove (5.41) we notice that, by Lemma 5.6.1

$$\sup_{x}\rho^{\epsilon}(\epsilon y,s|\eta^{(k)})\leq\sup_{x}\rho^{\epsilon}(\epsilon x,\epsilon^{1/4}|\eta^{(k)})$$

We then have

$$\rho^{\epsilon}(\epsilon y,\epsilon^{1/4}|\eta^{(k)})\leq h^{\epsilon}_{kr\epsilon^{\beta}}+\epsilon^{\gamma}+\|\rho_0\|_{\infty}+2\epsilon^{1/4-2\zeta}$$

and using the induction hypothesis

$$\leq c^{\star}k\epsilon^{\gamma}+\epsilon^{\gamma}+\|\rho_0\|_{\infty}+2\epsilon^{1/4-2\zeta}\leq 2\|\rho_0\|_{\infty}$$

by (5.39b). This proves (5.41).

We have therefore completed the proof of (5.40). By iterating (5.40) we get for $kr\epsilon^{\beta}+\epsilon^{1/4}\le t\le (k+1)r\epsilon^{\beta}$, $k\le k^*$

$$h^{\epsilon}_t \le (3+c_1+\|\rho_0\|_\infty)k\epsilon^{\gamma}+6\|\rho_0\|_\infty\int_0^t ds\chi_s h^{\epsilon}_s, \tag{5.42}$$

where

$$\chi_s=\begin{cases}1 & \text{if } s\in\cup_{k\ge 0}[kr\epsilon^{\beta}+\epsilon^{\frac14},(k+1)r\epsilon^{\beta}]\\ 0 & \text{otherwise}\end{cases}$$

By (5.39a)

$$h^{\epsilon}_{(k^*+1)r\epsilon^{\beta}}\le e^{6\|\rho_0\|_\infty T}(3+c_1+\|\rho_0\|_\infty)(k^*+1)\epsilon^{\gamma}\le c^*(k^*+1)\epsilon^{\gamma}$$

Proposition 5.5.3 is therefore proven. □

5.5.5 Proof of Proposition 5.5.2.

Call H^{ϵ}_k the subset in H^{ϵ} where the conditions 1), 2) and 3) are imposed only for $h\le k$, so that H^{ϵ}_k depends only on the configurations at the times $hr\epsilon^{\beta}$ with $h\le k$. Let χ^{ϵ}_k be the characteristic function of H^{ϵ}_k. Then, setting $\eta'\equiv\eta_{(k-1)r\epsilon^{\beta}}$,

$$\mathbb{P}^{\epsilon}_{\mu^{\epsilon}}(H^{\epsilon}_k)=\mathbb{E}^{\epsilon}_{\mu^{\epsilon}}\left(\chi^{\epsilon}_{k-1}\mathbb{P}^{\epsilon}_{\eta'}\left(\{\sup_x \eta(x,r\epsilon^{\beta})\le\epsilon^{-\zeta},\|\eta_{r\epsilon^{\beta}}-\rho^{\epsilon}(\cdot,r\epsilon^{\beta}|\eta')\|\le\epsilon^{\gamma}\}\right)\right) \tag{5.43}$$

Because of the characteristic function χ^{ϵ}_{k-1}, η' satisfies condition 2) so that for all x: $\eta'(x)\le\epsilon^{-\zeta}$. Then by Lemma 5.4.5 for any N there is c such that

$$\mathbb{P}^{\epsilon}_{\eta'}(\|\eta_{r\epsilon^{\beta}}-\rho^{\epsilon}(\cdot,r\epsilon^{\beta}|\eta')\|\ge\epsilon^{-\gamma})\le c\epsilon^{N}$$

By (5.36)

$$|\rho^{\epsilon}(q,\sigma,r\epsilon^{\beta}|\eta')-\rho(\epsilon q,\sigma,kr\epsilon^{\beta})|\le\bar{c}\epsilon^{\gamma}T\epsilon^{-\beta}$$

hence, since $\gamma>\beta$,

$$\sup_x \rho^{\epsilon}(x,\epsilon^{\beta}|\eta')\le\|\rho_0\|_\infty+c'\epsilon^{\gamma-\beta}\le d,\quad d\ge 1 \tag{5.44}$$

Using the Chebyshev inequality we get that

$$\mathbb{P}^{\epsilon}_{\eta'}(\sup_x\eta(x,r\epsilon^{\beta})>\epsilon^{-\zeta})\le\epsilon^{-1}\epsilon^{\zeta n}\mathbb{E}^{\epsilon}_{\eta'}(\eta(x,r\epsilon^{\beta})^n)$$

By (5.33a), (5.44) and Theorem 5.4.3, we have

$$\mathbb{E}^{\epsilon}_{\eta'}(\eta(x,r\epsilon^{\beta})^n)\le cd^n$$

Inserting the above estimates in (5.43) we get

$$\mathbb{P}^{\epsilon}_{\mu^{\epsilon}}(H^{\epsilon}_k)\ge\mathbb{P}^{\epsilon}_{\mu^{\epsilon}}(H^{\epsilon}_{k-1})\quad c\epsilon^{N}$$

By iterating the above inequality and by the arbitrarity of N, we prove the same bound as in (5.35) for the probability that conditions 2) and 3) are satisfied. A similar bound for condition 1) can be proved recalling that μ^{ϵ} is a product of Poisson measures and then using arguments similar to those in the proof of Lemma 5.4.5. □

§5.6 Proof of Theorem 5.4.3.

We have already mentioned that the a-priori bound valid for the Carleman equation (5.1), hold also for (5.20):

5.6.1 Lemma. *For any* $0 < \epsilon \leq 1$ *and any* $\eta \in \Omega_\epsilon$

$$\sup_{x,t} \rho^\epsilon(x,t|\eta) \leq \sup_x \eta(x)$$

Proof. For each ϵ, x takes finitely many values. Having fixed t, let then x_0 be the state where $\rho^\epsilon(x,t|\eta)$ reaches its maximum. The right hand side of (5.20a) computed at $(q,\sigma) = x_0$ is non positive hence the maximum of $\rho^\epsilon(x,t|\eta)$ is reached at time 0. □

Next we express the time derivative of the v-functions as linear combinations of v-functions. We need a definition

5.6.2 Definition: The operators a^ϵ and A^ϵ.

For any $\epsilon > 0$, $x = (q,\sigma)$ and $t \geq 0$ we define $a^\epsilon_t(\xi(x) \to \xi'(x))$ as follows: setting

$$\xi(q,1) = h, \quad \xi(q,-1) = k \;\; \xi'(q,1) = h' \;\; \xi(q,-1) = k'$$

we write

$$\begin{aligned}
a^\epsilon_t\big((h,k) \to (h-2,k)\big) &= \frac{1}{2}[\rho^\epsilon(q,-1,t)^2 - \rho^\epsilon(q,1,t)^2]h(h-1)\\
a^\epsilon_t\big((h,k) \to (h-2,k+1)\big) &= \rho^\epsilon(q,-1,t)h(h-1)\\
a^\epsilon_t\big((h,k) \to (h-1,k)\big) &= -\rho^\epsilon(q,1,t)h(h-1)\\
a^\epsilon_t\big((h,k) \to (h-2,k+2)\big) &= \frac{1}{2}h(h-1)\\
a^\epsilon_t\big((h,k) \to (h,k)\big) &= -\frac{1}{2}h(h-1) - 2\rho^\epsilon(q,1,t)h\\
a^\epsilon_t\big((h,k) \to (h-1,k+1)\big) &= 2\rho^\epsilon(q,-1,t)h\\
a^\epsilon_t\big((h,k) \to (h-1,k+2)\big) &= h\\
a^\epsilon_t\big((h,k) \to (h+1,k)\big) &= -h
\end{aligned}$$

The definition of a^ϵ is completed by defining $a^\epsilon\big((h_0,k_0) \to (h_0',k_0')\big)$ when the transition $(k_0,h_0) \to (k_0',h_0')$ is one of those considered above, as equal to the expression given above for $a^\epsilon\big((k_0,h_0) \to (k_0',h_0')\big)$ after having changed σ in $-\sigma$ inside the argument of the ρ's. For all the other transitions $a^\epsilon = 0$. In particular $a^\epsilon((0,0) \to (h,k)) = 0$ for all (h,k).

We then define

$$A^\epsilon(\xi \to \xi') = \sum_q a^\epsilon\big((\xi(q,1),\xi(q-1)) \to (\xi'(q,1),\xi'(q-1))\big) \prod_{q' \neq q,\sigma} 1(\xi(q',\sigma) = \xi'(q',\sigma)) \quad (5.45)$$

5.6.3 Remarks. There are just a few features to remember in the expression for a^ϵ:

(1) There are "deaths", i.e. $h' + k' < h + k$, only if $h \geq 2$ and/or $k \geq 2$.

(2) At most two particles die: $h' + k' \geq h + k - 2$

(3) At most one particle is born: $h' + k' \leq h + k + 1$

We shall often use the following notational convention:

$$A_t^\epsilon f(\xi) = \sum_{\xi'} A_t^\epsilon(\xi \to \xi') f(\xi') \tag{5.46}$$

In analogous way we define $P_t^{\epsilon,*} f(\xi)$.

5.6.4 Proposition. *Defining $v_t^\epsilon(\emptyset) \equiv 1$, we have*

$$\frac{d}{dt} v_t^\epsilon(\xi|\eta_0) = \epsilon^{-1} L_0^\star v_t^\epsilon(\xi|\eta_0) + A_t^\epsilon v_t^\epsilon(\xi|\eta_0) \tag{5.47}$$

where $L_0^\star$, defined in (5.8), acts on v_t^ϵ as a function of ξ.

Proof. The proof of the Proposition is based on the following heuristic idea. Since the interactions involve pairs of particles, they can affect at most two particles at the time: if these are among those in ξ, then the degree of the v-function cannot increase and, at most, decrease by 2. If the two particles are not in ξ their interaction should not affect the v-function, and, when one is in ξ and the other is not the degree of the v-function may vary at most by one.

Unfortunately we have not been able to translate in a simple way this kind of argument, and our proof remains essentially lengthy and computational, (even though some tricks are used to make the computations simpler). We therefore suggest to skip it at a first reading and to proceed in the proof of Theorem 5.4.3.

We have, recalling (5.31),

$$\frac{d}{dt} v_t^\epsilon(\xi|\eta_0) = \mathbf{E}_{\eta_0}^\epsilon \Big(\sum_q \{ \prod_{\substack{x'=(q',\sigma') \\ q' \neq q}} w_t^\epsilon(\xi(x'), \eta(x')|\eta_0) \} \{ (L^\epsilon + \frac{\partial}{\partial t}) \prod_{\sigma = \pm 1} w_t^\epsilon(\xi(x'), \eta(q,\sigma)|\eta_0) \} \Big) \tag{5.48}$$

where the time derivative acts on w^ϵ through ρ^ϵ. We set

$$\frac{\partial}{\partial t} \rho^\epsilon = \epsilon^{-1} (\frac{\partial}{\partial t})_0 \, \rho^\epsilon + (\frac{\partial}{\partial t})_c \, \rho^\epsilon \tag{5.49}$$

where

$$(\frac{\partial}{\partial t})_0 \, \rho^\epsilon = \rho^\epsilon(q - \sigma, \sigma) - \rho^\epsilon(q, \sigma) \tag{5.50a}$$

and

$$(\frac{\partial}{\partial t})_c \, \rho^\epsilon = \rho^\epsilon(q, -\sigma)^2 - \rho^\epsilon(q, \sigma)^2 \tag{5.50b}$$

The action of $\epsilon^{-1} L_0 + (\partial/\partial t)_0$ can be singled out in (5.48), we then get

$$\frac{d}{dt} v_t^\epsilon(\xi|\eta_0) = \epsilon^{-1} L_0 v_t^\epsilon(\xi|\eta_0) + \mathbf{E}_{\eta_0}^\epsilon \Big(\sum_q \{ \prod_{\substack{x'=(q',\sigma') \\ q' \neq q}} w_t^\epsilon(\xi(x'), \eta(x')|\eta_0) \} C_t^\epsilon(q) \Big) \tag{5.51}$$

where

$$C_i^\epsilon(q) = L_c(V_k^+(n)V_h^-(m)) + (\frac{\partial}{\partial t})_c(V_k^+(n)V_h^-(m)) \tag{5.52}$$

In the above equation

$$k = \xi(q,1), \quad h = \xi(q,-1), \quad n = \eta_t(q,1), \quad m = \eta_t(q,-1) \tag{5.53}$$

and

$$V_k^+(n)V_h^-(m) = w_i^\epsilon(\xi(q,1),\eta(q,1)|\eta_0)w_i^\epsilon(\xi(q,-1),\eta(q,-1)|\eta_0) \tag{5.53b}$$

In order to compute $C^\epsilon(q,\sigma)$ we observe that

$$\begin{aligned} L_c V_k^+(n)V_h^-(m) = & \frac{1}{2}n(n-1)[V_k^+(n-2)V_h^-(m+2) - V_k^+(n)V_h^-(m)] \\ & + \frac{1}{2}m(m-1)[V_k^+(n+2)V_h^-(m-2) - V_k^+(n)V_h^-(m)] \end{aligned}$$

It is convenient to interpret the above formula in terms of labelled particles, thinking of the factors $n(n-1)$ and $m(m-1)$ as a sum over distinct pairs of labels ranging in the set $\{1,\cdots,n\}$ and in the set $\{1,\cdots,m\}$ respectively. With this in mind we give the following definitions.

For any integers $k \geq 0$ and $n > 0$ we define the following sets.

$$\aleph(n,k) = \{(i_1,\cdots,i_k) \in \{0,\ldots,n\}^k : \text{ if } s \neq s' \text{ and } i_s = i_{s'} \text{ then } i_s = i_{s'} = 0\} \tag{5.54a}$$

Furthermore for any integers $i,j \in \{1,\cdots,n\}, i \neq j$ we let

$$\aleph_{i,j}(n,k) = \{(i_1,\cdots,i_k) \in \{0,1,\cdots,n\}^k : i_s \notin \{i,j\} \ \forall s = 1,\cdots,k\} \tag{5.54b}$$

Finally for any $(q,\sigma) \in Z \times \{-1,1\}$ and $i \in \{1,\cdots,n\}$ we set $\chi^\pm(i)(q,\sigma) \equiv \chi^\pm(i)$, and

$$\begin{aligned} \chi^\pm(i) &= 1 \quad if \quad i \neq 0 \\ \chi^\pm(i) &= -\rho_\pm \quad if \quad i = 0 \end{aligned} \tag{5.55a}$$

where

$$\rho_\pm \equiv \rho^\epsilon(q,\pm\sigma),t|\eta_0) \tag{5.55b}$$

Then from the definitions we have that

$$V_k^\pm(n) = \sum_{\aleph(n,k)} \prod_{s=1}^{k} \chi^\pm(i_s) \tag{5.56}$$

where a set A as a subscript to a sum means the sum over all the elements in A.

We also define for all $(i_1,\cdots,i_k) \in \aleph(n,k)$the functions $\bar{\chi}(i_s)$, as

$$\bar{\chi}(i_s) = \begin{cases} 1 & \text{if } i \neq 0 \\ 0 & \text{if } i = 0 \end{cases} \tag{5.57}$$

From the above definitions it follows that

$$L_c V_k^+(n)V_h^-(m) = \frac{1}{2}\sum_{i\neq j=1}^{n}[\sum_{\aleph_{i,j}(n,k)}\prod_{s=1}^{k}\chi^+(i_s)V_h^-(m+2) - V_k^+(n)V_h^-(m)]$$
$$+\frac{1}{2}\sum_{i\neq j=1}^{m}[\sum_{\aleph_{i,j}(m,h)}\prod_{s=1}^{h}\chi^-(i_s)V_k^+(n+2) - V_k^+(n)V_h^-(m)] \tag{5.58}$$

We next rewrite the right hand side of (5.58) in terms of $V^{\pm}$-functions evaluated at n and m respectively. We start from the following identity true for all integers $r \geq 0$ and $\ell > 0$.

$$V_r^{\pm}(\ell+2) = V_r^{\pm}(\ell)$$
$$+\sum_{s\neq s'=1}^{r}\sum_{\aleph(\ell+2,r)}\prod_{\bar{s}=1}^{r}\chi^{\pm}(i_{\bar{s}})1\{i_s=\ell+1, i_{s'}=\ell+2\}$$
$$+\sum_{s=1}^{r}\sum_{\aleph(\ell+2,r)}\prod_{\bar{s}=1}^{r}\chi^{\pm}(i_{\bar{s}})[1\{i_s=\ell+1, i_{s'}\neq\ell+2\forall s'\neq s\}$$
$$+1\{i_s=\ell+2, i_{s'}\neq\ell+1\ \forall s'\neq s\}] \tag{5.59a}$$

where $1\{\cdot\}$ denotes the characteristic function of $\{\cdot\}$. From (5.59a) it easily follows that

$$V_r^{\pm}(\ell+2) = V_r^{\pm}(\ell) + 2rV_{r-1}^{\pm}(\ell) + r(r-1)V_{r-2}^{\pm}(\ell) \tag{5.59b}$$

We use (5.59b) to rewrite the terms $V_h^-(m+2)$ and $V_k^+(n+2)$ which appear in (5.58). We obtain the following expression.

$$L_c V_k^+(n)V_h^-(m) = V_h^-(m)\frac{1}{2}\sum_{i\neq j=1}^{n}[\sum_{\aleph_{i,j}(n,k)}\prod_{s=1}^{k}\chi^+(i_s) - V_k^+(n)]$$
$$+[2hV_{h-1}^-(m) + h(h-1)V_{h-2}^-(m)]\frac{1}{2}\sum_{i\neq j=1}^{n}\sum_{\aleph_{i,j}(n,k)}\prod_{s=1}^{k}\chi^+(i_s)$$
$$+V_k^+(n)\frac{1}{2}\sum_{i\neq j=1}^{m}[\sum_{\aleph_{i,j}(m,h)}\prod_{s=1}^{h}\chi^-(i_s) - V_h^-(m)]$$
$$+[2kV_{k-1}^+(n) + k(k-1)V_{k-2}(n)]\frac{1}{2}\sum_{i\neq j}^{m}\sum_{\aleph_{i,j}(m,h)}\prod_{s=1}^{h}\chi^-(i_s) \tag{5.60}$$

We now use the following identity true for all integers $r \geq 0$ and $\ell > 0$ derived by writing explicitely the terms in the sum which defines $V_r^{\pm}(\ell)$ and which are not present in $\sum_{\aleph_{\ell,r}(m,h)}\prod_{s=1}^{h}\chi^{\pm}(i_s)$

$$\frac{1}{2}\sum_{i\neq j=1}^{\ell}[\sum_{\aleph_{i,j}(\ell,r)}\prod_{s=1}^{r}\chi^{\pm}(i_s) - V_r^{\pm}(\ell)] = -r\sum_{\aleph(\ell,r+1)}\prod_{s=1}^{r-1}\chi^{\pm}(i_s)\bar{\chi}(i_r)\bar{\chi}(i_{r+1})$$
$$-\frac{1}{2}r(r-1)\sum_{\aleph(\ell,r)}\prod_{s=1}^{r-2}\chi^{\pm}(i_s)\bar{\chi}(i_{r-1})\bar{\chi}(i_r) \tag{5.61}$$

We add and subctract $\chi^{\pm}$ to all the $\bar{\chi}$'s appearing in (5.61). Noticing that

$$\sum_{i_s=0}^{\ell}[\bar{\chi}(i_s)-\chi^{\pm}(i_s)]=\rho_{\pm} \tag{5.62}$$

we then get

$$\frac{1}{2}\sum_{i\neq j=1}^{\ell}[\sum_{\aleph_{i,j}(\ell,r)}\prod_{s=1}^{r}\chi^{\pm}(i_s)-V_r^{\pm}(\ell)]=-r[V_{r+1}^{\pm}(\ell)+2\rho_{\pm}V_r^{\pm}(\ell)+\rho_{\pm}^2V_{r-1}^{\pm}(\ell)]$$
$$-\frac{1}{2}r(r-1)[V_r^{\pm}(\ell)+2\rho_{\pm}V_{r-1}^{\pm}(\ell)+\rho_{\pm}^2V_{r-2}^{\pm}(\ell)] \tag{5.63}$$

Notice that (5.63) contains V-functions with degrees ranging from $r-2$ to $r+1$, as in the definition of a^{ϵ}. There is however a dangerous term: $-r\rho_{\pm}^2V_{r-1}^{\pm}(\ell)$. This means that the degree of the V-functions may decrease by 1 even when there are not two (or more) particles of the configuration ξ in the same state. This term will however be exactly cancelled by one arising from $(\frac{\partial}{\partial t})_c V^{\pm}$.

In order to compute the second and fourth term on the right hand side of (5.60) we use the following identity, true for all integers $r\geq 0$ and $\ell>0$.

$$\frac{1}{2}\sum_{i\neq j=1}^{\ell}\sum_{\aleph_{i,j}(\ell,r)}\prod_{s=1}^{r}\chi^{\pm}(i_s)=\frac{1}{2}\sum_{\aleph(\ell,r+2)}\prod_{s=1}^{r}\chi^{\pm}(i_s)\bar{\chi}(i_{r+1})\bar{\chi}(i_{r+2})$$
$$=\frac{1}{2}[V_{r+2}^{\pm}(\ell)+2\rho_{\pm}V_{r+1}^{\pm}(\ell)+\rho_{\pm}^2V_r^{\pm}(\ell)] \tag{5.64}$$

By (5.63) and (5.64) the sum of the first two terms on the right hand side of (5.60) gives (we write below $V_{k'}^{+}\equiv V_{k'}^{+}(n)$ and $V_{h'}^{-}\equiv V_{h'}^{-}(m)$ for any k' and h')

$$-kV_h^-[V_{k+1}^{+}+2\rho_+V_k^{+}+\rho_+^2V_{k-1}^{+}]-\frac{1}{2}k(k-1)V_h^-[V_k^{+}+2\rho_+V_{k-1}^{+}+\rho_+^2V_{k-2}^{+}]$$
$$+[2hV_{h-1}^{-}+h(h-1)V_{h-2}^{-}]\frac{1}{2}[V_{k+2}^{+}+2\rho_+V_{k+1}^{+}+\rho_+^2V_k^{+}]$$
$$=[-\frac{1}{2}k(k-1)\rho_+^2V_h^-V_{k-2}^{+}+\frac{1}{2}h(h-1)\rho_+^2V_{h-2}^{-}V_k^{+}]$$
$$+[-k\rho_+^2V_h^-V_{k-1}^{+}-\frac{1}{2}k(k-1)2\rho_+V_h^-V_{k-1}^{+}$$
$$+h\rho_+^2V_{h-1}^{-}V_k^{+}+\frac{1}{2}h(h-1)2\rho_+V_{h-2}^{-}V_{k+1}^{+}]$$
$$+[-k2\rho_+V_h^-V_k^{+}-\frac{1}{2}k(k-1)V_h^-V_k^{+}$$
$$+h2\rho_+V_{h-1}^{-}V_{k+1}^{+}+\frac{1}{2}h(h-1)V_{h-2}^{-}V_{k+2}^{+}]$$
$$+[-kV_h^-V_{k+1}^{+}+hV_{h-1}^{-}V_{k+2}^{+}]$$

On the right hand side of the above identity, we have grouped together terms with the factors $V_{k'}^{+}V_{h'}^{-}$ when $k'+h'$ have the same value, so we have 4 square brackets terms each corresponding to a value $k'+h'=-2,-1,0,1$. By (5.63) and (5.64) the sum of the last two terms on the right hand side of (5.60) is equal to the expression written above after changing k into h and superscripts and suberscripts plus into minus.

Summing up the quantities in all the square brackets we get (5.65) below. [In (5.65) we singled out the term multiplying $V_{k-1}^+ V_h^-$ and $V_k^+ V_{h-1}^-$ (note that this is linear in k and h) because it will cancel out with the last term in (5.52), as we shall see].

$$L_c V_k^+(n) V_h^-(m) = \sum_{i=-2}^{2} G_i \tag{5.65}$$

where

$$\begin{aligned}
G_{-2} &= \frac{1}{2}k(k-1)V_{k-2}^+(n)V_h^-(m)[\rho_-^2 - \rho_+^2] + \frac{1}{2}h(h-1)V_k^+(n)V_{h-2}^-(m)[\rho_+^2 - \rho_-^2] \\
G_{-1} &= k(k-1)[\rho_- V_{k-2}^+(n)V_{h+1}^-(m) - \rho_+ V_{k-1}^+(n)V_h^-(m)] \\
&\quad + h(h-1)[\rho_+ V_{k+1}^+(n)V_{h-2}^-(m) - \rho_- V_k^+(n)V_{h-1}^-(m)] \\
G_0 &= \frac{1}{2}k(k-1)[V_{k-2}^+(n)V_{h+2}^-(m) - V_k^+(n)V_h^-(m)] \\
&\quad + \frac{1}{2}h(h-1)[V_{k+2}^+(n)V_{h-2}^-(m) - V_k^+(n)V_h^-(m)] \\
&\quad + k[2\rho_- V_{k-1}^+(n)V_{h+1}^-(m) - 2\rho_+ V_k^+(n)V_h^-(m)] \\
&\quad + h[2\rho_+ V_{k+1}^+(n)V_{h-1}^-(m) - 2\rho_- V_k^+(n)V_h^-(m)] \\
G_1 &= k[V_{k-1}^+(n)V_{h+2}^-(m) - V_{k+1}^+(n)V_h^-(m)] \\
&\quad + h[V_{k+2}^+(n)V_{h-1}^-(m) - V_k^+(n)V_{h+1}^-(m)] \\
G_2 &= kV_{k-1}^+(n)V_h^-(m)[\rho_-^2 - \rho_+^2] + hV_k^+(n)V_{h-1}^-[\rho_+^2 - \rho_-^2]
\end{aligned} \tag{5.66}$$

It is easy to see that the sum of the last two terms on the right hand side of eq.(5.65) is equal to $-(\frac{\partial}{\partial t})_c(V_k^+(n)V_h^-(m))$, therefore $C^\epsilon(q,\sigma)$, defined in eq.(5.60), is equal to the sum in the right hand side of (5.65) without the last two terms. Substituting into (5.59) we get (5.56). This completes the proof of Proposition 5.6.4. □

5.6.5 The hierarchy of equations for the v-functions.

To simplify notation, sometimes in the sequel we drop the initial configuration η from the argument of the v and of the ρ functions. We can write (5.47) in integral form:

$$v_t^\epsilon(\xi) = P_t^{\epsilon,*} v_0^\epsilon(\xi) + \int_0^t ds\, P_{t-s}^{\epsilon,*} A_s^\epsilon v_s^\epsilon(\xi) \tag{5.67}$$

which can be iterated N times giving

$$\begin{aligned}
v_t^\epsilon(\xi) &- \int_0^t ds_1 \cdots \int_0^{s_{N-2}} ds_{N-1} P_{t-s_1}^{\epsilon,*} A_{s_1}^\epsilon \ldots A_{s_{N-1}}^\epsilon P_{s_{N-1}}^{\epsilon,*} v_0^\epsilon(\xi) \\
&+ \sum_{k=1}^{N-1} \int_0^t ds_1 \cdots \int_0^{s_{k-1}} ds_k P_{t-s_1}^{\epsilon,*} A_{s_1}^\epsilon \ldots A_{s_k}^\epsilon 1(\xi) \\
&+ \int_0^t ds_1 \cdots \int_0^{s_{N-1}} ds_N P_{t-s_1}^{\epsilon,*} A_{s_1}^\epsilon \ldots A_{s_N}^\epsilon v_{s_N}^\epsilon(\xi)
\end{aligned} \tag{5.68}$$

where $1(\xi) \equiv 1$ in the second term, recall that in our convention $v_t^\epsilon(\emptyset) \equiv 1$. We call respectively $R_t^{\epsilon,i}(\xi)$, $i = 1, 2, 3$, the three terms on the right hand side of (5.68).

The last term, $R_t^{\epsilon,3}(\xi)$ is the remainder, without it we would have a finite explicit expression for the v-function, its presence makes the hierarchy of equations for the v-functions infinite, as for the correlation functions u^ϵ. This was the origin of all the difficulties met in §5.4, but here the time interval when we study the process is very short, $t \le \epsilon^\beta$ and the remainder is really small, as seen in the following lemma.

5.6.6 Lemma. *Given any n and N positive, there is c which depends on n and N, such that for all $0 < \epsilon \le 1$*

$$\sup_{t \le \epsilon^\beta} \sup_{|\xi|=n} |R_t^{\epsilon,3}(\xi)| \le c\epsilon^{(\beta-3\zeta)N-\zeta n} \tag{5.69}$$

Proof. By Definition 5.6.2, using Lemma 5.6.1, we easily see that there is a function $d(m)$, $m \ge 1$, such that for all ϵ and t as above and $m \ge 1$

$$\|A_t^\epsilon f(\xi)\| \le d(m)\epsilon^{-2\zeta}\|f\|$$

where $\|\cdot\|$ is the sup norm with $|\xi| = m$. The v-function in $R_t^{\epsilon,3}(\xi)$ has order $\le n+N$. We have an a-priori bound on the v-function, based on the expression (5.22), using Lemma 5.3.3 to estimate the correlation functions and Lemma 5.6.1 to bound the ρ-functions. From this we have that there exist coefficents $d'(m)$, $m \ge 1$, such that for all t and ϵ as above and $m \ge 1$

$$\|v_t^\epsilon(\xi)\| \le d'(m)\epsilon^{-\zeta m}$$

By inserting all these bounds in (5.68) and performing the integrals we then obtain (5.69). □

Since $\beta > 3\zeta$, cf. Lemma 5.4.5, we can and will choose N so that

$$\beta N - 3\zeta N - \zeta n > \frac{1}{4}n \tag{5.70a}$$

namely

$$N = [\frac{1/4+\zeta}{\beta - 3\zeta} + 2]n \tag{5.70b}$$

Therefore the remainder term $R_t^{\epsilon,3}(\xi)$ is bounded by the right hand side of (5.24).

The crucial point for estimating the other terms are: 1) the degree of the v-function may decrease under the action of A^ϵ only if there are more particles in the same state; 2) the v-function at time 0 vanishes if there are states with only one particle. To illustrate these points we just examine two particular cases before giving the proof in general.

5.6.7 Heuristic analysis of (5.68).

The term $R_t^{\epsilon,1}$. The prototype of these terms is the first term on the right hand side of (5.67). As already mentioned,

$$v_0^\epsilon(\xi) = 0 \quad \text{if there is } x : \xi(x) = 1 \tag{5.71}$$

therefore

$$|\sum_{\xi} P_t^{\epsilon,\star}(\xi' \to \xi) v_0^\epsilon(\xi)| \leq [(\epsilon^{-1}t)^{-1/2}]^{n/2}[\epsilon^{-\zeta n}] \tag{5.72}$$

The first factor estimates the $P_t^{\epsilon,\star}$ probability that for no x, $\xi(x) = 1$. The second one bounds the sup over ξ of $v_0^\epsilon(\xi)$, by (5.22).

Let us now consider $R_t^{\epsilon,2}(\xi)$. The prototype of these terms, when $|\xi| = 2k$, is:

$$\int_0^t ds_1 \cdots \int_0^{s_{k-1}} ds_k \sum_{\xi_1,\bar{\xi}_1,\cdots,\xi_k,\bar{\xi}_k} \prod_{i=1}^k [1(|\bar{\xi}_i| = |\xi_i| - 2) A_{s_i}^\epsilon(\xi_i \to \bar{\xi}_i) P_{s_{i-1}-s_i}^{\epsilon,\star}(\bar{\xi}_{i-1} \to \xi_i)] \tag{5.73}$$

Recall that since $\rho^\epsilon \leq \epsilon^{-\zeta}$, each factor $|a^\epsilon|$ is bounded by $c\epsilon^{-2\zeta}$, then the absolute value of (5.73) is bounded by ($s_0 \equiv t$),

$$c \int_0^t ds_1 \cdots \int_0^{s_{k-1}} ds_k \prod_{i=1}^k (\epsilon^{-1}(s_{i-1} - s_i))^{-1/2} \epsilon^{-2\zeta k} \tag{5.74}$$

where c is a suitable constant. The factors $(\epsilon^{-1}(s_{i-1} - s_i))^{-1/2}$ in (5.74) estimate the probability that two asymmetric random walks are at same site after a time $\epsilon^{-1}(s_{i-1} - s_i)$ uniformly on their initial positions. By performing the integrals in (5.74) we then get a bound which goes like $\epsilon^{k/2} t^{k/2} \epsilon^{-2\zeta k}$, which is smaller than the right hand side of (5.24).

We have so far examined only *extreme cases* in the iteration of (5.67). There are other terms where there are deaths, creations and displacements of particles all mixed with each other. One might hope to prove (5.24) from (5.67) using an inductive procedure which involves the sup norms of $v_t^\epsilon(\xi)$ over all ξ with same number of particles. Let us call

$$\bar{v}_t^\epsilon(n) = \sup_{\xi:|\xi| \leq n} \sup_{s \leq t} |v_s^\epsilon(\xi)| \tag{5.75}$$

then from (5.67) and the above discussions we have that for a suitable constant c

$$\begin{aligned} \bar{v}_t^\epsilon(n) \leq c[(\epsilon^{-1}t)^{-1/2}]^{n/2}\epsilon^{-2\zeta n} + c \int_0^t ds \Bigg(& n[\bar{v}_s^\epsilon(n+1) + \epsilon^{-\zeta}\bar{v}_s^\epsilon(n)] \\ & + \frac{\epsilon^{-\zeta} n^2}{\sqrt{\epsilon^{-1}(t-s)}} [\bar{v}_s^\epsilon(n) + \bar{v}_s^\epsilon(n-1) + \epsilon^{-\zeta}\bar{v}_s^\epsilon(n-2)] \Bigg) \end{aligned} \tag{5.76}$$

The quantity

$$\int_0^\epsilon ds \frac{1}{\sqrt{\epsilon^{-1}(t-s)}} [\bar{v}_s^\epsilon(n-1) + \bar{v}_s^\epsilon(n-2)] \tag{5.77}$$

is smaller than the right hand side of (5.76), but it is larger (for large values of n) than $(\epsilon^{-1}t)^{-n/4}$ (at least when $t = \epsilon^\beta$), because for $t = \epsilon$, $\bar{v}_t^\epsilon(m)$ is of the order of the unity. In fact $v_s^\epsilon(\xi)$ for all $s \leq \epsilon$ is bounded away from 0 if for instance all the particles in ξ have same velocity and position.

Such a term arises because we have taken sup norms, otherwise we would have a factor related to the probability that the particles are in the same state, which gives the right contribution for obtaining the bound (5.24). This will require a careful analysis of the branching process associated to the iteration of (5.67).

§5.7 The branching process.

The equation (5.68) will be interpreted in terms of a branching process with births and deaths, which will then be studied by an iterative scheme.

5.7.1 Definition: branching trajectories.

Given $t > 0$ and $0 < \epsilon \leq 1$, let

$$\mathcal{X} = \bigcup_{m \geq 0} \mathcal{X}_m, \quad \mathcal{X}_m = D([0,t], \Omega_\epsilon^0) \times [0,t]^m \tag{5.78}$$

where Ω_ϵ^0 is the set of all the labelled configurations $\underline{x} = \{x_{i_1}, \ldots, x_{i_p}\}$, $x_{i_j} \in \mathbb{Z}_\epsilon \times \{-1, 1\}$, p being any positive integer and $(i_1, \ldots, i_p)$ any subset of the positive integers. $(i_1, \ldots, i_p)$ are the labels of the particles and $\underline{x}$ their states. The elements of $[0,t]^m$ are denoted by $\underline{t} = (t_1, \ldots, t_m)$, t_i will be referred to as the i-th interaction time, because in the branching process that we are going to define $\underline{t}$ will be such that $t_0 = 0 < t_i < t_{i+1} < t_{n+1} = t$ and at the t_i there will be branching events.

5.7.2 The labelled branching process.

Given $t > 0$, $\epsilon \in (0,1]$ and $\underline{x} = (x_1, \ldots, x_n) \in \Omega_\epsilon^0$, the labelled branching process is a σ-finite signed measure $\mathcal{P}^\epsilon$ on $\mathcal{X}$ determined on each $\mathcal{X}_m$ by the values

$$\int d\mathcal{P}^\epsilon f = \int_0^t dt_1 \cdots \int_0^{t_{m-1}} dt_m \sum_{\underline{x}^1, \ldots, \underline{x}^m} \mathbf{E}^\epsilon_{\underline{x}, \ldots, \underline{x}^m} \Big(\prod_{i=1}^m B^\epsilon_{t-t_i}(\underline{x}(t_i^-) \to \underline{x}^i) f\Big) \tag{5.79}$$

f being any bounded measurable function on $\mathcal{X}_m$. In (5.79) $\mathbf{E}^\epsilon_{\underline{x}, \ldots, \underline{x}^m}$ is the expectation with respect to a process which, in any time interval $[t_{i-1}, t_i)$, $1 \leq i \leq m+1$, is the independent process with initial condition $\underline{x}^i$, $\underline{x}^0 = \underline{x}$, where each particle jumps with intensity 1 by $-\sigma$, if σ is its velocity. Each particle keeps its label and preserves its velocity during these time intervals, while at the interaction times t_i, particles may disappear, or flip their velocities (keeping their positions) or new particles may be created, as determined by the transition $\underline{x}(t_i^-) \to \underline{x}^i$. This has weight $B^\epsilon_{t-t_i}(\underline{x}(t_i^-) \to \underline{x}^i)$, if the weight is 0 the transition is forbidden. It remains therefore to specify these weights: we set, recalling (2.2) for notation and 5.6.2,

$$B^\epsilon_{t-s}(\underline{x} \to \underline{x}') = A^\epsilon_{t-s}(U(\underline{x}) \to U(\underline{x}'))\phi_s(\underline{x}(\cdot)) \tag{5.80}$$

where ϕ_s is a characteristic function which imposes some constraints on the labelling of the configuration after the transition: if these are not met it equals 0, otherwise 1. The constraints, which depend on the trajectory up to time s included, allow to reconstruct uniquely $\underline{x}'$ from $U(\underline{x}')$ once $\underline{x}(s')$ is given for all $s' < s$. The rules are rules 1), 2) and 3) below. Before stating them we recall that: if $U(\underline{x}') \neq U(\underline{x})$ and $A^\epsilon_{t-s}(U(\underline{x}) \to U(\underline{x}')) \neq 0$, then there is a site q where the configuration $U(\underline{x})$ differs from $U(\underline{x}')$ while at all the other sites the configurations are the same. There are two subcases: (a) at (q, σ) the number of particles decreases, say by $h > 0$ and at $(q, -\sigma)$ increases by $k \geq 0$; (b) at (q, σ) the number of particles increases by 1, while at $(q, -\sigma)$ does not change.
We are now ready for stating the rules specifying ϕ_s:

1) If $U(\underline{x}') = U(\underline{x})$ then $\underline{x}' = \underline{x}$.

If $U(\underline{x}') \neq U(\underline{x})$ and the change is in q, then

2) In case (a), if $h \geq k$, the $h - k$ highest-labelled particles at (q, σ) disappear and the next k flip their velocities. If $h < k$ the h highest labelled particles at (q, σ) flip their velocities and $k - h$ particles are added in the state $(q, -\sigma)$ with labels, $p + 1, \ldots, p + k - h$, if p is the highest label ever appeared till the time of the interaction under consideration.

3) In case (b), one particle is added, its state is (q, σ) and its label $p + 1$, p being as in 2).

5.7.3 Definition: The ancestor of a particle.

By looking at 5.6.2, in Definition 5.7.2 if $k > h$ then $k = h + 1$. The particle with the lowest label in (q, σ), is then the *ancestor* of the new particle, we are using the same notation as in 2) and 3). Same convention is used to define the ancestor in case (b).

Remarks. Since the operator $B_i^\epsilon(\underline{x} \to \underline{x}')$ relates configurations with possibly different particles numbers, (5.79) describes birth-death processes at the interaction times t_i, with independent evolution of the particles between branchings.

Consider a function f on $\mathcal{X}$ defined as $v_0^\epsilon(\xi(t))1(\xi(t) \neq \emptyset)$ on $\mathcal{X}_m$ for $m < N$ and equal to 0 when $m \geq N$. The integral of this f gives $R_i^{\epsilon,1}(\xi)$, see (5.68). $R_i^{\epsilon,2}(\xi)$ is the integral of the function which vanishes on all $\mathcal{X}_m$, $m \geq N$, while, in the complement, it equals the characteristic function of extinction, namely of the event that there is an interaction time so that, after the transition, the resulting configuration is the empty one.

5.7.4 Definition: The skeleton of an element of $\mathcal{X}$.

Let $(\underline{x}(\cdot), \underline{t}) \in \mathcal{X}_m$, we then denote by $\mathcal{L}^{i,b}$ and $\mathcal{C}^{i,b}$, $1 \leq i \leq m$, $b = \pm$, respectively the labels of the particles in $\underline{x}(t_i^b)$ and the following partition of these labels: $h, k \in \mathcal{L}^{i,b}$ are in the same atom of $\mathcal{C}^{i,b}$ if and only if $x_h(t_i^b) = x_k(t_i^b)$. The cluster structure of $(\underline{x}(\cdot), \underline{t})$ is then the set of all $\mathcal{L}^{i,b}, \mathcal{C}^{i,b}$, while the cluster structure at (before/after) the i-th interaction time is $\mathcal{L}^{i,b}, \mathcal{C}^{i,b}$, with b equal respectively to $-$ or to $+$.

Two elements $(\underline{x}(\cdot), \underline{t})$ and $(\underline{y}(\cdot), \underline{t}')$ in $\mathcal{X}_m$ are equivalent if they have the same cluster structure. The equivalence class under the above relation defines a partition in $\mathcal{X}_m$, which will be called a skeleton, and will be denoted by π. π is also called the skeleton of any of its elements. $(\mathcal{C}^{i,\pm}(\pi), \mathcal{L}^{i,\pm}(\pi))$ is the cluster structure common to all the elements in the skeleton π.

Remarks. In $\mathcal{X}_m$ there are finitely many skeletons, recall in fact that the skeleton does not specify the values of the interaction times.

Notice further that: given π the configuration at t_i^+ is uniquely determined from the configuration at t_i^-; all the elements of π have the same number of particles with same labels and velocities in each interval $[t_i, t_{i+1})$; if a labelled particle in $\underline{x}(\cdot)$ dies or flips its velocity at the i-th interaction time, the same happens in all the elements of the skeleton π containing $\underline{x}(\cdot)$, and each particle has the same ancestor in any element of π.

5.7.5 The independent branching process.

Given $\underline{x}$, a skeleton π in $\mathcal{X}_m$ and $\underline{t} \in [0,t]^m$ we define a process $\underline{x}^0(s)$, $0 \le s \le t$, whose law is denoted by $\mathbb{P}^\epsilon_{\underline{x},\underline{t},\pi}$, as follows. $\underline{x}^0(0) = \underline{x}$; in the time intervals $[t_i, t_{i+1})$ the particles move independently as before; the initial condition at t_i, $i \ge 1$, is determined by $\underline{x}^0(t_i^-)$ and by π by saying that if at the i-th interaction time in π a particle dies or flips its velocity the same particle is present in $\underline{x}^0(t_i^-)$ and does the same. If a new particle is born in π at the i-th interaction time, say with label h and velocity σ, then a particle with label h is present in $\underline{x}^0(t_i^+)$ with velocity σ at the site q of the state $x^0_\ell(t_i^+)$, if ℓ is the ancestor of h.

We call this process independent because the evolution between the interaction times is independent and at the interaction times death and flips are specified by π and not by the position of the particles. Also births are specified by π, but the site where a particle is born is determined by the position of its ancestor at the time of the birth.

5.7.6 The bound on $R_t^{\epsilon,2}(\xi)$.

We denote by p^0_m the finite collection of all the skeleton in $\mathcal{X}_m$ where all the particles disappear after the m-th interaction time (and not before). Then there is a constant c, which only depends on N, such that

$$|R_t^{\epsilon,2}(\xi)| \le c \sum_{m=1}^{N-1} \sum_{\pi \in p^0_m} \int_0^t dt_1 \cdots \int_0^{t_{m-1}} dt_m \, \mathbb{E}^\epsilon_{\underline{x},\underline{t},\pi}\Big(\prod_{i,b} \chi^{i,b}\Big) \epsilon^{\zeta(q_-(\pi)+q_0(\pi))} \tag{5.81}$$

where $\chi^{i,b}$ is the characteristic function that the particles configuration at t_i^b has the cluster structure $C^{i,b}$. $q_-(\pi)$ is the overall number of particles in the skeleton, while $q_0(\pi)$ is the number of interaction times when no particle dies. One can in fact verify that in each of these events a function ρ^ϵ at most might be added, we then bound it by $\epsilon^{-\zeta}$, using Lemma 5.6.1, and obtain (5.81). Some other numerical and combinatorial factors are taken into account by means of the constant c.

We shall now bound (5.81). Call k the highest label in π, t'_k the time when k is born (equal to 0 if no particle is created in π) and t''_k when it dies. Let $C_k^{i,b}$ the cluster structure obtained after dropping the particle k and let $\chi_k^{i,b}$ be its characteristic function, which therefore does not have constraint on the state of particle k, for all i. We then have

$$\mathbb{E}^\epsilon_{\underline{x},\underline{t},\pi}\Big(\prod_{i,b} \chi^{i,b}\Big) \le \mathbb{E}^\epsilon_{\underline{x},\underline{t},\pi}\Big(1(x^0_k(t''^-_k) = x^0_h(t''^-_k)) \prod_{i,b} \chi_k^{i,b}\Big) \tag{5.82}$$

where h is the label of a particle which is in the same state as particle k at the i-th interaction time in any of the elements of π, recall that a particle may die only if another one is in the same state. We condition on the σ-algebra $\mathcal{F}$ determined by the positions of all the particles $x^0_\ell(s)$, $0 \le s \le t$, $1 \le \ell < k$. We have

$$\mathbb{E}^\epsilon_{\underline{x},\underline{t},\pi}\Big(1(x^0_k(t''^-_k) = x^0_h(t''^-_k)) \prod_{i,b} \chi_k^{i,b} | \mathcal{F}\Big) = \prod_{i,b} \chi_k^{i,b} \mathcal{E}^\epsilon_{x,\underline{t},\pi}(\{x^0(t''^-_k) = (q,\sigma)\}) \tag{5.83}$$

where $x^0_h(t''^-_k) = (q,\sigma)$ is fixed by $\mathcal{F}$, $x^0(t'^+_k) = x$ is also fixed by π which specifies the ancestor of particle k, and by $\mathcal{F}$ which fixes its state. $\mathcal{E}^\epsilon_{x,\underline{t},\pi}$ denotes then the process of a single particle which in the intervals $[t_i, t_{i+1})$ jumps with intensity 1 by $-\sigma$, if σ is the velocity of the particle k after the i-th interaction time in π, while at the interaction times t_i the particle k does not move, but may change its velocity, if this happens in π; finally, this process starts at t'^{+}_k from x. To derive (5.83) we have used that after particle k no other particle is created; in fact, if k were the ancestor of another particle, since the positions of this latter are fixed by the conditioning, then also the position of particle k would be fixed, at the time when it creates the new particle.

We can now use the local central limit theorem to get

$$\mathbf{E}^{\epsilon}_{\underline{t},\pi}\left(\prod_{i,b} \chi^{i,b}\right) \leq \frac{c}{\sqrt{\epsilon^{-1}(t''_k - t'_k)}} \mathbf{E}^{\epsilon}_{\underline{t},\pi}\left(\prod_{i,b} \chi^{i,b}_k\right) \tag{5.84}$$

We can iterate the above procedure starting from $k-1$. Call $L(\pi)$ the labels h of the particles in π such that when h dies there is another particle with smaller label in the same state. We then have

$$\mathbf{E}^{\epsilon}_{\underline{t},\pi}\left(\prod_{i,b} \chi^{i,b}\right) \leq \prod_{h\in L(\pi)} \frac{c}{\sqrt{\epsilon^{-1}(t''_h - t'_k)}} \tag{5.85}$$

and from (5.81)

$$|R^{\epsilon,2}_t(\xi)| \leq c \sum_{m=1}^{N-1} \sum_{\pi \in p^0_m} \int_0^t dt_1 \cdots \int_0^{t_{m-1}} dt_m\, \epsilon^{-\zeta(q_-(\pi)+q_0(\pi))} \prod_{h\in L(\pi)} \frac{c}{\sqrt{\epsilon^{-1}(t''_h - t'_h)}} \tag{5.86}$$

We start by integrating the death times of the particles with label in $L(\pi)$, beginning with the highest ones: each one produces a factor bounded by $\sqrt{t}$. Then we integrate the times when the particles are born: each one produces a factor t. The same contribution arises from each of the integrals when no particle dies or is born. We therefore get

$$|R^{\epsilon,2}_t(\xi)| \leq c \sum_{m=1}^{N-1} \sum_{\pi \in p^0_m} \epsilon^{(1/4+\beta/2-\zeta)q_-(\pi)+(\beta-\zeta)q_0(\pi)} \tag{5.87}$$

This is smaller than the right hand side of (5.24), therefore, to complete the proof it is enough to prove the bound (5.24) for the term $R^{\epsilon,1}_t(\xi)$.

5.7.7 The bound of $R^{\epsilon,1}_t(\xi)$.

We have to consider the case when after the last interaction time there are particles alive: we then have to compute the v-function at time 0 on the state occupied by these particles. As already remarked, this vanishes if there is a state occupied only by one particle. We can then intepret it as a new kind of death, we call it f-death. Also the f-death has in fact a characteristic function which imposes that the particles cannot f-die alone. As a difference with the previous case, arbitrarily many particles may f-die together. We obtain the same expression as (5.86), including also the f-deaths. As before we start by integrating the death times in $L(\pi)$ which are not f-deaths. From each of this we get a factor $\sqrt{t}$, as before. We then integrate over the birth times: if the birth refers to a particle which f-dies, then we do not get t as before but $\sqrt{t}$, while we get t from the integral of an interaction time with no birth or death. Finally we are left with a factor $t^{-1/2}$ for each particle with label in $L(\pi)$ and present at time 0. In fact neither the initial time nor the final one are integrated. Their number is denoted by $q_1(\pi)$. Calling again $q_-(\pi)$ the total number of particles, we have

$$|R_t^{\epsilon,1}(\xi)| \leq c \sum_{m=1}^{N-1} \sum_{\pi \in p_m^0} \epsilon^{-\zeta q_-(\pi)+(1/4+\beta/2)(q_-(\pi)-q_1(\pi))+(\beta-\zeta)q_0(\pi)} \left(\frac{\epsilon}{t}\right)^{q_1(\pi)} \tag{5.88}$$

From this it follows that the worst case is that considered in 5.7.7 which gives the same bound as in (5.24) which is therefore proven. □

§5.8 Bibliographical notes.

The Carleman equation has been derived in [25] from a model of particles which are in $\mathbb{R}$. They move like brownian motions with drift ± 1 and collide when they are suitably close with rules similar to those considered here. Convergence to (5.1) is proven in the usual kinetic scaling and letting then the diffusion of the brownians vanish. The methods we have presented here do not apply to such a case because the particles positions are not discrete, and the correlation functions of single configurations have delta functions for which we do not have the Lanford short time estimate. For some time we have considered this as a technical problem, maybe not too difficult to overcome, but we have not yet solved it.

The two dimensional analogue of the particle model that we have considered gives rise to the Broadwell equation, as it has been shown in [26]. Convergence is proven till times when the solution of the limiting equation is bounded. Convergence to the Broadwell equation is proven in [35] in the low density limit, for a HPP lattice gas, we refer for more details to the original paper and to Chapter VII.

CHAPTER VI

THE GLAUBER+KAWASAKI PROCESS

Many of the technical difficulties that we have found in the previous chapters were caused by the possibility of anomalous accumulations of particles. To some extent such effects are physically spurious, since in general there are other forces in nature, which, being repulsive, prevent the molecules from getting too close to each other. Taking these effects to the extreme, we study in this chapter hard core interactions, so that the occupation numbers are rigorously bounded: for a lattice gas with just one kind of particles, at each site x, $\eta(x) = 0, 1$. This is particularly useful in computer simulations where it is convenient to describe the configurations by means of boolean variables with 0-1 values. Instead of particles we can equivalently talk of spins, then the variables are denoted by $\sigma(x)$ and their values are ± 1; we often switch from one interpretation to the other. Going back to the particle language, in order to preserve the exclusion condition, we need a non zero range interaction, since we must suppress jumps of particles on sites which are already occupied. Quite surprisingly there is a lattice gas which is almost as simple as the independent system of Chapter II: this is the simple symmetric exclusion process, which is also called the stirring process. For such a process, very sharp probability estimates are known and there is a natural graphical representation which makes some of the proofs quite transparent. We shall discuss these features in the present chapter and use them to prove that, if we add "small perturbations" (in the sense of Chapters IV and V) to the stirring process, we then have convergence, in the macroscopic limit, to reaction-diffusion equations. In terms of spins, the stirring process becomes the Kawasaki dynamics at infinite temperature, which describes an evolution where spins exchange freely. Spin flips updatings, corresponding to a Glauber dynamics at finite temperature, are added as small perturbations in order to obtain, in the macroscopic limit, reaction-diffusion equations. In Chapter VIII we shall consider again Glauber and Kawasaki dynamics, when analysing phase separation phenomena. These will then be studied in details in Chapters IX and X in the context of the same Glauber+Kawasaki process that we consider here.

§6.1 The exclusion process.

We consider here the one dimensional symmetric simple exclusion process with nearest neighbor jumps, (for a general definition see [90]). This is the Markov process on $\{0,1\}^{\mathbf{Z}}$ with generator L_0, where

$$L_0 f(\eta) = \sum_x \sum_{b=\pm 1} \frac{1}{2}\eta(x)(1-\eta(x+b))[f(\eta^{(x,x+b)}) - f(\eta)] \tag{6.1}$$

where f is any cylindrical function and

$$\eta^{(x,y)}(z) = \begin{cases} \eta(z) & \text{if } z \neq x, y \\ \eta(y) & \text{if } z = x \\ \eta(x) & \text{if } z = y \end{cases} \tag{6.2}$$

The process may be defined by a limiting procedure, starting from a finite volume version, where the particles are in the circle $[-L, L]$, $L \in \mathbb{N}$, i.e. with periodic boundary conditions, and then letting $L \to \infty$. When L is finite, the process is a pure jump process: the particles, after independent exponential times of unit intensity, try to jump with probability $1/2$ to one of their two nearest neighbor sites. The jump does or does not take place if the chosen site is empty or, respectively, occupied. We do not discuss the convergence problem when $L \to \infty$, because we shall construct an explicit realization of the infinite volume process and we refer to [90] for a general existence proof.

6.1.1 The Stirring Process.

From (6.1) and (6.2), since $\eta^{(x,y)} = \eta$ when $\eta(x) = \eta(y)$, the right hand side of (6.1) simplifies and after a straightforward computation

$$L_0 f(\eta) = \frac{1}{2} \sum_x \left[f(\eta^{(x,x+1)}) - f(\eta) \right] \tag{6.3}$$

We can read (6.3) by saying that after independent exponential times of intensity $1/2$ nearest neighbor variables exchange their values and for this reason the process is also known as the stirring process. By interpreting the process in terms of stirring updatings, we have a natural and useful way for constructing it, as we are going to discuss.

6.1.2. Definition: The Marks Process.

This is a direct product of identically distributed processes on $\mathbb{R}_+$, one for each pair of nearest neighbor sites of $\mathbb{Z}$. Each single process is a Poisson process with intensity $1/2$. The events are called *marks*. Each mark is graphically represented by an arrow connecting the two (nearest neighbor) sites which specify where the mark has appeared.

The stirring process is defined as follows. Given an initial particle configuration, a realization ω of the marks process determines uniquely an evolution of the particles in the following way (graphical representation). We say that (x, t) and (y, s), $0 \leq s < t$, are ω-connected if there is a sequence $(x(0) = y, t(0))$, ..., $(x(n) = y, t(n))$, $t(0) = s < t(1) \cdots < t(n) = t$, such that at any $t(i)$, $i = 1, \cdots, n-1$, there is an arrow (in ω) connecting $x(i-1)$ and $x(i)$ (which necessarily are nearest neighbor sites), while no other arrow touches $x(i-1)$ in the time interval $(t(i-1), t(i))$. This is graphically visualized in Fig. 6.1, next page. Given ω the evolution starting from any given configuration η is defined by putting a particle at (x, t) if and only if (x, t) is ω-connected to $(y, 0)$ and $\eta(y) = 1$. It is easy to see that this evolution is with probability one well defined and that the corresponding process is a Markov process with generator (6.3).

Once ω is given, it is possible to identify the particles during their motion hence to define a process of labelled particles.

Remarks. Instead of particles we may consider any other kind of variables, which are then exchanged according to the marks: in particular we will consider spins, so that at each site there is a spin $\sigma(x)$, either up, $\sigma(x) = 1$, or down, $\sigma(x) = -1$. The process is then the Kawasaki process at infinite temperature, which, by its definition, evidently, leaves the "total magnetization" invariant. Notice, finally, that the above construction defines a universal coupling for processes starting from any initial condition, we simply let them evolve following the same marks.

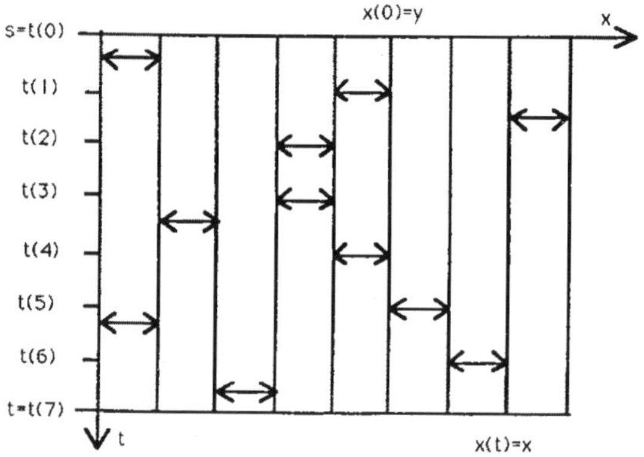

FIG. 6.1.

6.1.3. Invariant measures.

The only invariant measures for the stirring process are the exchangeable measures: μ is exchangeable if for any n-tuple $(x_1, \cdots, x_n)$ of mutually distinct sites and any $(a_1, \cdots, a_n) \in \{0,1\}^n$,

$$\mu(\{\eta(x_i) = a_i, i = 1, \cdots, n\}) = \mu(\{\eta(x_i) = a_{\pi(i)}, i = 1, \cdots, n\})$$

for any permutation π of $(1, \cdots, n)$. A proof of the above statement is essentially contained in 6.4.4 below.

By the De Finetti's theorem any exchangeable measure is an integral of Bernoulli measures: ν_ρ, $0 \le \rho \le 1$, is a Bernoulli measure if it is a product measure on $\{0,1\}^{\mathbf{Z}}$ and if $\mathbb{E}_{\nu_\rho}(\eta(x)) = \rho$, for all x, (ρ is the average density of particles in ν_ρ). Therefore the extremal invariant measures of the stirring process are the Bernoulli measures ν_ρ and any invariant measure is a convex combination (integral) of Bernoulli measures.

§6.2 The Glauber+Kawasaki process.

We shall introduce particle models for reaction-diffusion equations of the form:

$$\frac{\partial m}{\partial t} = \frac{1}{2}\frac{\partial^2 m}{\partial r^2} - V'(m), \qquad m = m(r), \quad r \in \mathbb{R} \tag{6.4a}$$

where $-1 \le m \le 1$; the reactive term $-V'(m)$ is a polynomial in m, we consider hereafter the case

$$-V'(m) = \alpha m - \beta m^3 \tag{6.4b}$$

6.2.1 The microscopic model.

The configuration space is $\{-1,1\}^{\mathbb{Z}}$, its elements are spins configuration denoted by

$$\sigma \equiv \{\sigma(x) \in \{-1,1\}, \quad x \in \mathbb{Z}\} \tag{6.5}$$

The evolution is a Markov process whose generator is

$$L^\epsilon = \epsilon^{-2} L_0 + L_G \tag{6.6}$$

where L_0 is defined in (6.3) (η being replaced by σ), while L_G, the generator of the Glauber dynamics, has the following form:

$$L_G f(\sigma) = \sum_x c(x,\sigma)[f(\sigma^{(x)}) - f(\sigma)] \tag{6.7}$$

where

$$\sigma^{(x)}(z) = \begin{cases} \sigma(z) & \text{if } z \neq x \\ -\sigma(x) & \text{if } z = x \end{cases} \tag{6.8a}$$

$$c(x,\sigma) = 1 - \gamma\sigma(x)[\sigma(x+1) + \sigma(x-1)] + \gamma^2[\sigma(x+1)\sigma(x-1)] \tag{6.8b}$$

with

$$0 \leq \gamma < 1$$

We denote by $\mathbb{E}^\epsilon$ the expectation with respect to the process with generator L^ϵ. For questions concerning the existence of the process, we refer to [90].

The initial measure μ^ϵ is a product measure with averages

$$\mathbb{E}_{\mu^\epsilon}(\sigma(x)) = m_0(\epsilon x) \tag{6.9}$$

where $|m_0(r)| \leq 1$ for all r, and $m_0 \in C^2(\mathbb{R})$. The macroscopic behavior of the Glauber+Kawasaki process is described by (6.4), more precisely

6.2.2 Theorem. *([36]). For any $n \geq 1$ and $L > 0$*

$$\lim_{\epsilon \to 0} \sup_{\substack{x_1 \neq .. \neq x_n \\ |x_i| \leq \epsilon^{-1} L}} \left|\mathbb{E}^\epsilon_{\mu^\epsilon}\left(\prod_{i=1}^n \sigma(x_i,t)\right) - \prod_{i=1}^n m(\epsilon x_i, t)\right| = 0 \tag{6.10}$$

where $m(r,t)$ solves (6.4) with

$$-V'(m) = \mathbb{E}_{\nu_m}(L_G \sigma(0)) \tag{6.11a}$$

and ν_m is the Bernoulli measure with spin average equal to m.

Remarks. The restriction to $|x_1| \leq \epsilon^{-1} L$ is unnecessary, but the proof would require some extra argument, and, for the sake of simplicity we have stated the theorem in the present weaker form. In [36] the fluctuations from the limiting behavior of the magnetization density are also studied. Using (6.7) and (6.8), the right hand side of (6.11a) is equal to the right hand side of (6.4b) with

$$\alpha = -2(1-2\gamma) \qquad \beta = 2\gamma^2 \tag{6.11b}$$

There are two different regimes when γ is smaller or larger than $1/2$. In the former case V ($\equiv$ the integral of $V'(m)$), has a single minimum, in the latter there are two minima; at the transition the minimum is quartic. We shall see some of the implications of this structure in Chapters IX and X, when studying the stability properties of the model; here we only derive (6.4), following the line of proof used in Chapter IV.

§6.3 Duality.

Like the system of independent symmetric random walks, also the stirring process is self-dual, as suggested by the following analogue of (2.51):

$$L_0\sigma(x) = \frac{1}{2}[\sigma(x+1) + \sigma(x-1) - 2\sigma(x)] \tag{6.12}$$

6.3.1 Duality for the Stirring Process.

For any $\xi \in \{0,1\}^{\mathbf{Z}}$, $|\xi| = \sum_x \xi(x) < \infty$, and any $\sigma \in \{-1,1\}^{\mathbf{Z}}$ we define

$$\begin{aligned} D(\xi,\sigma) &= \prod_{x:\xi(x)=1} \sigma(x) \equiv \prod_x D_{\xi(x)}(\sigma(x)) \\ D_0(\pm 1) &= 1 \qquad D_1(\pm 1) = \pm 1 \end{aligned} \tag{6.13}$$

By direct inspection we can easily verify that $L_0 D$ is the same both if L_0 acts on ξ and on σ, by an abuse of notation we use the same symbol L_0 both for the operator acting on $\{-1,1\}^{\mathbf{Z}}$ and for that on $\{0,1\}^{\mathbf{Z}}$. Denote by $\mathcal{E}^0$ the expectation of the process on $\{0,1\}^{\mathbf{Z}} \times \{-1,1\}^{\mathbf{Z}}$ defined as the product of the ξ-process with generator $\epsilon^{-2}L_0$ starting from $\xi \in \{0,1\}^{\mathbf{Z}}$ and the σ-process also with generator $\epsilon^{-2}L_0$ and initial measure μ^ϵ. Then for $0 \le s \le t$

$$\frac{d}{ds}\mathcal{E}^0(D(\xi_{t-s},\sigma_s)) = 0 \tag{6.14}$$

which, integrated over s, implies that

$$\mathbf{E}^0_{\mu^\epsilon}(D(\xi,\sigma_t)) = \mathbf{E}^0_\xi\Big(\mathbf{E}_{\mu^\epsilon}(D(\xi_t,\sigma_0))\Big) = \sum_{\xi'} P^\epsilon_t(\xi \to \xi')\mathbf{E}_{\mu^\epsilon}(D(\xi',\sigma_0)) \tag{6.15}$$

$P^\epsilon_t(\xi \to \xi')$ being the transition probability from ξ to ξ' in a time t, when the generator is $\epsilon^{-2}L_0$.

(6.15) has a simple interpretation in terms of the graphical representation of the stirring process. In fact the occupation numbers at time t and at $\underline{x} = (x_1, \cdots, x_n)$, if these are the sites where ξ has value 1, are determined by knowing the sites at time 0 connected to them. The connection is established by following backwards the arrows in the marks process for a time t. The law of the final positions is the same as that of n stirring particles at time t which initially were at $\underline{x}$, hence (6.15).

Observe that even with finitely many particles, i.e. when

$$|\xi| \equiv \sum_x \xi(x) < \infty$$

the stirring process with generator $\epsilon^{-2}L_0$ has a non trivial structure, in fact the exclusion rule builds up correlations among the $|\xi|$ particles. The crucial point is to prove that the stirring process with finitely many particles becomes close to the independent process in the limit $\epsilon \to 0$, we will come back to this problem in the sequel. Here, before ending this paragraph, we introduce some more notation.

6.3.2 Notation.

Let

$$\mathcal{M}_n \equiv \{\underline{x} = (x_1, \cdots, x_n) \in \mathbb{Z}^n : x_i \neq x_j \ \forall i \neq j\} \tag{6.16}$$

Denote by $\mathcal{P}_{\underline{x}}$ the law induced on $\mathcal{M}_n$ by the labelled stirring process starting from $\underline{x}$, see Definition 6.1.2. Let $\underline{x}(t) = (x_i(t), i = 1, \cdots, n)$, be the position variables at time t in this process. Let finally $U(\underline{x})$ be the map from $\mathcal{M}_n$ onto $\mathcal{N}_n \equiv \{\xi \in \{0,1\}^{\mathbb{Z}} : |\xi| = n\}$ which simply ignores the labels of the particles in $\mathcal{M}_n$.

6.3.3 Duality for the Glauber+Kawasaki process.

Let $\mathcal{E}$ be the expectation with respect to the product of the ξ-process with generator $\epsilon^{-2}L_0$ and the σ-process with generator L^ϵ. The ξ-process starts from a given ξ, the σ-process from μ^ϵ as in (6.9). We define the correlation function

$$u^\epsilon(\xi, t|\mu^\epsilon) = \mathbb{E}^\epsilon_{\mu^\epsilon}(D(\xi, \sigma_t)) \tag{6.17}$$

where $D(\cdot,\cdot)$ is defined in (6.13). Observe that

$$\frac{d}{ds}\mathcal{E}(D(\xi_{t-s}, \sigma_s)) = \mathcal{E}(L_G D(\xi_{t-s}, \sigma_s))$$

where L_G acts on the σ variable. Integrating over s and using (6.15) we have

$$u^\epsilon(\xi, t|\mu^\epsilon) = \sum_{\xi'} P^\epsilon_t(\xi \to \xi') u^\epsilon(\xi', 0|\mu^\epsilon) + \int_0^t ds \sum_{\xi'} P^\epsilon_{t-s}(\xi \to \xi') \mathbb{E}^\epsilon_{\mu^\epsilon}(L_G D(\xi', \sigma_s)) \tag{6.18}$$

Notice that the last expectation can be written as a linear combination of correlation functions so that (6.18) defines an integral relation between correlation functions, known as the BBGKY hierarchy (for the Glauber+Kawasaki process).

§6.4 Equicontinuity of the correlation functions.

The proof of Theorem 6.2.2 is based on the analysis of the BBGKY hierarchy (6.18), analogously to the proof of Theorem 4.3.1. We first compute $L_G D$. Let $\underline{x} \in \mathcal{M}_n$, $\xi = U(\underline{x})$, see 6.3.2, and

$$D(\xi, \sigma) = \prod_{i=1}^n \sigma(x_i) \tag{6.19}$$

Then

$$L_G D(\xi, \sigma) = \chi\big[n\alpha D(\xi, \sigma) + 2\gamma \sum_{i=1}^n (\Delta_i D(\xi, \sigma) - \gamma^2 D(\xi + \delta_{x_i+1} + \delta_{x_i-1}, \sigma))\big] + (1-\chi)\Gamma(\xi, \sigma) \tag{6.20a}$$

where

$$\chi = 1(|x_i - x_j| > 1, \text{ for all } i \neq j) \tag{6.20b}$$

$$\chi\Delta_i D(\xi,\sigma) = \chi[\sum_{b=\pm 1} D(\xi - \delta_{x_i} + \delta_{x_i+b},\sigma) - 2D(\xi,\sigma)] \tag{6.20c}$$

$$\Gamma(\xi,\sigma) = (1-\chi)L_G D(\xi,\sigma) \tag{6.20d}$$

Since $\sigma(x)^2 = 1$,

$$|L_G D(\xi,\sigma)| \leq c, \quad |\Gamma(\xi,\sigma)| \leq c, \qquad c = c(|\xi|) \tag{6.21}$$

For any $n \geq 1$, $\underline{x} \in \mathcal{M}_n$, we set $\underline{r} = \epsilon\underline{x}$ and, writing $\underline{x}$ instead of $U(\underline{x})$ in the argument of the correlation functions, see 6.3.2, we define

$$\gamma^\epsilon(\underline{r},t) = u^\epsilon(\underline{x},t|\mu^\epsilon) \tag{6.22}$$

By linear interpolation $\gamma^\epsilon(\underline{r},t)$ is then defined on the whole $\mathbb{R}^n$. Since $|D(\xi,\sigma)| \leq 1$, also $|\gamma^\epsilon(\underline{r},t)| \leq 1$. Next we prove that for each n the family $\gamma^\epsilon(\underline{r},t)$ is equicontinuos.

6.4.1 Proposition. *Given $\underline{x}$ and $\underline{y}$ in $\mathcal{M}_n$, see (6.16), let*

$$\|\underline{x} - \underline{y}\| = \max_{i=1,\cdots,n} |x_i - y_i| \tag{6.23}$$

Then given any $T > 0$ and $\zeta > 0$, there is $\delta > 0$ such that, writing $u^\epsilon(\underline{x},t|\mu^\epsilon)$ for $u^\epsilon(U(\underline{x}),t|\mu^\epsilon)$,

$$\sup_{\substack{|t-t'|\leq\delta \\ t,t'\leq T}} \sup_{\|\underline{x}-\underline{y}\|\leq\epsilon^{-1}\delta} |u^\epsilon(\underline{x},t|\mu^\epsilon) - u^\epsilon(\underline{y},t|\mu^\epsilon)| \leq \zeta \tag{6.24}$$

Proof . In the course of the proof we shall use the symbol $P_t^\epsilon(\underline{x} \to \underline{z})$ also for the transition probabilities in the labelled processes.

The proof is very similar to that of Theorem 4.4.1 given in §4.4.3. There is here one more difficulty, coming from the stirring evolution replacing the independent evolution of Chapter IV. We shall prove estimates which relate the two: in particular (4.32) and (4.33) are replaced by the following two properties:

1) For any positive ℓ^* and ζ^* there is $\tau^* > 0$ so that for all $t \leq \tau^*$, $\underline{x} \in \mathcal{M}_n$

$$\sum_{\underline{z}} P_t^\epsilon(\underline{x} \to \underline{z}) 1(\|\underline{z} - \underline{x}\| > \epsilon^{-1}\ell^*) < \zeta^* \tag{6.25}$$

2) There exists a constant c_2 so that uniformly on $\underline{x},\underline{y} \in \mathcal{M}_n$, $t > 0$ and $\epsilon > 0$

$$\sum_{\underline{z}} |P_t^\epsilon(\underline{x} \to \underline{z}) - P_t^\epsilon(\underline{y} \to \underline{z})| \leq c_2 \frac{1}{\sqrt{\epsilon^{-2}t}} \|\underline{y} - \underline{x}\| \tag{6.26}$$

We postpone the proof of 1) and 2) and prove equicontinuity.

6.4.2 Details of the proof.*

The proof of equicontinuity is now essentially the same as in 4.4.3. For brevity we treat simultaneously the two terms on the right hand side of of (6.18), while in 4.4.3 we have considered separately the two analogous terms in (4.25).

We choose arbitrarily the two positive numbers ζ and T, we then want to determine $\delta > 0$ so that (6.24) holds. We set $\zeta' = 10^{-2}\zeta$ and introduce $\ell > 0$ so that

$$\sup_{\|\underline{z}-\underline{z}'\|\le\epsilon^{-1}\ell} |u^\epsilon(\underline{z},0|\mu^\epsilon) - u^\epsilon(\underline{z}',0|\mu^\epsilon)| \le \zeta' \tag{6.27}$$

for all $\epsilon > 0$, (this can be done because of (6.9) and of the assumption that $m_0(r)$ has uniformly bounded derivative). We then choose τ_0 in the interval $(0, \zeta' c_1^{-1})$, $c_1 = \max\{1, c(n)\}$, where $c(n) = c(|\xi|)$ is defined in (6.21). Another request on τ_0 is that (6.25) holds with $\ell^\star = \ell$, $\zeta^\star = \zeta'$ and all $t \le \tau_0$.

We now choose $\delta > 0$ in such a way that

i) $\delta < \tau_0/2$, $\delta < \ell$, $c_1\delta < \zeta'$ and

$$(1 + Tc_1)\frac{c_2\delta}{\sqrt{\tau_0/2}} < \zeta'$$

ii) (6.25) holds with $t \le \delta$, $\zeta^\star = \zeta'$ and $\ell^\star = \ell'$, where

$$\ell' \equiv [c_2(1 + 2Tc_1)]^{-1}\zeta'\sqrt{\tau_0/2}$$

Since $\delta < \tau_0/2$ and in (6.24) $|t - t'| \le \delta$, it follows that either both $t, t' \le \tau_0$ or both $t, t' \ge \tau_0/2$. We start by considering the case $t, t' \le \tau_0$. By (6.18) and (6.21), dropping μ^ϵ from the argument of the correlation functions,

$$|u^\epsilon(\underline{x},t) - u^\epsilon(\underline{y},t)| \le 2c_1\tau_0 + |\sum_{\underline{z}} P_t^\epsilon(\underline{x} \to \underline{z})u^\epsilon(\underline{z},0) - P_{t'}^\epsilon(\underline{y} \to \underline{z})u^\epsilon(\underline{z},0)|$$

By our choice of τ_0, $c_1\tau_0 \le \zeta'$. Using (6.25) with $\ell^\star = \ell$ and $\zeta^\star = \zeta'$ and then (6.27) we get

$$\begin{aligned}|u^\epsilon(\underline{x},t) - u^\epsilon(\underline{y},t')| &\le 4\zeta' + |\sum_{\underline{z}} P_t^\epsilon(\underline{x} \to \underline{z})[u^\epsilon(\underline{z},0) - u^\epsilon(\underline{x},0) + u^\epsilon(\underline{x},0)]1(\|\underline{z} - \underline{x}\| \le \epsilon^{-1}\ell) \\ &\quad - \sum_{\underline{z}} P_{t'}^\epsilon(\underline{y} \to \underline{z})[u^\epsilon(\underline{z},0) - u^\epsilon(\underline{y},0) + u^\epsilon(\underline{y},0)]1(\|\underline{z} - \underline{y}\| \le \epsilon^{-1}\ell)| \\ &\le 6\zeta' + |u^\epsilon(\underline{x},0) - u^\epsilon(\underline{y},0)|\end{aligned} \tag{6.28}$$

Since $\delta \le \ell$ by (6.27) we then get the bound (6.24) for $t, t' \le \tau_0$.

Case $t, t' \ge \tau_0/2$. Let $0 < t' - t \le \delta$. Then by (6.18) and (6.21)

$$|u^\epsilon(\underline{y},t') - u^\epsilon(\underline{y},t)| \le \sum_{\underline{z}} |P_{t'-t}^\epsilon(\underline{y} \to \underline{z})|u^\epsilon(\underline{z},t) - u^\epsilon(\underline{y},t)| + c_1\delta$$

Recalling that $c_1\delta \le \zeta'$, by ii) we get

$$|u^\epsilon(\underline{y},t') - u^\epsilon(\underline{y},t)| \le \sup_{|\underline{z}-\underline{y}|\le\epsilon^{-1}\ell'} |u^\epsilon(\underline{z},t) - u^\epsilon(\underline{y},t)| + 2\zeta' \tag{6.29}$$

hence

$$|u^\epsilon(\underline{y},t') - u^\epsilon(\underline{x},t)| \le 2 \sup_{|\underline{x}-\underline{y}|\le\epsilon^{-1}(\ell'+\delta)} |u^\epsilon(\underline{x},t|\mu^\epsilon) - u^\epsilon(\underline{y},t|\mu^\epsilon)| + 2\zeta' \tag{6.30}$$

From (6.30), using (6.18) we get

$$|u^\epsilon(\underline{y},t') - u^\epsilon(\underline{x},t)| \le +2\zeta' + 2 \sup_{|\underline{x}-\underline{y}|\le\epsilon^{-1}(\ell'+\delta)} A_t^\epsilon(x,y)$$

where, writing $\underline{z}$ for $U(\underline{z})$ in the argument of the D functions

$$\begin{aligned} A_t^\epsilon(x,y) = &\sum_{\underline{z}} |P_t^\epsilon(\underline{x}\to\underline{z}) - P_t^\epsilon(\underline{y}\to\underline{z})| \\ &+ \int_{t-\tau_0/2}^{t} ds \sum_{\underline{z}} |P_{t-s}^\epsilon(\underline{x}\to\underline{z}) - P_{t-s}^\epsilon(\underline{y}\to\underline{z})| |\mathbf{E}_{\mu^\epsilon}^\epsilon(L_G D(\underline{z},\sigma_s))| \\ &+ \int_0^{t-\tau_0/2} ds \sum_{\underline{z},\underline{z}'} |P_{\tau_0/2}^\epsilon(\underline{x}\to\underline{z}) - P_{\tau_0/2}^\epsilon(\underline{y}\to\underline{z})| P_{t-\tau_0/2-s}^\epsilon(\underline{z}\to\underline{z}') |\mathbf{E}_{\mu^\epsilon}^\epsilon(L_G D(\underline{z}',\sigma_s))| \end{aligned}$$

(we have used that $|u^\epsilon| \le 1$). By (6.26) and (6.21) we then get

$$\begin{aligned} |u^\epsilon(\underline{y},t') - u^\epsilon(\underline{x},t)| &\le 2\zeta' + 2c_2\frac{1}{\sqrt{\epsilon^{-2}\tau_0/2}}\epsilon^{-1}(\ell'+\delta) + 2c_1\tau_0/2 + 2Tc_1c_2\frac{1}{\sqrt{\epsilon^{-2}\tau_0/2}}\epsilon^{-1}(\ell'+\delta) \\ &\le 7\zeta' \end{aligned}$$

The Proposition is therefore proven, modulo (6.25) and (6.26). □

6.4.3 Proof of (6.25).

It is enough to prove that for all $\underline{x}$

$$\sum_{\underline{z}} P_t^\epsilon(\underline{x}\to\underline{z}) 1(|z_1 - x_1| > \epsilon^{-1}\ell^*) \le \frac{\zeta^*}{n}$$

The left hand side equals

$$\sum_{z} P_t^\epsilon(x_1 \to z_1) 1(|z_1 - x_1| > \epsilon^{-1}\ell^*)$$

hence the proof of (6.25) is the same as the proof of (4.32), to which we refer. □

6.4.4 Proof of (6.26).

By repeated use of the triangular inequality we can reduce (6.26) to the case when $\underline{x}$ and $\underline{y}$ differ by one entry, for instance the last one, i.e. $x_i = y_i$ for $i < n$. We let the $\underline{x}$ and $\underline{y}$ particles move according to the same realization of the mark process, which therefore defines a natural coupling between the process starting from $\underline{x}$ and $\underline{y}$, see Definition 6.1.2 and the Remarks after it. Before doing that it is however convenient to consider a slightly different realization of the marks process.

6.4.5 Definition: The active-passive marks process.

This is the same as the marks process defined in 6.1.2 except that the exponential times have intensity one (twice as much the one they had before) and each mark is either ***active or passive*** with equal probability, independently of anything else. The exchanges, however, are determined only by the active marks, the passive ones are completely ignored: since the law of the active marks is the same as before, the particle process is unchanged.

In the sequel we consider the process as realized in the active-passive marks process and, for ease of future references, we consider below $\epsilon = 1$: notice that (6.26) depends on ϵ and t through $\epsilon^{-2}t$, so that the general case can be readily recoverd from the case $\epsilon = 1$. Given $\underline{x} = (x_1, \cdots, x_n)$ and $\underline{y} = (x_1, \cdots, x_{n-1}, y_n)$, we define $\hat{x} = (x_1, \cdots, x_n, y_n) \in \mathcal{M}_{n+1}$. We therefore have for any $\underline{z} \in \mathcal{M}_n$

$$P_t(\underline{x} \to \underline{z}) = \sum_{\underline{\hat{z}} \in \mathcal{M}_{n+1}} P_t(\underline{\hat{x}} \to \underline{\hat{z}}) 1(\hat{z}_i = z_i, i = 1, \ldots, n-1) 1(\hat{z}_n = z_n)$$

$$P_t(\underline{y} \to \underline{z}) = \sum_{\underline{\hat{z}} \mathcal{M}_{n+1}} P_t(\underline{\hat{x}} \to \underline{\hat{z}}) 1(\hat{z}_i = z_i, i = 1, \ldots, n-1) 1(\hat{z}_{n+1} = z_n)$$

We need to show that

$$\sum_{\underline{z} \in \mathcal{M}_n} \Big| \sum_{\underline{\hat{z}} \in \mathcal{M}_{n+1}} P_t(\underline{\hat{x}} \to \underline{\hat{z}}) f_{\underline{z}}(\underline{\hat{z}}) \Big| \le c \frac{|x_n - y_n|}{\sqrt{t}} \tag{6.31a}$$

where

$$f_{\underline{z}}(\underline{\hat{z}}) = 1(\hat{z}_i = z_i,\ 1 \le i \le n-1) \big[1(\hat{z}_n = z_n) - 1(\hat{z}_{n+1} = z_n) \big] \tag{6.31b}$$

We next introduce the stopping times

$$\tau_{n,n+1} = \sup \{ t : \text{there is no mark at time } t \text{ between } x_n(t) \text{ and } x_{n+1}(t) \} \tag{6.32}$$

We then condition on $\tau_{n,n+1}$ and use the strong Markov property: since the probability that the mark at $\tau_{n,n+1}$ is active or passive is the same, the process $\underline{\hat{x}}(s) \in \mathcal{M}_{n+1}$, $s \ge \tau_{n,n+1}$, is symmetric under the exchange of $\hat{x}_n(s)$ and $\hat{x}_{n+1}(s)$. On the other hand $f_{\underline{z}}(\underline{\hat{z}})$ is antisymmetric, so that, in order to prove (6.31), it is enough to show that

$$\mathbb{P}(\tau_{n,n+1} > t) \le c \frac{|x_n - y_n|}{\sqrt{t}} \tag{6.33}$$

$\mathbb{P}$ being the law of the active-passive marks process. The event in (6.33) only refers to the particles n and $n+1$. We let $w(t) = x_n(t) - x_{n+1}(t)$, so that $w(0) = x_n - y_n \equiv a$, and we observe that $w(t)$ for $t < \tau_{n,n+1}$ is a symmetric random walk, starting from a and moving on the lattice $\mathbb{Z}$ after exponential times of mean one. Then the stopping time $\tau_{n,n+1}$ has the same law as the first time τ_a the walk hits the origin. In Feller's book, see eq.(7.5) in Chapter III of [58], one can find the formula

$$\tilde{P}(\tilde{\tau}_a = m) = \frac{a}{m} \left(\frac{1}{2}\right)^m \binom{m}{\frac{m+a}{2}}$$

which gives the probability that a discrete time random walk starting from a hits the origin at time m. Then, by the Stirling's formula, there is a constant c such that

$$\tilde{P}(\tilde{\tau}_a = m) \le c \frac{a}{m^{3/2}}$$

The continuous time analogue is then obtained by writing

$$P(\tau_a > t) = \sum_{n=0}^{\infty} \sum_{m=n}^{\infty} \frac{t^n}{n!} e^{-t} \tilde{P}(\tilde{\tau}_a = m) \le c\, a \sum_{n=0}^{\infty} \frac{t^n}{n!} e^{-t} \sum_{m=n}^{\infty} \frac{1}{m^{3/2}} \le \bar{c} \frac{a}{\sqrt{t}}$$

From this (6.33) follows. □

§6.5 Proof of Theorem 6.2.2.

The proof of tightness for the correlation functions is now completed, we therefore know that the correlation functions converge by subsequences as $\epsilon \to 0$: we need now to identify their limiting values. We use (6.18) and (6.20a). The main point is the following probability estimate: for all $n \ge 1$, all $\underline{x} = (x_1, \cdots, x_n)$ and all $t > 0$

$$\lim_{\epsilon \to 0} \sup_{\underline{x} \in \mathcal{M}_n} \sum_{\underline{y}} \left| P_t^{\epsilon}(\underline{x} \to \underline{y}) - \prod_{i=1}^{n} [G_{\epsilon^{-2}t}((x_i - y_i))] \right| = 0 \tag{6.34a}$$

where

$$G_t(r) = e^{-r^2/2t} / \sqrt{2\pi t} \tag{6.34b}$$

We postpone the proof of (6.34) to the end of this chapter.

We fix $s < t$. Then by (6.34) and (6.21), we can neglect, when $\epsilon \to 0$, the contribution of Γ in (6.20a), as well as the contribution of the terms with the discrete laplacian Δ_i. Call $\gamma_s(\underline{r})$ the limit of (6.22) along a converging subsequence. Then by (6.34), we can replace by 1 the characteristic function (6.20b) appearing in (6.20a), so that, by the equicontinuity of the correlation functions,

$$\begin{aligned} \gamma_t(\underline{r}) = & \prod_{i=1}^{n} [\int dr G_t(r_i - r) m_0(r)] \\ & + \int_0^t ds \int [\prod_{i=1}^{n} dr_i' G_{t-s}(r_i - r_i')] \sum_{j=1}^{n} \sum_{\ell \in \{1,3\}} (A_{j,\ell}\gamma_s)(r_1', \ldots, r_n') \end{aligned} \tag{6.35a}$$

where

$$\begin{aligned} (A_{j,1}\gamma_s)(\underline{r}) &= (4\gamma - 2)\gamma_s(\underline{r}) \\ (A_{j,3}\gamma_s)(\underline{r}) &= -2\gamma^2 \gamma_s(r_1, \ldots, r_{j-1}, r_j, r_j, r_j, r_{j+1}, \ldots, r_n) \end{aligned} \tag{6.35b}$$

This equation has the same structure as (4.42). In the paragraph §4.5.3 it is proven that (4.42) has a unique solution in the class of correlation functions satisfying $|\gamma(\underline{r})| \le c^n$. Since $|\gamma_s(\underline{r})| \le 1$, the same proof applies to this case, hence the correlation functions converge as $\epsilon \to 0$ to the unique solution γ to (6.35). We can then check, as in the paragraph §4.5.1 that

$$\bar{\gamma}_t(\underline{r}) \equiv \prod_{i=1}^{n} m_t(r_i)$$

solves (6.35) if $m_t(r)$ solves (6.4) with initial datum $m_0(r)$. Theorem 6.2.2 is therefore proven, modulo the proof of (6.34). □

In order to prove (6.34) we need to study the relation between the stirring and the independent processes, this is done in the next paragraph.

§6.6 Coupling the stirring and the independent processes.

We consider n, $n \geq 2$, labelled particles moving according to the stirring process and n labelled particles moving as independent random walks. A coupling of these two processes is a process of $2n$ particles whose marginal over the first n particles has the law of the stirring process while the marginal over the others is the independent process. We shall use same labels, $1, \cdots, n$, for the stirring and the independent particles. The coupling we are going to define is such that the displacements of the particles with same labels are as far as possible the same. The coupled process is a jump Markov process on $\mathcal{M}_n \times \mathbf{Z}^n$ with generator $\mathcal{L}$ which will be defined below after setting some new notation. We denote by $\underline{x}$ and $\underline{x}^0$ the configurations of n particles in $\mathcal{M}_n$, respectively $\mathbf{Z}^n$. We then denote by $\underline{e}_i \in \mathcal{M}_n$ the unit vector in the i-th direction, i.e. the element whose entries are all 0 except the i-th one which has value 1. Sums or differences of configurations should be read componentwise, so that, for instance, $\underline{x} + \underline{e}_i$ equals $(x_1, \cdots, x_i + 1, \cdots, x_n)$, and this is well defined in $\mathcal{M}_n$ if $x_j \neq x_i + 1$ for all j. We then introduce the operators $\partial_{\pm,i}$, $i = 1, \cdots, n$, on $\mathcal{M}_n$ as

$$\partial_{\pm,i}\, \underline{x} = \begin{cases} \underline{x} \pm \underline{e}_i & \text{if for all } j,\ x_j \neq x_i \pm 1 \\ \underline{x} \pm \underline{e}_i \mp \underline{e}_j & \text{if } x_j = x_i \pm 1 \text{ and } j > i \\ \underline{x} & \text{if } x_j = x_i \pm 1 \text{ and } j < i \end{cases} \tag{6.36}$$

and finally we set

$$\mathcal{L}f(\underline{x}, \underline{x}^0) = \frac{1}{2} \sum_{i=1}^{n} \sum_{b=\pm} [f(\partial_{b,i}\underline{x},\ \underline{x}^0 + b\underline{e}_i) - f(\underline{x}, \underline{x}^0)] \tag{6.37}$$

6.6.1. A first realization of the coupling.

We shall consider different realizations of the process. The first one uses the same space $\mathbf{Z}^n \times \mathbf{R}_+$ where the independent process lives. Assume the process starts from $\underline{x}, \underline{x}^0 \in \mathcal{M}_n \times \mathbf{Z}^n$. When an independent particle, say particle i, jumps by $b = \pm 1$, then the stirring particle i tries to do the same jump. If this leads to an empty site, i.e. $x_i + b \neq x_j$, for all j, then this jump is done. If $x_i + b = x_j$, then we distinguish: if j is a label with lower priority, namely $j > i$, then the two particles, i and j, exchange their positions. If on the contrary $j < i$, then j has priority and the exchange (attempted by i) does not take place. In this way it is seen that (6.37) gives the same intensities of exchanges in the $\underline{x}$-process as in the stirring process. Notice also that the marginal on $\underline{x}^0$ is obviously the independent process.

The priority given to the smaller labels is clearly arbitrary, any other list of priorities would again define a coupling with essentially the same properties. Notice that the stirring particle 1 moves the same as the independent particle 1. We shall say that all the other stirring particles are second class with respect to particle 1. Furthermore the motion of the first k stirring particles is completely determined by the motion of the first k independent particles.

As we shall see below the main feature of the coupling is that with large probability the independent and the stirring particles are close at time t at least by $ct^{\frac{1}{4}+a}$, for any fixed arbitrary $a > 0$, c depending on a, we have assumed that $\underline{x} = \underline{x}^0$ at time 0. Since typically the independent particles have distance of the order of $\sqrt{t}$, the above is a significant estimate, which will actually imply (6.34). To understand the origin of this estimate remember that typically the total time in the time interval $[0,t]$ when two random walks are at nearest neighbor sites is of the order of $\sqrt{t}$. When they are not nearest neighbor [assume for the moment that there are just two particles] they have same displacement as the independent particles, according to (6.37). When they are nearest neighbor then the difference between the positions of the stirring and of the corresponding independent particles might jump by ± 1, after an exponential time of mean 1. Since the sign is chosen with equal probability and independently of the previous discrepancies the total difference will grow only as the square root of the total time spent *together*, hence overall as $t^{1/4}$. For a formal proof it is convienent to introduce a new realization of the process.

6.6.2 A second realization of the coupling.

The previous coupling can be realized in the active-passive marks process defined before. The rule is simple: whenever an active mark makes one or two stirring particles move, then the independent particle [the one with minimal label, if there are two stirring particles involved] makes the same jump as the corresponding stirring particle. If on the other hand there appears a passive mark between two stirring particles, then the independent particle with larger label moves. In all the other cases no particle moves. It is easy to see that this has indeed the same law as the previous coupling.

6.6.3 Proposition. *Let $\{\underline{x}(t), \underline{x}^0(t),\ t \geq 0\}$, be the process defined by (6.37) starting from $\underline{x}, \underline{x}^0 \in \mathcal{M}_n \times \mathbb{Z}^n$, $n > 1$, and let $\mathbb{P}_{\underline{x},\underline{x}^0}$ be its law. Then for any positive a and k there is c (which depends on a, k and n) such that uniformly on $t > 0$*

$$\mathbb{P}_{\underline{x},\underline{x}^0}\left(\|[\underline{x}(t) - \underline{x}] - [\underline{x}^0(t) - \underline{x}^0]\| \geq t^{\frac{1}{4}+a}\right) \leq ct^{-k} \tag{6.38}$$

where $\|\underline{x}\|$ is defined in (6.23). Furthermore, $\mathbb{P}_{\underline{x},\underline{x}^0}$ a.s., $x_i(t)$, for any t, is determined by $\{x_j^0(s)\}$, for all $s \leq t$ and all j which have priority on i and including i itself.

Proof. The statement about the measurability of $x_i(t)$ has been already proven, so we need only show the validity of (6.38).

First of all notice that it is enough to prove the bound (6.38) for each single difference $x_i(t) - x_i^0(t)$. This quantity may vary only at times t when $|x_i(t) - x_j(t)| = 1$ for some $j < i$ and an active or passive mark appears between $x_i(t)$ and $x_j(t)$. Call $\partial D_{i,j}(t) = \pm 1$ the jump at time t of the variable $x_i(t) - x_i^0(t)$ due to the presence of particle j, namely

$$\partial D_{i,j}(t) = \begin{cases} \operatorname{sign}(x_i(t^-) - x_j(t^-)) & \text{if the mark is passive} \\ -\operatorname{sign}(x_i(t^-) - x_j(t^-)) & \text{if the mark is active} \end{cases}$$

Then

$$[x_i(t) - x_i] - [x_i^0(t) - x_i^0] = \sum_{j<i} D_{i,j}(t)$$

$$D_{i,j}(t) = \sum_{s \leq t} \partial D_{i,j}(s)$$

(6.38) follows then from proving that the probability that $|D_{i,j}(t)| > t^{\frac{1}{4}+a}$ is bounded by ct^{-k}. It is convenient to use again the graphical representation which we reformulate in terms of the dual graph Γ. Γ is the set contained in the upper half plane made by all vertical lines (i.e. lines parallel to the y-axis) which intersect the x-axis at the points $x+\frac{1}{2}$, $x \in \mathbf{Z}$. On Γ we define two independent Poisson processes each of intensity 1/2, the points in the two processes are called respectively active and passive marks. A mark (active-passive) at $(x+\frac{1}{2},t)$ corresponds in the previous picture to an active-passive arrow connecting x and $x+1$ at time t. To compute $D_{i,j}(t)$ we first need to know the times when the particles i and j are nearest neighbour. For this purpose we introduce the notion of a-set. $A \subset \Gamma$ is a a-set if it is a finite union of bounded vertical segments in Γ whose projections on the y axis have empty intersection. The random set

$$\mathcal{A}(t) = \{(x+\frac{1}{2},s) \; : \; 0 \le s \le t, \; |x_i(s) - x_j(s)| = 1, \; x = \min(x_i(s), x_j(s))\}$$

is therefore (with probability 1) an a-set. The value of $D_{i,j}(t)$ depends on the marks in $\mathcal{A}(t)$, in fact when a mark appears in $\mathcal{A}(t)$, and only then, $D_{i,j}$ jumps by ± 1; it only matters for the following to observe that if the increment is 1 for a given value of the mark, then it would be -1 for the opposite one. We therefore need to know the distribution of the active and passive marks in $\mathcal{A}(t)$. If this set were not random, then the marks would be put in $\mathcal{A}(t)$ with the same law as in Γ. This is not true in general if $\mathcal{A}(t)$ is random, but it holds for our particular $\mathcal{A}(t)$. The crucial point is the following: assume $\mathcal{A}(t)$ is a measurable a-set valued function such that for any a-set A

$$\{\mathcal{A}(t) = A\} \in \Sigma(A^c)$$

almost surely, (where $\Sigma(A^c)$ is the σ-algebra of the events depending only on the marks in the complement of A). Then the conditional distribution of the marks in A is still given by two independent Poisson processes of intensity 1/2. The above measurability condition is indeed fulfilled by our specific $\mathcal{A}(t)$, we can then conclude, using the above criterion, that the marks are distibuited in $\mathcal{A}(t)$ as in the whole Γ.

The reader may convince himself of the validity of the criterion by looking at the similar statement for the discrete analogue of our process, namely when each vertical line is replaced by a lattice of given spacing. At each lattice site we may have either zero occupation or a mark, either active or passive. The distribution of all these variables is Bernoulli with given parameters. In this case the previous statement can be easily proven. To recover from this the continuous case, we need to let the spacing go to zero and at the same time to choose suitably the parameters of the Bernoulli measure. We leave the details to the reader.

Denote by T the total lenght of the intervals in $\mathcal{A}(t)$, then the conditional distribution of $D_{i,j}(t)$ given $\mathcal{A}(t)$ equals the distribution at time T of a symmetric random walk of intensity 1 which starts from the origin. On the other hand the probability that $T > t^{\frac{1}{2}+a}$, vanishes faster than any power of t for any given $a > 0$. In fact the time interval between two consecutive returns has the distribution of the return to the origin of a symmetric random walk which moves with intensity 2 and starts from site 1. The probability that the first $N \equiv [t^{\frac{1}{2}+a}]$ returns occur all in a time shorter than t goes like $(1-c/\sqrt{t})^N$ which vanishes faster than any power of t. Collecting all these estimates we then prove Proposition 6.6.3. □

6.6.4 Corollary. *For any a and k there is c so that uniformly on $\underline{x}, \underline{x}^0$*

$$\mathbb{P}_{\underline{x},\underline{x}^0}\Big(\sup_{s\geq t} \frac{\|[\underline{x}(s)-\underline{x}]-[\underline{x}^0(s)-\underline{x}^0]\|}{s^{\frac{1}{4}+a}} > 1\Big) \leq ct^{-k} \tag{6.39}$$

The Corollary is a simple consequence of Proposition 6.6.3, we omit the details.

§6.7 Distance between stirring and independent particles.

We denote by P_t^0 the transition probability of the independent process. Then, by (4.34), it is enough to prove that for any n, and any $\underline{x}$, $\underline{y} \in \mathcal{M}_n$,

$$\lim_{t\to\infty} \sum_{\underline{y}} |P_t(\underline{x}\to\underline{y}) - P_t^0(\underline{x}\to\underline{y})| = 0$$

Given $t>0$ let $T=t^{\frac{1}{2}+b}$, $b>0$ and small enough for what follows to hold. Let $\pi(\underline{y},\underline{y}^0)$ be the joint representation of $P_{t-T}(\underline{x}\to\underline{y})$ and $P_{t-T}(\underline{x}^0\to\underline{y}^0)$, $\underline{x}^0\equiv\underline{x}$, realized by the above coupling. By Proposition 6.5.3 we can neglect the pairs $\|\underline{y}-\underline{y}^0\| > t^{\frac{1}{4}+a}$. On the other hand we can also neglect those $\underline{y}^0$ for which there are $i\neq j$ such that $|y_i^0 - y_j^0| \leq t^{\frac{1}{2}-a}$, this is a classical estimate on random walks, see (4.34). We choose $a<1/8$, so that $t^{1/2-a}-t^{1/4+a}\to\infty$. If $\underline{y}$ and $\underline{y}^0$ are as above,

$$\sum_{\underline{z}} |P_T(\underline{y}\to\underline{z}) - P_T^0(\underline{y}\to\underline{z})|$$

is infinitesimal when $t\to\infty$ (hence when $T\to\infty$), if a and b are small enough. In fact $|y_i - y_j| \geq t^{1/2-a} - 2t^{1/4+a}$ and the two probability that two random walks travel such a distance in a time $T=t^{1/2+b}$ is infinitesimal if $1/4+b/2<1/2-a$. Finally if $2a<b$ then $T^{1/2}-t^{1/4+a}\to\infty$, so that also

$$\sum_{\underline{z}} |P_T^0(\underline{y}\to\underline{z}) - P_T^0(\underline{y}^0\to\underline{z})|$$

is infinitesimal when $t\to\infty$ for all $(\underline{y},\underline{y}^0)$ as above, and this completes the proof of (6.34). □

§6.8 Bibliographical Notes.

The derivation of the reaction-diffusion equation for the Glauber+Kawasaki dynamics has been obtained in [36], as well as the convergence of the fluctuation fields to a generalized Ornstein Uhlenbeck process. The proof that we have presented is simpler than that in [36] because we do not need here the more refined estimates proven in [36] for studying the fluctuations. In [36] a probabilistic interpretation of the integral equation for the correlation functions is discussed and used. The GPV method of Chapter III (in the finite volume case) can also be used to derive (6.4), see [79].

A general reference for the exclusion process is Liggett's book, [90], to which we refer for a more complete list of references. A two particles version of the coupling between independent and exclusion particles can be found in [12]. The coupling we have described was first introduced in [60], but such a publication is not easily accessible, so we also refer to Chapter III of [38]. More properties of the coupling can be found in [39], [41], in the Appendix A of [44] and in [62a].

There are inequalities relating the exclusion and the independent particles which have been extensively used in the analysis of the symmetric exclusion process, cf. [2], [9] and [90].

CHAPTER VII

HYDRODYNAMIC LIMITS IN KINETIC MODELS

We have used extensively, when deriving the macroscopic equations, that the macroscopic fields are smooth and that the macroscopic times are finite. If such conditions fail, we are in the presence of critical phenomena: it might happen, for instance, that the solutions of the macroscopic equations develop singularities and even that at long times the qualitative behavior of the system is no longer described by the macroscopic equations. There is not, at the moment, a general theory on these aspects, but just a few results for particular models: shock waves have been studied in the asymmetric simple exclusion and zero range processes, in the Boghosian-Levermore model and in the contact process. We do not have much to add to what Herbert Spohn says in his book, [125], to which the reader is addressed. For references see also the Bibliographical Notes at the end of Chapter I.

Singularities may also appear when there are phase transitions, since some values of the macroscopic fields are thermodynamically forbidden, and this may force the appearence of discontinuities. In Chapter VIII we discuss at a qualitative level these issues, while, in the last two chapters, we examine in detail how phases separate in the context of the Glauber+Kawasaki process. In the present chapter, on the other hand, we study the long time behavior of models for kinetic equations, deriving Euler and Navier-Stokes like equations, no proofs are given, but only some heuristic considerations.

§7.1 The Carleman equation.

The macroscopic equations describe a system in some space-time limit, but if we observe it through other space-time windows we often see new phenomena, even different macroscopic equations. Such a phenomenon is particularly evident in models for kinetic equations, since the interactions in the kinetic regime have not yet fully developed. One of the main successes of the Boltzmann equation, has in fact been the realization of a a scheme for deriving hydrodynamics. We shall follow the same scheme for determining the hydrodynamical equations associates to stochastic particle models which simulate kinetic equations.

We start with the model of Chapter V, where the analysis is simpler, and then we discuss generalizations to more interesting systems. An important feature of all of these is the existence of two characteristic lengths: the first one is the mean free path of a particle, namely the average space a particle travels between two successive collisions. It is determined by the relative strength of the generators of the streaming and the collision processes. If anomalous density fluctuations can be neglected, as in Chapter V, the mean free path is of the order of ϵ^{-1}, since this is the factor which multiplies L_0 in (5.3). The second typical length is the macroscopic length which measures the average number of sites for the particles' density to vary significantly. The ratio of these two, namely the mean free path and the macroscopic length, is called the Knudsen number.

In the macroscopic limits both lengths go to infinity, but there are different behaviors according to the value of the Knudsen number. If it converges to a finite limit, then we are in the case considered in §5.1, and, for the model of Chapter V, to which we restrict from now on in this paragraph, the limiting equation is the Carleman equation. Less interesting is when the Knudsen number goes to infinity: namely when, having fixed the generator as in (5.3), we choose an initial measure which scales as δ^{-1}, $\delta \to 0$ and $\delta^{-1}\epsilon \to 0$. Since the mean free path is ϵ^{-1}, the density profile changes before there is a significant number of collisions. By adapting to the present case the analysis of Chapter V, we can prove that at times $\delta^{-1}\tau$, τ finite, the limiting density profile evolves freely, namely as in (5.1), but with the right hand side set equal to 0.

The opposite case, when the Knudsen number goes to 0, corresponds to "the hydrodynamical limit", as we shall see. The initial measure μ^ϵ is a product of Poisson measures on $\mathbb{N}^{\mathbb{Z}\times\{-1,1\}}$ with averages:

$$\mathbb{E}_{\mu^\epsilon}(\eta(q,\sigma)) = \rho_0(\delta q,\sigma), \quad \delta=\delta(\epsilon),\ \lim_{\epsilon\to 0}\delta^{-1}\epsilon = \infty \tag{7.1a}$$

where ρ_0 is a regular function, as in Chapter V, while δ is the macroscopic length.

We denote by r and R the coordinate values of the lattice site q respectively in the new and in the old (as in Chapter V) macroscopic space units:

$$r = \delta q, \quad R = \epsilon q \tag{7.1b}$$

Notice that the average occupation number at (q,σ) which, by (7.1a) is $\rho_0(r,\sigma)$, when written in terms of the old macroscopic space scale does depend on ϵ: $\rho^\epsilon(R,\sigma) = \rho_0(\delta\epsilon^{-1}R,\sigma)$. Therefore we cannot apply the results of Chapter V to study the limit as $\epsilon \to 0$, also because the system is defined in the whole $\mathbb{Z}$ and not in $\mathbb{Z}_\epsilon$. It is however not too difficult to extend to this case the analysis of Chapter V (to which we refer for notation). We have the following result: (calling $\rho^\epsilon(R,\sigma,t)$ the solution to (5.1) with initial datum $\rho^\epsilon(R,\sigma)$)

$$\lim_{\epsilon\to 0}|\mathbb{E}^\epsilon_{\mu^\epsilon}(D(\xi,\eta_t)) - \prod_{i=1}^n \rho^\epsilon(\epsilon q_i,\sigma,t)| = 0, \quad \xi = \sum_{i=1}^n \delta_{(q_i,\sigma_i)} \tag{7.2}$$

uniformly on $|q_i| \le \epsilon^{-k}$, for any given k: we assume hereafter that $\delta^{-1} \le \epsilon^{-k}$ for some k.

However $\rho^\epsilon(R,\sigma,t)$ does not converge, when $\epsilon \to 0$, in fact it does not even converge at $t = 0$. However if we use the proper space scale and choose $R = R(\epsilon) = \epsilon\delta^{-1}r$ with r fixed as $\epsilon \to 0$, then, recalling (7.1b),

$$\rho_\epsilon(r,\sigma,t) = \rho^\epsilon(R,\sigma,t) = \rho^\epsilon(\epsilon\delta^{-1}r,\sigma,t) \to \hat\rho(r,\sigma,t)$$

where $\hat\rho$ is the solution of the new macroscopic equation

$$\frac{\partial\hat\rho(r,\sigma,t)}{\partial t} = \hat\rho(r,-\sigma,t)^2 - \hat\rho(r,\sigma,t)^2 \tag{7.3a}$$

$$\hat\rho(r,\sigma,0) = \rho_0(r,\sigma) \tag{7.3b}$$

Furthermore, uniformly on $|q_i| \le \epsilon^{-k}$,

$$\lim_{\epsilon\to 0}|\mathbb{E}^\epsilon_{\mu^\epsilon}(D(\xi,\eta_t)) - \prod_{i=1}^n \hat\rho(\delta q_i,\sigma,t)| = 0 \tag{7.4}$$

This result is far from unexpected: since the Knudsen number is now vanishingly small, the mean free path of a particle is much smaller than δ^{-1}, the new macroscopic space unit. Therefore the local densities cannot significantly change after finite times, (but the velocities can and do indeed change in this time scale). The space variable r in (7.3) in fact enters only as a parameter: for each r there is an autonomous evolution of the local velocity densities independently of what happens elsewhere.

To recover a non trivial space structure, we have to scale the time in such a way that a particle may travel by finite (in δ^{-1} units) distances in the new unit time. But, during this time, each particle has infinitely many collisions, as $\epsilon \to 0$, the velocity changes too often and it is no longer a good macroscopic variable. The Young measure, however, keeps a meaning, since it gives, for each value of the velocity, the fraction of time when that value is observed. The relevant Young measures, if local equilibrium holds, are determined by the equilibrium distribution of the velocities: from (7.3) it is natural to guess that at equilibrium ± 1 are equally distributed. In fact it is easy to check that the product of Poisson distributions with the same parameter, one for the particles with velocity 1, the other for velocity -1, is invariant under the process with generator (5.3).

By local equilibrium, the macroscopic variables reduce to only the local particle density, the velocities are in fact thermalized and have equiprobable values. Since local equilibrium characterizes the hydrodynamical regime, the limit when the Knudsen number vanishes is a hydrodynamical limit. The Carleman model is in this respect atypical, in general there are several equilibrium velocity disributions, for the true Boltzmann equation, for instance, the equilibrium distributions are the Maxwellians, indexed by their mean and variance, i.e. by the mean velocity and the temperature.

Denote again by $\rho^\epsilon(R,\sigma,t)$ the solution of (5.1) where $\rho^\epsilon(R,\sigma,0) = \rho_0(\delta\epsilon^{-1}R,\sigma,0) = \rho_0(r,0)$, (see (7.1b)) with ρ_0 fixed. We assume local equilibrium at time 0, i.e. that $\rho_0(r,\sigma) = \rho_0(r,-\sigma)$, and, to make notation simpler, we set

$$\gamma = \delta\epsilon^{-1}, \quad \tilde\rho_\gamma(r,\sigma,t) = \rho^\epsilon(\gamma^{-1}r,\sigma,\gamma^{-2}t), \;\; \tilde\rho_\gamma(r,\sigma,0) = \rho_0(r,\sigma) \tag{7.5}$$

Kurtz, [81], has proved that for all (r,σ,t)

$$\lim_{\gamma\to 0} \tilde\rho_\gamma(r,\sigma,t) = \tilde\rho(r,t) \tag{7.6}$$

where $\tilde\rho(r,t)$ solves the diffusion equation

$$\frac{\partial}{\partial t}\tilde\rho(r,t) = \frac{1}{2}\frac{\partial}{\partial r}(D(\tilde\rho(r,t))\frac{\partial}{\partial r}\tilde\rho(r,t)), \quad D(x) = \frac{1}{x} \tag{7.7}$$

with initial condition $\tilde\rho(r,\sigma,0) = \rho_0(r,\sigma)$. We therefore conjecture that uniformly on $|q_i| \le \epsilon^{-k}$, $k > 1$,

$$\lim_{\epsilon\to 0} |\mathbb{E}^\epsilon_{\mu^\epsilon}(D(\xi,\eta_{\gamma^{-2}t}) - \prod_{i=1}^n \tilde\rho^\epsilon(\delta q_i,\sigma,t)| = 0, \tag{7.8}$$

The methods used in [34] combined with the results of Chapter V allow to prove (7.8) when $\gamma = \delta\epsilon^{-1} = |\ln\epsilon|^{-u}$, $0 < u < 1/2$. The proof essentially consists in showing that 1) the particle system is well described by the Carleman equation till time $a|\ln\epsilon|$, for $a > 0$ small enough, and 2) the solution of the kinetic equation converges at these times to the solution of the hydrodynamic equation (7.7). The first step requires the extension of Proposition 5.5.3, and more generally of the analysis of §5.5, to times $T = a|\ln \epsilon|$, we shall discuss this aspect in Chapter X, in the context of the Glauber+Kawasaki process. In the proof one faces the problem of controlling errors which have initially the order of ϵ to some positive power, but then they increase exponentially with the macroscopic time. It is then clearly possible to extend to times $T \approx a|\ln \epsilon|$ the whole analysis of §5.5. For the step 2) mentioned above, one can use the results proven by Kurtz which relate the solution of (5.1) to that of (7.7).

We believe that (7.8) holds also for other choices of $\delta(\epsilon)$, but this requires a more careful analysis of the discretized version of (5.1) and of its convergence to (7.7), there are no results of this kind, as far as we know. An unpleasant feature of the above scheme is the assumption of local equilibrium at time 0, namely that $\rho_0(r,\sigma) = \rho_0(r,-\sigma)$. This however can be easily removed and then $\tilde{\rho}_\gamma$ converges again to the solution of (7.7), but with initial condition $\tilde{\rho}(r,0) = [\rho_0(r,1) + \rho_0(r,-1)]/2$. There is therefore an initial layer, i.e. a discontinuity at time 0. All this can be easily understood by examining (7.3) which describes the evolution on the previous time scale. Time γ^{-1} in (7.3) corresponds to time 1 in (7.7) and since by (7.3a) the difference $\hat{\rho}(r,\sigma,t) - \hat{\rho}(r,-\sigma,t)$ goes to 0 exponentially fast. In an infinitesimal time, as measures in time units (7.7), the memory of the initial velocity distribution is completely lost.

A much more serious problem is when the initial condition does depend on γ, for instance if we choose the same initial condition as in Chapter V, we have

$$\tilde{\rho}_\gamma(r,\sigma,0) = \rho_0(R,\sigma) = \rho_0(\gamma^{-1}r,\sigma) \tag{7.9}$$

ρ_0 being a fixed function. There are indications, see [96], that again, in the limit $\gamma \to 0$, there is convergence to (7.7), with some suitable initial condition. But a complete proof does not seem to exist in the literature nor the convergence of the particle system, which would prove the validity of (7.8) also in the present case. Physically such a result implies that the system is intrinsically diffusive at large times, while, in the frame of (7.5), it is the spatial structure of the initial datum which determines the time scale on which the diffusive behavior is established: whether this is really necessary remains therefore an open and interesting question.

7.2 The Broadwell equation.

The next model we consider is definitely more interesting from a physical point of view: it is two dimensional and the collisions preserve momentum. The corresponding kinetic equation is the Broadwell equation, but convergence is proven only at short times, or better till when the Broadwell equation has a bounded solution, [26].

The particles' state space is $\mathbb{N}^{\mathbb{Z}^2 \times \mathcal{V}}$, where $\mathcal{V} = \{e_i, i = 1, \ldots, 4\}$ is the "velocity" space: $e_1 = -e_3$ is the unit vector along the positive x-axis, while $e_2 = -e_4$ is the unit vector along the positive y-axis. The evolution is a Markov process whose generator has the form (5.3), namely $L^\epsilon = \epsilon^{-1}L_0 + L_c$

where

$$L_0 f(\eta) = \sum_{(q,e)\in \mathbf{Z}^2\times\mathcal{V}} \eta(q,e)[f(\eta - \delta_{q,e} + \delta_{q+e,e}) - f(\eta)] \tag{7.10}$$

$$L_c f(\eta) = \frac{1}{2} \sum_{(q,e)\in \mathbf{Z}^2\times\mathcal{V}} \eta(q,e)\eta(q,-e)[f(\eta + \sum_{b=\in\{-1,1\}} [-\delta_{q,be} + \delta_{q,be^\perp}]) - f(\eta)] \tag{7.11}$$

(for the existence of the process we refer generically to [90]), L_0 therefore generates an independent process where the particles jump after exponential times of intensity one in the direction of their velocities. The collisions described by L_c are local, at each site q any pair of particles with opposite velocities after independent exponential times of intensity one rotates both velocities by $\pi/2$.

As in Chapter V the initial measure μ^ϵ is a product of Poisson measures with averages

$$\mathbb{E}_{\mu^\epsilon}(\eta(q,e)) = \rho_0(\epsilon q, e) \tag{7.12}$$

ρ_0 being a suitably regular non negative function. The analogue of (5.7) is proven in [26] for the model defined on a torus of side $\epsilon^{-1}L$, $L>0$, and with $\rho_0(r,e) \in C^1$. In this case the limiting function ρ satisfies the equation:

$$\frac{\partial \rho(r,e,t)}{\partial t} + e\nabla\rho(r,e,t) = \rho(r,e^\perp,t)\rho(r,-e^\perp,t) - \rho(r,e,t)\rho(r,-e,t) \tag{7.13a}$$

$$\rho(r,e,0) = \rho_0(r,e) \tag{7.13b}$$

but the convergence is proven only till when the solution of (7.13) is bounded: there is a local existence theorem for (7.13), but the solution might become singular after a finite time. As in Chapter V the smoothness of the limiting profile is an input in the analysis and the method, as it stands, is not useful for establishing existence for the limiting Broadwell equation (7.13). Nonetheless the behavior of the particle model might, or better, should clarify the nature of the possible discontinuities arising in the Broadwell equation and how the solution should be continued past these times. Unfortunately there are no results in this direction for this model.

We can repeat here the analysis done earlier in 7.1. By letting the Knudsen number go to zero with the time fixed, we derive (7.13) without the streaming term $e\cdot\nabla\rho$. Calling $\hat\rho(r,e,t)$ the solution of this latter equation, we readily see that when $t\to\infty$ it converges to $\hat\rho(r,e)$ which satisfies the local equilibrium relations: ($i=1,2$ below)

$$\hat\rho(r,e)\hat\rho(r,-e) = \hat\rho(r,e^\perp)\hat\rho(r,-e^\perp), \quad \hat\rho(r,e_i) - \hat\rho(r,-e_i) = \hat\rho(r,e_i,0) - \hat\rho(r,-e_i,0) \tag{7.14a}$$

To derive an equation with a non trivial space structure, we change, as before, the time scale. To be precise, we first choose μ^ϵ as a product of Poisson measures with averages

$$\mathbb{E}_{\mu^\epsilon}(\eta(q,e)) = \rho_0(\delta q, e) \tag{7.14b}$$

and then assume that ρ_0 satisfies the first condition in (7.14a). Then we set $\delta = \delta(\epsilon)$ so that $\gamma = \delta\epsilon^{-1} \to 0$ as $\epsilon\to 0$ and denote by $\rho^\epsilon(R,e,t)$ the solution to (7.13) with initial condition $\rho^\epsilon(R,e,0) = \rho_0(\gamma R, e)$. We finally define $\tilde\rho_\gamma(r,e,t) = \rho^\epsilon(\gamma^{-1}r, e, \gamma^{-1}t)$: this is the analogue of

(7.5), notice however that times here are scaled by γ^{-1} and not by γ^{-2}, due to the fact the equilibrium condition (first condition in (7.14a)) does not impose that the average velocity is 0, as in the Carleman equation. When $\gamma \to 0$ we have ($r = (r_1, r_2) \in \mathbb{R}^2$)

$$\lim_{\gamma\to 0} \sum_{e\in\mathcal{V}} \tilde{\rho}_\gamma(r,e,t) = \rho(r,t) \tag{7.15a}$$

$$\lim_{\gamma\to 0} \tilde{\rho}_\gamma(r,e_i,t) - \tilde{\rho}_\gamma(r,-e_i,t) = j_i(r,t), \quad i = 1,2 \tag{7.15b}$$

where the density ρ and the current j, satisfy the Euler-like equations which, written in divergence form, are

$$\frac{\partial}{\partial t}\rho(r,t) + \nabla\cdot j(r,t) = 0, \quad j = (j_1, j_2) \tag{7.16a}$$

$$\frac{\partial}{\partial t} j_i(r,t) + \frac{\partial}{\partial r_i}\pi_i(r,t) = 0, \quad i = 1,2 \tag{7.16b}$$

$$\pi_1 = \frac{1}{2\rho}(1 + j_1^2 - j_2^2), \quad \pi_2 = \frac{1}{2\rho}(1 + j_2^2 - j_1^2) \tag{7.16c}$$

Notice that π_1 and π_2 are respectively equal to $\rho_1 + \rho_3$ and $\rho_2 + \rho_4$, where the ρ_i solve the system:

$$\rho_1 - \rho_3 = j_1, \quad \rho_2 - \rho_4 = j_2, \quad \rho_1 + \rho_2 + \rho_3 + \rho_4 = \rho, \quad \rho_1\rho_3 = \rho_2\rho_4 \tag{7.17}$$

The validity of (7.15) is restricted to times t for which (7.16) has a regular solution. The proof can be obtained by adapting to this case the analysis of [21], we omit the details. We can then prove the analogue of (7.8) by choosing $\gamma = \delta\epsilon^{-1} = |\ln\epsilon|^u$, $u < 1$, arguing as below (7.8).

We have so far studied the Euler regime, but the Broadwell model there are also "Navier-Stokes effects". We shall rather examine these aspects in the HPP model where results have already been obtained and the analysis is technically simpler.

7.3 The HPP model.

The HPP model was introduced by Hardy, Pomeau and de Pazzis in 1976, [72]. It is a variant of the system defined in §7.2 where an exclusion condition is imposed. The original model is completely deterministic, its state space is $\{0,1\}^{\mathbb{Z}^2\times\mathcal{V}}$ and the time is discrete. At each unit time the configuration is updated by means of two successive sub-updatings. The first one is collisional: at each site q with 0,1,3,4 particles nothing changes. At the others, if the two particles have opposite velocities, e and $-e$, then they both rotate by $\pi/2$, becoming $e^\perp$ and $-e^\perp$. If they do not have opposite velocities, then nothing happens. The second sub-updating is just streaming, each particle moves one step in the direction of its velocity.

The model is the ancestor of the modern cellular automata and it was introduced for simulating on a computer hydrodynamical equations, as we shall discuss in the sequel, but before doing it, we study the kinetic limit. This is characterized by a limit in which the Knudsen number remains finite. There are several ways to achieve this as we shall see, they are all based on depressing the collisions with respect to the free motion. The simplest way to achieve this is to make the system random by assuming that at each collision updating each site is given independently a variable which may be 0 or 1: if it equals 1, the HPP rule is applied, otherwise it is not. We choose 1 with probability ϵ and the initial measure μ^ϵ as the product probability on $\{0,1\}^{\mathbb{Z}^2\times\mathcal{V}}$ with averages

$$\mathbb{E}_{\mu^\epsilon}\big(\eta(q,e)\big) = \rho_0(\epsilon q, e) \tag{7.18}$$

where ρ_0 is a regular function with values in $[0,1]$. In [35] it is proven that for all $n \geq 1$ and for all $t > 0$

$$\lim_{\epsilon\to 0} \sup_{q_1,e_1,\dots,q_n,e_n} \left|\mathbb{E}_{\mu^\epsilon}\big(\prod_{i=1}^{n}\eta(q_i,e_i,t)\big) - \prod_{i=1}^{n}\rho(\epsilon q_i,e_i,t)\right| = 0 \tag{7.19}$$

where the sup is over all the $(q_1,e_1,\dots,q_n,e_n)$ which are mutually distinct and $\rho(r,e,t)$ solves the equation

$$\frac{\partial\rho(r,e,t)}{\partial t} + e\nabla\cdot\rho(r,e,t) = Q(\rho)(r,e,t) \tag{7.20a}$$

where the collision operator Q is, for simplicity we drop the dependence on r and t:

$$Q = \rho(e^\perp)\rho(-e^\perp)(1-\rho(e))(1-\rho(-e)) - \rho(e)\rho(-e)(1-\rho(e^\perp))(1-\rho(-e^\perp)) \tag{7.20b}$$

The proof is based on Lanford's perturbative technique, which converges at short times, see §5.4. However since the variables are a-priori bounded the result extends to arbitrary times.

Since the system is not one dimensional, by a low density limit we can also have finite Knudsen numbers. Let us start by some general considerations on the kinetic equations of the form:

$$\frac{\partial f(r,e,t)}{\partial t} + e\cdot\nabla f(r,e,t) = Q(f)(r,e,t) \tag{7.21}$$

where Q is an operator acting on f as function of e for each r fixed. Let $f_\epsilon(r,e,t) = f(\epsilon^{-1}r,e,\epsilon^{-1}t)$, then f_ϵ solves (7.21) with Q replaced by $\epsilon^{-1}Q$. Therefore if we choose $Q = Q_\epsilon$ with $Q_\epsilon = \epsilon Q_1$, Q_1 fixed, we then see that the corresponding equations become invariant under the above space-time scaling. This choice, $Q = Q_\epsilon$, corresponds in the particle model to making collisions with probability ϵ. However if Q depends quadratically on f, we may not change Q, but just the initial condition, setting $f^\epsilon(r,e) = \epsilon g_0(\epsilon r,e)$, with g_0 fixed. Calling $f^\epsilon(r,e,t)$ the solution of (7.21) with initial condition $f^\epsilon(r,e)$, then

$$g(r,e,t) = \epsilon^{-1}f^\epsilon(\epsilon^{-1}r,e,\epsilon^{-1}t) \tag{7.22}$$

is independent of ϵ and solves (7.21) with initial condition g_0. The density associated to f^ϵ is proportional to ϵ, therefore the particle model for (7.21) should be studied when the particles density is proportional to ϵ.

The true Boltzmann equation is derived in this same limit, the low density limit, also called the Boltzmann-Grad limit, and for this reason, the Boltzmann equation is usually emploied for studying rarefied gases, like the atmosphere at high altitudes. For hard spheres the low density limit corresponds to the following. Each particle has radius R and mass m and the particles interact by elastic collisions. The particles are in a box of diameter $\epsilon^{-1}L$, $L > 0$, in a space with d dimensions, and the system is studied for times of the order of $\epsilon^{-1}t$. To have the density proportional to ϵ, we put N_ϵ particles, where $N_\epsilon \approx \epsilon[\epsilon^{-1}L]^d$. Using the scale invariance of Newton's laws, we can as well consider N_ϵ spheres of radius $R_\epsilon = \epsilon R$ in a box of diameter L for a time of the order of unity. The relation between N_ϵ and R_ϵ is then

$$N_\epsilon \approx R_\epsilon^{-d+1}, \quad R_\epsilon^{d-1} N_\epsilon \approx 1$$

which is the usual formula for the Boltzmann-Grad limit, see for instance [83].

In the HPP model, the collision kernel (7.20) is not quadratic, therefore the low density limit will not reproduce it, but, at best, only its quadratic part, which coincides with the collision operator in the Broadwell equation. By this scaling argument one would therefore conjecture that in the regular (deterministic) HPP model with initial measure μ^ϵ, which is a product measure with averages

$$\mathbb{E}_{\mu^\epsilon}\big(\eta(q,e)\big) = \epsilon\rho_0(r,e) \tag{7.23}$$

the averages of the variables $\eta(q,e,t)$, $t = \epsilon^{-1}\tau$, τ fixed, should be close to $\epsilon\rho(\epsilon q, e, \tau)$, where ρ solves the Broadwell equation with initial condition ρ_0. This is not what happens: the limit exists, propagation of chaos holds (in some weaker sense than as in Chapter V) but the limiting averages do not obey the Broadwell equation nor any kinetic-like equation. This is the Uchyiama phenomenon, [129], first recognized by Uchyiama in a mechanical model for the Broadwell equation. The phenomenon is essentially due to recollisions among particles, whose probability remains finite even in the Boltzmann-Grad limit, because the velocities have finitely many values. The same happens in the HPP model, see [35], and, in general, in the deterministic kinetic models with discrete velocities.

Unfortunately it is not infrequent to find in the literature papers where the authors assume at some stage good factorization properties, ignoring completely that, by the Uchyiama result, they are not fulfilled, and then pretending to derive the kinetic equations: see for instance Remark 2.1, p.222 in

E. Gabetta, *On a Broadwell like lattice in a box*, Discrete models of fluid dynamics (A.S. Alves, eds.), World Scientific, Singapore, 1991, pp. 218–229.

To derive a kinetic equation we need to add some stochasticity to the purely deterministic HPP model. The Carleman model, see Chapter V, has two sources of randomness, the first one is in the collisions mechanism, each collision has in fact an exponential waiting time (of intensity 1). It is essentially the same as saying that collisions have probability ϵ, because the streaming is speeded up by a factor ϵ^{-1}, so that two particles, typically, have to "decide whether to collide" in a time interval of the order of ϵ, otherwise they may be separated by the streaming. The other randomness comes from the streaming, the exponential waiting times before jumps make the motion of each particle similar to a random walk. The stochasticity in the collisions was used for establishing the right value of the Knudsen number. However, in the low density regime a particle has a finite probability of not colliding at all in a finite macroscopic time, therefore the randomness of collisions is not enough to avoid the Uchyiama phenomenon, we can as well consider deterministic collisions, without loosing much. On the other hand, since we want to keep the exclusion rule, we have to let the particles move in the streaming updatings all at the same time (in the direction of their velocities). It is therefore not completely evident how a brownian-like motion could be added, this will be discussed in the following subparagraph.

7.4 The Boghosian-Levermore cellular automata.

The particles will move at discrete times with the updating rules described in [34], which are based on those introduced by Boghosian and Levermore, [14], to simulate the Burgers equation. They describe random walks which move simultaneously at discrete times and whose only interaction is an exclusion rule: in this sense these automata are similar to the exclusion processes considered in the previous chapter and in fact are used as more convenient alternatives to the exclusion processes for computer simulations, since they have the economical advantage of simultaneous updatings. To understand the underlying idea let us first consider a single random walk moving at the integer times. It can be thought of as a particle with a fictitious velocity which is updated each time before jumping. In two dimensions the velocity is $\sigma \in \mathcal{V}$ and the particle moves one step in the direction of σ, after the displacement σ is updated and the previous procedure applied again. If the values of σ are independent and symmetric, then the particles move like a symmetric random walk (at integer times). If desired we can bias the motion by suitably changing the law of σ. It is convenient to interpret the four values of σ as corresponding to four parallel planes $\mathbb{Z}^2$, so that when updating σ the particle changes plane (moving vertically). Each plane σ has a specified horizontal direction determined by the same value σ when thought of as an element of $\mathcal{V}$. After having updated σ the particle moves one step along the characteristic direction of the plane where it is. Since we are interested in the HPP model, the particles have also a "real" velocity $e \in \mathcal{V}$, so that we have 4 parallel planes $\mathbb{Z}^2 \times \mathcal{V}$, namely each "physical state" (q, e) may be placed in each of these four planes, the corresponding state being (q, e, σ). We then define the evolution by composing HPP updatings (independently on each plane) with vertical updatings, where particles move from one plane to the other. In doing this we must preserve the exclusion rule (at most one particle in each state in each plane). A way to achieve this is to choose for each state (q, e) independently a variable $a_{(q,e)} \in \{1, \dots, 4\}$: then the content of the cell (q, e, σ), $\sigma = e_i$, is put in the state (q, e, σ'), $\sigma' = e_j$, with $j = i + a_{(q,e)} \bmod 4$. In such a way the exclusion rule is preserved. We call this updating a "stirring updating" and the evolution is then defined by alternating a given number of stirring updatings to HPP updatings. For a precise definition of the model we refer to [34], where it is proven that in the low density limit, if the number of stirring updatings in between two consecutive HPP updatings diverges as $\epsilon^{-\nu}$ for any $0 < \nu < 1$, then the expected occupation numbers multiplied by the normalization factor ϵ^{-1} converge when $\epsilon \to 0$ to the solution of the Broadwell equation (till when its solution stays bounded). Propagation of chaos is also proven, in the strong form stated in Chapter V and earlier in this paragraph.

On the other hand if we make the density finite, even if we keep, as we do, the stirring updatings, to have a finite Knudsen number we need to randomize the collisions, for instance by assuming like before, that they have probability ϵ. By the Lanford technique, which applies again, we have convergence to the kinetic equation (7.13). But the presence of the stirring updatings makes it possible to prove estimates of the type proven in Chapter V, the analysis is however much simpler because the occupation numbers are a-priori bounded. Using these estimates, we can show that the particle system behaves as predicted by the kinetic equation (7.20), at least till times of the order of $a|\ln \epsilon|$, with $a > 0$ small enough. As explained in §7.1 this can be exploited for establishing hydrodynamic behavior. The Euler equation for the model differs from the Euler equation associated to the Broadwell equation because of the exclusion rule. We choose the initial measure as in (7.14b) with $0 \leq \rho_0 \leq 1$ assuming that it satisfies the analogue of the first condition in (7.14a), namely, see (7.20),

$$Q(\rho_0, \rho_0)(r, e) = 0, \text{ for all } r, e \tag{7.24}$$

The solutions of (7.24) can be written as

$$M(e|h(r), c(r)) = \frac{1}{1 + \exp\left[h(r) + c(r) \cdot e\right]}$$

and we choose $\rho_0(r, e)$ as the above expression with $h = h_0$ and $c = c_0$, h_0 and c_0 being smooth functions.

Calling $\rho^\epsilon(R, e, t)$ the solution of (7.20) with initial condition $\rho^\epsilon(R, e, 0) = \rho_0(\gamma R, e)$, $\gamma = \delta\epsilon^{-1}$ and defining $\tilde{\rho}_\gamma(r, e, t) = \rho^\epsilon(\gamma^{-1}r, e, \gamma^{-1}t)$, see below (7.14b), we have, as proven in [34], that

$$\lim_{\gamma \to 0} \left|\tilde{\rho}_\gamma(r, e, t) - M(e|h_t(r), c_t(r))\right| = 0$$

Furthermore, having defined

$$n(r, t) = \sum_e M(e|h_t(r), c_t(r))$$
$$v(r, t) = \sum_e eM(e|h_t(r), c_t(r))$$

the density n and the velocity v satisfy the "Euler equations"

$$\frac{\partial n}{\partial t} + \nabla \cdot (nv) = 0 \tag{7.25a}$$
$$\frac{\partial (nv_i)}{\partial t} + \sum_{j=1}^{2} \frac{\partial \pi_{i,j}}{\partial r_j} = 0 \tag{7.25b}$$
$$\pi_{i,j}(r, t) = \delta_{i,j}\left[M(e_i|h_t(r), c_t(r)) + M(-e_i|h_t(r), c_t(r))\right], \quad i, j = 1, 2 \tag{7.25c}$$

In [34] it is proven that the particle model behaves at the Euler times as predicted by (7.25), in the same sense as for the Carleman and the Broadwell models discussed before.

7.5 The Boltzmann, Euler and Navier-Stokes equations.

We next discuss the Navier-Stokes equation and the appearence, at the hydrodynamical scale, of dissipative effects, like viscous and diffusive phenomena. At least from a rigorous point of view, the validity problem for the Navier-Stokes equation is far from settled. There is a wide agreement on the Euler equation, it should describe the behavior of the system when space and time are scaled in the same way (the Euler scaling): it might be difficult, even extremely difficult, to get an actual proof for specific systems, but what to be done is at least clear. The situation is different for the Navier-Stokes equation, which, being not scale invariant, cannot be derived by just a space-time scaling. Something else has to be done, but what to do is in general not known. We start by the Boltzamnn equation, where things are more clear.

7.5.1 The Boltzmann equation and the derivation of hydrodynamics.

We consider the Boltzmann equation (7.21), with the collision kernel corresponding to the hard sphere model, but, since we shall not need its explicit form, we just refer to the literature, [28], for its expression. The compressible Euler equations for a gas whose state equation is that of a perfect gas (hence describing rarefied gases) are the following: (below ρ is the density, u the velocity and T the temperature)

$$\frac{\partial}{\partial t}\rho + \nabla \cdot (\rho u) = 0 \tag{7.26a}$$

$$\frac{\partial}{\partial t}u + (u \cdot \nabla)u + \frac{1}{\rho}\nabla(\rho T) = 0 \tag{7.26b}$$

$$\frac{\partial}{\partial t}\left[\rho(\frac{1}{2}u^2 + \frac{3}{2}T)\right] + \nabla \cdot \left[\rho(\frac{1}{2}u^2 + \frac{3}{2}T) + \rho u T\right] = 0 \tag{7.26c}$$

The Navier-Stokes equations for the same system are:

$$\frac{\partial}{\partial t}\rho + \nabla \cdot (\rho u) = 0 \tag{7.27a}$$

$$\frac{\partial}{\partial t}u + (u \cdot \nabla)u + \frac{1}{\rho}\nabla(\rho T) = \frac{4}{3}\eta \Delta u \tag{7.27b}$$

$$\frac{\partial}{\partial t}\left[\rho(\frac{1}{2}u^2 + \frac{3}{2}T)\right] + \nabla \cdot \left[\rho(\frac{1}{2}u^2 + \frac{3}{2}T) + \rho u T\right] = \nabla \cdot (\lambda \nabla T) + 2\eta \sum_{i,j} \frac{\partial}{\partial r_j}(u_i S_{i,j}) \tag{7.27c}$$

$$S_{i,j} = \frac{1}{2}(\frac{\partial}{\partial r_i}u_j + \frac{\partial}{\partial r_j}u_i - \frac{1}{3}\nabla \cdot u \delta_{i,j}) \tag{7.27d}$$

where η and λ are respectively the viscosity and the heat diffusion coefficients.

The validity problem starting from the Boltzmann equation consists in relating (7.21) to (7.26)-(7.27). This goes back to Hilbert and Grad, the first complete proof is due to Caflisch, [20], who has proven the following.

Assume the initial datum $f_0(r,v)$ for (7.21) depends on the scaling parameter ϵ as:

$$f_0(r,v) = M(\rho_0^\epsilon(r), u_0^\epsilon(r), T_0^\epsilon(r); v) \tag{7.28}$$

where $\rho_0^\epsilon(r) = \rho_0(\epsilon r)$, $u_0^\epsilon(r) = u_0(\epsilon r)$ and $T_0^\epsilon(r) = T_0(\epsilon r)$, with ρ_0, u_0 and T_0 given smooth functions and

$$M(\rho, u, T : v) = \rho(2\pi T)^{-3/2} \exp\left[-(v-u)^2/2T\right]$$

(this amounts to assuming local equilibrium at the inital time). There is a solution f_t of (7.21) with initial condition (7.28) such that there is $\tau > 0$ and for $t \le \epsilon^{-1}\tau$, all $R > 0$ and all v

$$\lim_{\epsilon \to 0} \sup_{|r| \le \epsilon^{-1}R} \left| f_t(r,v) - M(\rho_t^\epsilon(r), u_t^\epsilon(r), T_t^\epsilon(r) : v) \right| = 0 \tag{7.29a}$$

ρ_t^ϵ, u_t^ϵ and T_t^ϵ are the solutions of (7.26) with initial datum ρ_0^ϵ, u_0^ϵ and T_0^ϵ. The time $\epsilon^{-1}\tau$ is any time up to which the solution of (7.26) remains smooth.

Due to the scale invariance of (7.26), $\rho_t(r) = \rho_t^\epsilon(\epsilon^{-1}r)$, $u_t(r) = u_t^\epsilon(\epsilon^{-1}r)$ and $T_t(r) = T_t^\epsilon(\epsilon^{-1}r)$ solve (7.26) with initial datum $\rho_0(r)$, $u_0(r)$ and $T_0(r)$ so that we can rewrite (7.29a) in the more familiar form

$$\lim_{\epsilon \to 0} \sup_{|r| \le R} \left| f_{\epsilon^{-1}t}(\epsilon r, v) - M(\rho_t(r), u_t(r), T_t(r) : v) \right| = 0 \tag{7.29b}$$

The time T is any time up to which the solution of (7.26) remains bounded and smooth.

Instead of the Euler solution we might have taken as well in (7.29) the solution of the Navier-Stokes equation (7.27). In fact it is known that its solution with initial datum $\rho_0(\epsilon r)$, $u_0(\epsilon r)$ and $T_0(\epsilon r)$ till $t < \epsilon^{-1}\tau$ has the same limiting behavior when $\epsilon \to 0$ as the corresponding solution to (7.26), so that in this limit (7.27) cannot be distinguished from (7.26). To see the relevance of (7.27) in the description of the evolution (7.21) we have to study other features than the pure Euler hydrodynamical limit. One possibility is to look at higher order corrections in ϵ at fixed Euler times, i.e. at times $\epsilon^{-1}t$ in (7.21). It has been proven in [20] that the Maxwellian with parameters chosen as the solution of (7.27) with initial datum $\rho_0(\epsilon r)$, $u_0(\epsilon r)$ and $T_0(\epsilon r)$ correctly reproduces some of the terms appearing in the asymptotic expansion (in ϵ) of (7.21), for a suitable choice of the viscosity and the diffusion coefficients in (7.27). There are other terms of the expansion, "orthogonal" to the Maxwellian, which are not obtained in this way, but they are in some sense unimportant: this is purposedly vague, just like our understanding of the matter. The unpleasant feature of this approach is that we get the same terms starting from the linearization of (7.27) around the solution of the Euler equation, so that, in this way one never recognizes the full non linear Navier-Stokes equation.

The other possibility, which we better understand from a physical point of view, is to look at cases where the Euler and the Navier-Stokes equations have qualitatively different behaviors and see which one is followed by the solution of the Boltzmann equation. This occurs in the steady heat transfer between reservoirs at different temperatures. Assume that the system is in a cubic region of side $\epsilon^{-1}L$, two opposite faces are connected to reservoirs at temperatures T_- and $T_+ > T_-$, the other opposite faces are, for instance, identified, putting therefore periodic boundary conditions. Such conditions are imposed to the Boltzmann equation by suitable boundary conditions. Caflisch, [20], has proven that in the limit as $\epsilon \to 0$ the steady state solution of (7.21), rescaled in space, converges to the stationary solution (where no convection is present) of (7.27) in a box of side L and with a suitable choice of the viscosity and the diffusion coefficients. The result shows that at least in the case considered, (7.27) describes (7.21) in a better way than the Euler equation (7.26).

In a sense the above is an extreme case: we have set conditions for which the velocities are absent (here we talk of hydrodynamical velocities u, the velocities v in the Boltzmann equation are of course always present) so that the Euler equation in this case is trivial. A natural question, therefore, is whether (7.27) is relevant also when there are non zero velocities. De Masi, Esposito and Lebowitz, [33], have attacked this question by studying the case when the initial density and temperature are constant while the velocity is $\epsilon u_0^\epsilon(r)$, where $u_0^\epsilon(r) = u_0(\epsilon r)$, u_0 being a smooth function. Besides the usual scaling properties they also assume that the macroscopic velocity is vanishingly small, as $\epsilon \to 0$. Of course at times $\epsilon^{-1}t$ there is no significant difference, as $\epsilon \to 0$, between (7.26) and (7.27) so that, by Caflisch's theorem, there is convergence to (7.26) starting from the Boltzmann equation. At times $\epsilon^{-2}t$, Euler and Navier-Stokes equation are still identical at the leading order in ϵ, i.e. ϵ^0, the density and the temperature have the same value they had initially, the velocity is equal to 0. To first order in ϵ there is a difference: the Navier-Stokes equation to first order converges in the diffusive scaling to the incompressible Navier-Stokes equation

$$\rho\frac{\partial}{\partial t}u + \rho u \cdot \nabla u = -\nabla p + \eta\Delta u, \qquad \nabla \cdot u = 0 \tag{7.30}$$

(where ρ is a constant density and p, the pressure, is determined by the equation itself) In [33] it is shown that in the same diffusive scaling, the parameters of the Maxwellian which match correctly the first order terms of the expansion in ϵ of the solution of (7.21), (neglecting the terms "orthogonal to the Maxwellian") have to be chosen as in (7.30). We refer to [33] for a precise statement of the result, we only want to stress here that (1) "the relevant" terms of the expansion in ϵ of (7.21) are described by the non-linear incompressible Navier-Stokes equation and that (2) to this order, the non-linear incompressible Navier-Stokes equation is the same as (7.27); (7.27) therefore describes (7.21) in a more accurate way than (7.26), at least for the present choice of the initial data.

Also this result in a sense is not completely satisfactory, since one needs to look at first order corrections in ϵ. Next we consider another physically interesting case, the motion of a fluid in a channel. Assume the initial velocity is directed along the x-axis, that of the channel, but that it varies along the y-axis. We consider for simplicity the two dimensional case, and assume periodic boundary conditions in the y-direction. Temperature and density are initially constant. For the Euler equation this state is stationary, but it is not for the Navier-Stokes equation. The presence of a viscosity and the non-homogeneity of the initial velocity causes a gradient of temperature, hence a gradient of the pressure which, in turns, determines a vertical motion. There are indications that the Navier-Stokes equation has a definite limit in the diffusive scaling, but a proof is still missing: such a result and its analogue at the Boltzmann level would be a clear step forward in the derivation of the Navier-Stokes equation. The situation looks better for a stationary version of the previous problem. Assume that there is an external, constant (in space and time) force, directed along the x-axis and impose at the boundaries along the y-axis thermal boundary conditions: namely fix the temperature equal to some given value T_0. Esposito, Lebowitz and Marra have a proof (work in progress) that the stationary state of (7.27) in the diffusive scaling (namely scaling space like ϵ^{-1} and force as ϵ^2) converges to some given profile as $\epsilon \to 0$ and that the stationary solution of (7.21) has the same limit as the Maxwellian with parameters given by those obtained from (7.27).

There is still much to understand about hydrodynamics for the Boltzmann equation, but we think the above results present convincing evidence that (7.27) and (7.26) describe correctly (7.21) in the hydrodynamic limits. We next turn to the same problem for other evolutions, starting from the HPP model.

7.5.2 The Navier-Stokes behavior of the HPP model.

The previous considerations applied to the Carleman model become in a sense trivial, because the average velocity at equilibrium is zero and the Euler equation for this model is trivial. If instead the time scales like the square of the space, the limiting equation is a non linear diffusion, see (7.7). Same circumstance was present in Chapter III where we have considered the symmetric zero range processes: the first non trivial scaling is the diffusive one, since no average velocity exists at equilibrium, being the system symmetric.

For the HPP model instead there may be a net velocity at equilibrium and in fact the Euler scaling gives rise to a non trivial equation. Navier-Stokes effects appear in HPP in the incompressible limit discussed earlier, namely when choosing the average velocity of the order of ϵ. This is done in [34] by letting $h_0(r) = h_0$, a constant, and $c(r) = \gamma c_0(r)$ (actually, in [34] all the terms up to γ^2 have a definite form). In the hydrodynamic limit when time is scaled by γ^{-2}, instead of (7.25) we find the "incompressible Navier-Stokes equations", $n(r,t) = n_0$ (the initial density) while the velocity v (normalized by the factor γ^{-1}) satisfies

$$n_0 \frac{\partial u_i}{\partial t} + \frac{1}{2} n_0 g(n_0) \frac{\partial u_i^2}{\partial r_i} = -\frac{\partial p}{\partial r_i} + \nu(n_0) \frac{\partial u_i}{\partial r_i^2} \qquad i = 1,2 \tag{7.31}$$

where $\nabla \cdot u = 0$

$$g(n) = \frac{2-n}{4-n}$$
$$\nu(n) = \frac{4}{4-n}$$

and $\nabla \cdot p$ is uniquely determined by the equation itself.

7.5.3 The Navier-Stokes correction in other particle systems.

For particle systems whose evolution does not depend on the scaling parameter ϵ, the situation is much less clear. The Navier-Stokes corrections to local equilibrium at the Euler times have been studied in [52], [126], [112], (other papers on the subject are in preparation) with reference to two, purely mechanical, systems, the one dimensional hard rods system and the chain of linear oscillators. In these systems equilibium is degenerate due to the lack of ergodicity and to the presence of infinitely many first integral of motions. However introducing state parameters which label these extra conserved quantities, one can derive the analogue (a caricature, as the authors say) of the Euler equations, which consist of a set of infinitely many coupled equations. It is possible to compute in these cases the next order correction in ϵ and to find viscosity-like terms which rule their evolution.

There are essentially two types of arguments for relating the first order corrections in ϵ to a Navier-Stokes behavior. One is obvious, in nature there is no limit as $\epsilon \to 0$, so that the corrections are finite, maybe even not too small at realistic values of ϵ. Another argument is based on the long time corrections to the evolution: after times which, in the Euler scale, are of the order of ϵ^{-1}, also corrections of the order of ϵ may lead to finite effects. Both arguments have their shortcomings, the first order corrections to the state may spuriously depend on the choice of the initial state: states which give rise to the same macroscopic profile in the limit as $\epsilon \to 0$ may differ at first order in ϵ and such a difference may persist at the Euler times. Like for the Boltzmann equation, it is believed that not all the terms which appear in the expansion are all important, but there is no general prescription for singling out the relevant ones. The underlying idea, we think, is that only those which produce relevant effects at longer times do really matter, and this leads to the second argument presented as a justification for this procedure. But here the difficulty is more serious: in general there is no limit as $\epsilon \to 0$ of the state at the Euler times ϵ^{-1}, due to the non zero velocities, and one needs to specify what to look at, at these longer times. In principle, the same procedure used in §7.5 should apply as well, but the analysis is much harder.

So far we have seen examples where the Navier-Stokes corrections are present and effective, next we discuss cases where this is not so. The best counterexample is the asymmetric simple exclusion process, namely the process of Chapter VI, but where the jump probability to the right, $p > 1/2$, differs from that to the left, $q = 1 - p$. The hydrodynamic equation which describes the system in the Euler scaling, see [125] for references, is the non viscous Burgers equation

$$\frac{\partial}{\partial t}\rho + (p - q)\frac{\partial}{\partial r}[\rho(1 - \rho)] = 0 \tag{7.32}$$

There are solutions of (7.32) which are interpreted as "shock waves". They are step functions with density ρ_- and $\rho_+ > \rho_-$ before, respectively after, the discontinuity. The velocity of the wave is $c = (p - q)(1 - \rho_- - \rho_+)$. If the longer time behavior of the system is described by Navier-Stokes corrections to (7.32), we should add to (7.32) a viscosity term, which changes this equation into the viscid Burgers equation

$$\frac{\partial}{\partial t}\rho + (p - q)\frac{\partial}{\partial r}[\rho(1 - \rho)] = D\epsilon\frac{\partial^2}{\partial r^2}\rho \tag{7.33}$$

where D is some positive coefficient. This equation has still traveling waves, with the same relation between the velocity of the wave and its asymptotic densities: the shape of the wave is however modified, it is described by a smooth function which varies on the scale ϵ^{-1}, so that ϵ^{-1} is the order of the effective broadening of the shock in (7.33). In contrast with this picture, the shape of the shock in the true particle system is rigorously stationary and (7.33) gives a wrong prediction on the behavior of the system at times longer than the Euler ones.

While the shape of the shock is stationary, its position varies randomly. Taking this into account, one finds an effective broadening of the shape: namely if we look at the average occupation numbers, we are essentially averaging over the random displacements of the shock, hence we see a broadening effect. Also this cannot be explained in terms of (7.33), since the actually observed broadening of the average occupation numbers increases indefinitely on time, while (7.33) predicts a finite broadening.

Something similar happens also for the traveling waves in the one dimensional purely mechanical hard rod system, [112]. We leave the problem of derivation of the Navier-Stokes equation at this stage with many questions still to be answered.

We conclude this paragraph by discussing some variants of the HPP model used for computer simulations. As we have seen the hydrodynamic equations related to the HPP model have several unpleasant features. The two most disturbing ones are the lack of Galilean invariance and the fact that the pressure term depends on the velocity. The streaming term in the HPP-Euler equation has not the correct form $(u \cdot \nabla)u$, u being the velocity field, and this is due to the fact that there are only two possible directions of the velocity, so that the e_i-component of the velocity is carried only by the particles which have exactly the e_i velocity. To overcome this difficulty people have considered more complex lattice structures, the first generalization being to a triangular lattice. In this way it has been possible to gain back the correct galilean symmetry for two dimensional systems. For three dimensions the problem is more difficult and much more complex structures are needed. We refer to the literature, see for instance [65]. The fact that the pressure term turns out to be unrealistic and in general that the equilibrium states for these lattice gas automata do not have a Gibbs structure, making the connection with thermodynamics not really straightforward, seems somehow intrinsic to the schematization in terms of such systems and a price that must be paid for the simplicity of the evolution.

A final remark on these cellular automata: in computer simulations the value of the analogue of our ϵ is something finite and not particularly small, whatever this might really mean; the random updatings are limited to a minimum if not totally absent, since they rapidly increase the cost of the simulation; the automata used to simulate hydrodynamics do not work as the kinetic systems we have discussed in this chapter: there is no intermediate kinetic step, streaming and collision updatings have roughly the same intensity. Our theory therefore does not explain why or if the computer experiments really simulate the equations they are claimed to simulate. The theoretical arguments usually given are based on assumptions of local equilibrium. Since the interaction are local, it often happens that the product measures are invariant, with average occupation numbers satisfying proper conditions. There is then the ansatz that even not at equilibrium the state at all times is a product state, so that one can write down an equation for the averages. Of course such an ansatz is literally wrong, since a product state which is not an equilibrium state does not remain such during the evolution. We have already discussed in §7.3 that such an assumption is inconsistent with the low density limit, because of the Uchyiama phenomenon, but here one is studying the hydrodynamical equation and the same argument does not apply. On the other hand the factorization assumption might be relaxed by just assuming local equilibrium, as usually done when discussing the derivation of the Euler equation from hamiltonian evolutions, [99]. Presumably the system does not have such good ergodic-like properties and local equilibrium does not hold; it is still possible, nonetheless, that the answers are not too far off from the real ones.

Much less clear is the validity of the Navier-Stokes equation for these systems. In general local equilibrium is not sufficient for deriving Navier-Stokes. As predicted by the Gren-Kubo theory, see [125], the transport coefficients are in fact expressed in terms of fluctuations at equilibrium, and the very slow decay of some correlation functions may give rise to anomalous effects, which have indeed been observed in some computer experiments. These questions are of great interest, both numerically and conceptually, and would deserve a much more careful analysis, which however goes far beyond the purpose of this book.

§7.6 Bibliographical notes.

The hydrodynamical behavior of the Carleman model is discussed in [81] and [96]. The hydrodynamical limits for the HPP automaton with noise are discussed in [34] and [35]. There is a paper by Fritz in preparation where the limit of vanishing noise is discussed. There is a large number of papers on cellular automata, see for instance [65] and the Vol I of the journal *Complex Systems*.

In the Bibliographical Notes to Chapter I, we have quoted some references on particle models for the Burgers equations.

CHAPTER VIII

PHASE SEPARATION AND INTERFACE DYNAMICS

In this chapter we study the macroscopic behavior of systems which at equilibrium have phase transitions. We start by recalling some basic facts about phase transitions in equilibrium statistical mechanics, that we present in the context of the nearest neighbor Ising model.

8.1 The Ising model.

The two dimensional ferrogmanetic Ising model in the region $\Lambda \subset \mathbb{Z}^2$ with boundary conditions $S_{\Lambda^c} = \{S_x, x \notin \Lambda\} \in \{-1,1\}^{\Lambda^c}$ is the measure $\mu^{\beta,h}_{\Lambda,S_{\Lambda^c}}$ on $\{-1,1\}^\Lambda$ which gives probability

$$\mu^{\beta,h}_{\Lambda,S_{\Lambda^c}}(S_\Lambda) = \frac{1}{Z^{\beta,h}_\Lambda(S_{\Lambda^c})} \exp\big(-\beta H^h_\Lambda(S_\Lambda|S_{\Lambda^c})\big) \tag{8.1}$$

to the configuration S_Λ. In (8.1) $\beta \geq 0$ is the inverse temperature ($1/kT$, k being the Boltzmann constant) and the energy H^h_Λ is given by

$$H^h_\Lambda(S_\Lambda|S_{\Lambda^c}) = H^h_\Lambda(S_\Lambda) - \sum_{\substack{x\in\Lambda \\ y\notin\Lambda}} J(x,y)S_xS_y \tag{8.2a}$$

$$H^h_\Lambda(S_\Lambda) = -\frac{1}{2}\sum_{\substack{x,y\in\Lambda \\ x\neq y}} J(x,y)S_xS_y - \sum_{x\in\Lambda} hS_x \tag{8.2b}$$

$$J(x,y) = \begin{cases} J>0 & \text{if } x \text{ and } y \text{ are nearest neighbor sites} \\ 0 & \text{otherwise} \end{cases} \tag{8.3}$$

If $J < 0$ the model is anti-ferromagnetic. In the sequel we will also consider Ising models with long range interactions. h is an external magnetic field and $Z^{\beta,h}_\Lambda(S_{\Lambda^c})$, the normalization constant in (8.1), is the (grand-canonical) partition function in the region Λ with boundary conditions S_{Λ^c}.

8.1.1 Theorem. *For $h \neq 0$ and for $h = 0$ and $\beta \leq \beta_c$, where $\sinh 2\beta_c J = 1$, there is a unique probability $\mu^{\beta,h}$, the Gibbs measure at the inverse temperature β and magnetic field h, such that for any increasing sequence $\Lambda \nearrow \mathbb{Z}^2$ and any choice of the configurations S_{Λ^c},*

$$\text{weak-}\lim \mu^{\beta,h}_{\Lambda,S_{\Lambda^c}} = \mu^{\beta,h} \tag{8.4a}$$

[weak convergence means that the integral of any continuous function converges]. If $h = 0$ and $\beta > \beta_c$ there is $m_\beta > 0$ and two probabilities $\mu^{\beta,0}_{\pm m_\beta}$ so that any weak limit point of $\mu^{\beta,h}_{\Lambda,S_{\Lambda^c}}$ is a convex combination of these two measures, namely for any converging sequence there is $\alpha \in [0,1]$ so that

$$\text{weak-}\lim \mu^{\beta,h}_{\Lambda,S_{\Lambda^c}} = \alpha\mu^{\beta,0}_{m_\beta} + (1-\alpha)\mu^{\beta,0}_{-m_\beta} \tag{8.4b}$$

The measures $\mu^{\beta,0}_{\pm m_\beta}$ are translationally invariant and the average spin is

$$\mathbf{E}_{\mu^{\beta,0}_{\pm m_\beta}}(S_0) = \pm m_\beta \tag{8.5a}$$

They may be obtained by taking the limit as $\Lambda \nearrow \mathbf{Z}^2$ having chosen S_{Λ^c} as the configuration with all the spins equal respectively to ± 1.

The set $\{(\beta, h) : \beta > \beta_c,\ h = 0\}$ is the phase transition region and the two Gibbs states $\mu^{\beta,0}_{\pm m_\beta}$ correspond to the two pure phases. Their magnetization, $\pm m_\beta$, is called the order parameter, it is the extra variable besides β and h needed to specify the state when the phase transition occurs. Of course the magnetization is also defined outside of the phase transition region:

$$m(\beta, h) = \mathbf{E}_{\mu^{\beta,h}}(S_0) \tag{8.5b}$$

The graph of the magnetization versus h for fixed β looks like

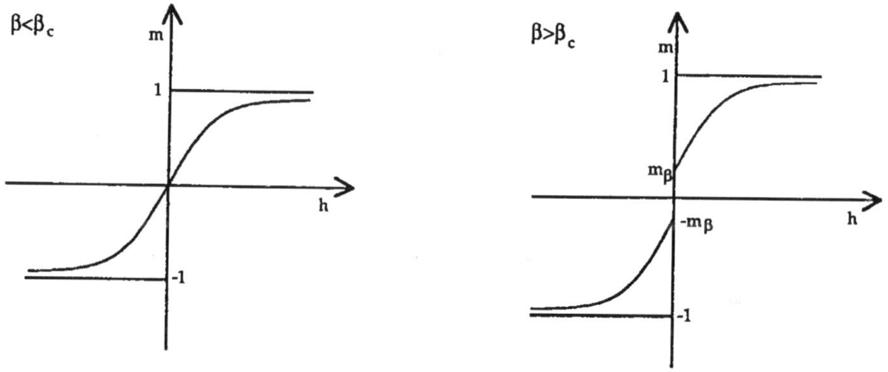

FIG. 8.1.

When $\beta > \beta_c$, the magnetization jumps from $-m_\beta$ to m_β as h passes through 0, so that there is a forbidden interval of magnetizations, $(-m_\beta, m_\beta)$: it is evident that this will have deep implications on the hydrodynamical behavior of the model. Of course convex combinations of $\mu^{\beta,0}_{\pm m_\beta}$, as in (8.4b), are still Gibbs measures and have average magnetizations in $(-m_\beta, m_\beta)$. Their typical configurations, though, being in the union of the supports of $\mu^{\beta,0}_{m_\beta}$ and $\mu^{\beta,0}_{-m_\beta}$, have magnetization densities equal either to m_β or to $-m_\beta$.

The same picture is obtained from the analysis of the partition function:

8.1.2 Theorem. *For any value of β and h there is a function $p(\beta, h)$, the thermodynamical pressure, such that for any sequence $\Lambda_n \nearrow \mathbf{Z}^2$ and any choice of $S_{\Lambda_n^c}$*

$$\lim_{n\to\infty} \frac{1}{\beta|\Lambda_n|} \log Z^{\beta,h}_{\Lambda_n}(S_{\Lambda_n^c}) = p(\beta, h) \tag{8.6}$$

$p(\beta, h)$, as a function of h for fixed β, is a convex function and it has continuous derivatives except in the phase transition region, $\{\beta > \beta_c,\ h = 0\}$, where

$$\frac{\partial}{\partial h} p(\beta, 0^\pm) = \pm m_\beta \tag{8.7}$$

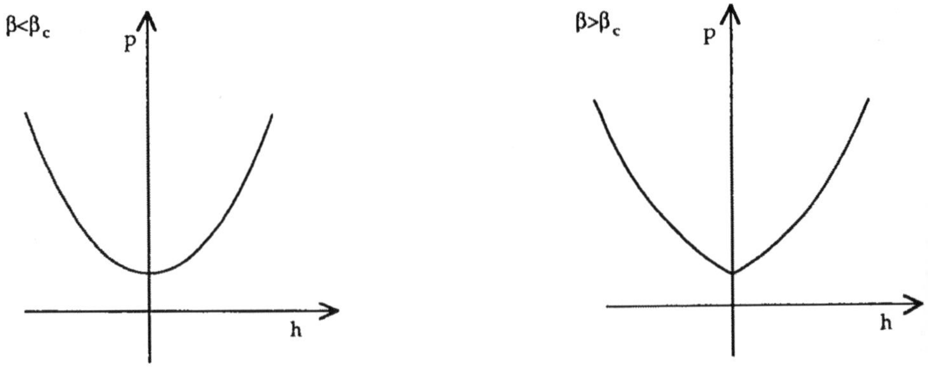

FIG. 8.2.

In agreement with thermodynamics, the pressure is not influenced by the boundary conditions in the thermodynamical limit. Its graph for fixed β is like in Fig.8.2.

The free energy $F(\beta, m)$ is the Legendre transform of $p(\beta, h)$, namely

$$F(\beta, m) = \sup_h \left[hm - p(\beta, h)\right] \tag{8.8a}$$

correspondingly

$$p(\beta, h) = \sup_m \left[hm - F(\beta, m)\right] \tag{8.8b}$$

The free energy F is the thermodynamic limit of $-(\beta|\Lambda|)^{-1} \log \hat{Z}_\Lambda^{\beta,m}(S_{\Lambda^c})$, where $\hat{Z}_\Lambda^{\beta,m}$ is the canonical partition function obtained from (8.1) by restricting the spin configurations to those whose sum equals $m|\Lambda|$.

The free energy as a function of m for fixed β is like as in Fig.8.3. The flat part in the graph of F corresponds to the phase transition region.

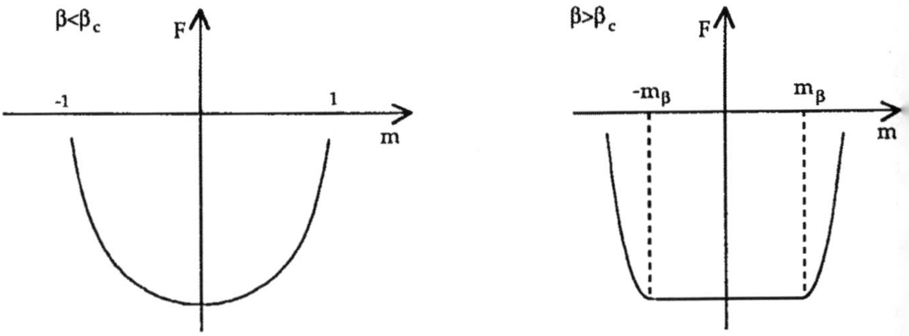

FIG. 8.3.

8.2 Time evolution.

Any Markov process for which the Gibbs measure is invariant is a good candidate for defining the time evolution. In finite regions the evolution is a jump process, the infinite volume dynamics is defined by a limiting procedure. There are two types of evolutions, whether they preserve or not the total magnetization, and are respectively called the Kawasaki and the Glauber dynamics. The generator L_β of the Kawasaki evolution acts, on the cylindrical functions, as

$$L_\beta f(S) = \frac{1}{2}\sum_{x,y} c_\beta(x,y;S)\big[f(S^{x,y}) - f(S)\big] \tag{8.9a}$$

$$c_\beta(x,y;S) = \exp\Big(-\frac{\beta}{2}\big[H^h(S^{x,y}) - H^h(S)\big]\Big) \tag{8.9b}$$

where $S^{x,y}$ is obtained from S by interchanging the spins at x and y, and x and y in (8.9) are nearest neighbor sites. Notice that the energy difference in (8.9b) is well defined also in infinite volumes and that it does not depend on h. For the existence of the process we refer to Liggett, [90].

L_β is self-adjoint in L^2 of any Gibbs measure at inverse temperature β and any magnetic field h. In the physics literature self-adjointness in this context is often referred to as the detailed balance condition. To prove such a property, consider a finite region Λ and any two sites x and y in Λ which are nearest neighbor. Then, in equilibirum, the rate at which the spins at x and y are exchanged when the configuration is S_Λ equals the probability of such a configuration times $c_\beta(x,y;S_\Lambda)$. Analogous expression holds for the exchange in the configuration $S_\Lambda^{x,y}$ and they compensate exactly because, by the choice of c_β,

$$e^{-\beta H_\Lambda^h(S_\Lambda|S_{\Lambda^c})} c_\beta(x,y;S) = e^{-\beta H_\Lambda^h(S_\Lambda^{x,y}|S_{\Lambda^c})} c_\beta(x,y;S^{x,y})$$

This proves the self-adjointness in finite volumes and, by a limiting procedure, the proof can be extended to infinite volumes.

The generator of the Glauber evolution is

$$L_{\beta,h} f(S) = \sum_x \exp\Big(-\frac{\beta}{2}\big[H^h(S^x) - H^h(S)\big]\Big)\big[f(S^x) - f(S)\big] \tag{8.10}$$

where S^x is obtained from S by flipping the spin at x; $L_{\beta,h}$ is self-adjoint with respect to any Gibbs measure at inverse temperature β and magnetic field h (recall that there are more Gibbs measures when there is phase transition).

We first consider the Kawasaki evolutions, which have a richer structure, and start by the simpler case $\beta < \beta_c$. Since the evolution preserves the magnetization we expect, in the continuum limit, a continuity equation for the magnetization density $m(r,t)$:

$$\frac{\partial m}{\partial t} + \nabla j = 0 \tag{8.11}$$

with j the magnetization flux. To have a closed equation, we need to express j in terms of the magnetic density field. If the Fourier law applies j is "proportional" to minus the gradient of the magnetic density, then

$$\frac{\partial m}{\partial t} = \frac{1}{2}\nabla(D\nabla m) \tag{8.12}$$

where D (which depends on $m(r,t)$), represents the diffusion matrix.

(8.12) is the true equation in the limiting case $\beta = 0$, i.e. at infinite temperature. In such a case in fact the generator L_β reduces to the stirring generator considered in Chapter VI and the hydrodynamic equation for this process is the heat equation with $D = 1$. However if we raise β to any positive value, no matter how small it is, the hydrodynamic limit becomes suddenly an extremely hard problem, which has not been solved so far. It is generally believed, however, that (8.12) holds and there is even a definite conjecture, based on the Green-Kubo theory, on the value of the diffusion matrix. First notice that there is a microscopic analogue of (8.11):

$$S(x,t) - S(x,0) = \sum_{i=1}^{2} \left[J(x-e_i, x, [0,t]) - J(x, x+e_i, [0,t])\right]$$

where $J(x, x+e_i, [s,t])$ is the magnetic flux in the time interval $[s,t]$ through the bond $(x, x+e_i)$, (e_i being the unit vector along the x-axis, y-axis, if i=1, respectively 2). This is twice the sum of the times counted as positive [negative] when the spins at x and $x+e_i$ are exchanged and the spin at x was positive [negative] before the exchange.

In the average, at equilibrium, the flux is zero, but it has fluctuations. Assuming its validity, the fluctuation dissipation theorem states that the variance, of the fluctuation fields equals the diffusion matrix D in (8.12). The variance is defined as

$$\lim_{t\to\infty} \frac{1}{t} \lim_{\Lambda \nearrow \mathbb{Z}^2} \frac{1}{|\Lambda|} \sum_{\substack{x\in\Lambda \\ y\notin\Lambda}} \mathbb{E}_{\mu_{\beta,m}}\Big(J(x, x+e_i, [0,t]) J(y, y+e_{i'}, [0,t])\Big) \tag{8.13}$$

where $\mu_{\beta,m}$ is the Gibbs measure with inverse temperature β and magnetization m, (if $m \in (-m_\beta, m_\beta)$ then the measure is a convex combination, as on the right hand side of (8.4b)). Spohn, [125], has proven that the limit in (8.13) exists, that it defines a non negative definite diffusion matrix, $D^{(GK)}$, the so called Green-Kubo diffusion matrix, supposedly equal to D in (8.12). There is a formula for $D^{(GK)}$, also derived by Spohn,

$$D^{(GK)}_{\ell,\ell'} = \frac{1}{2\chi}\Big(\frac{1}{2}\delta_{\ell,\ell'} \sum_{|e|=1} \mathbb{E}_{\mu_{\beta,m}}(c(0,e;S)) - 2\int_0^\infty ds \sum_x \mathbb{E}_{\mu_{\beta,m}}(j_\ell(x,s) j_{\ell'}(0,0))\Big) \tag{8.14a}$$

where

$$j_i(x,t) = [S(x,t) - S(x+e_i,t)]c(x, x+e_i; S) \tag{8.14b}$$

and χ is the magnetic susceptibility

$$\chi^{-1} = \frac{\partial F(\beta,m)}{\partial m} \tag{8.14c}$$

Spohn has also proved that the integral in (8.14a) is finite.

By (8.14c), the magnetic susceptibility diverges in the phase transition region, where we expect a degenerate, vanishing, diffusion. In particular, therefore, according to the Green-Kubo theory, an initial profile with values in $(-m_\beta, m_\beta)$ does not change, at least on the hydrodynamic space-time scale.

Computer simulations, [80], seem to agree with such a conjecture, in fact the macroscopic profile does not apparently change, we report in Fig. 8.5 some evidence of that, in the case of a long range interaction. But this does not imply that nothing happens. Assume first that the typical profile is constant, so that the initial configurations at $t = 0$ have a local magnetic density everywhere approximately equal to $m \in (-m_\beta, m_\beta)$. The system then changes this structure in a short time, since configurations with the above property do not belong to the support of any pure Gibbs measure, i.e. of any invariant measure. To reach "allowed values of the magnetic density" while preserving the total magnetization, the local magnetization should decrease in some regions and increase in others, thus forming clusters of the two pure phases. This is observed clearly in computer simulations, see for instance Fig. 8.6, which refers to a long range interaction, which also show that the size of these clusters has some well definite value.

We get essentially the same picture when the initial profile is not constant (and it has values in $(-m_\beta, m_\beta)$): as the parameter ϵ which defines the hydrodynamic limit becomes sufficiently small, the magnetic profile looks constant over regions significantly larger than the size of the clusters, so that the phases separate like in the constant case. The relative proportion of one phase to the other is adjusted so that the local (on the hydrodynamic scale) values of the initial magnetization are preserved. Such a picture is then stable at least on the hydrodynamic time scale. Therefore, on such a scale, the magnetic density has oscillations of very high frequency, infinitely fast in the hydrodynamic limit: the regions where this happens are called *mushy regions*. The oscillations disappear when integrated against functions which are smooth on the hydrodynamical scale, but are recorded by the Young measure which is concentrated on the two values $\pm m_\beta$.

Before completing this paragraph we discuss the Glauber evolution. Far from the phase transition region nothing particular is expected. There is only one Gibbs state invariant for the evolution and the system cannot but evolve toward it. If instead $h = 0$ and $\beta > \beta_c$, there are two Gibbs states corresponding to the two pure phases and the system has to choose among them. Assume initially that the magnetization is 0. Then, by the symmetry of the evolution for the change $S \to -S$, it will be a stochastic fluctuation which determines the final state. But such fluctuations in far apart regions are largely independent, so that again the phases will separate with a non trivial spatial pattern. The interface is the transition region which separates one phase from the other. The shape and the evolution of the interface is of great interest theoretically and in the applications, we shall be back on this in the sequel. Something non trivial may occur even outside of the phase transition region: if $\beta > \beta_c$ and $h > 0$ is very small, metastable effects are in fact possible. Assume the initial state is the pure phase with magnetization $-m_\beta$. This is not invariant for the Glauber evolution $L_{\beta,h}$, but the state is expected to change very slowly. In recent works Neves and Schonmann, [101], [121], and Martinelli, Olivieri and Scoppola, [93], have shown, roughly speaking, that the change occurs after a very long time, it is very abrupt and it has a definite pattern: a cluster of the stable phase grows somewhere with a definite shape and eventually it invades the whole space, we refer to the original papers for a precise formulation of the results and for limits of validity.

8.3 The Lebowitz-Penrose limit.

In this paragraph we consider Kac potentials: the strength and the range of the interaction are controlled by a parameter γ which eventually goes to 0; the energy is given by (8.2) with J

replaced by

$$J_\gamma(x,y) = \gamma^d J(\gamma|x-y|) \tag{8.15a}$$

where J is a smooth bounded continuous non negative function such that

$$0 < \int_{\mathbb{R}^d} dr\, J(|r|) = a < \infty \tag{8.15b}$$

We shall see that this model presents interesting features even when $d = 1$, as $\gamma \to 0$. Notice finally that the interaction of a spin with any other single one is vanishingly small as $\gamma \to 0$, yet its total interaction with all the other spins stays finite.

We assume that either J has compact support or that it decays exponentially at infinity, all these assumptions are not really necessary for what follows and may be relaxed. We first study equilibrium:

8.3.1 Theorem. *(Lebowitz-Penrose). Let $p_\gamma(\beta,h)$ be the pressure in the thermodynamical limit at given $\gamma > 0$. Then*

$$\lim_{\gamma\to 0} p_\gamma(\beta,h) = p(\beta,h) \equiv \sup_{|m|\le 1}\left(\frac{1}{2}am^2 + \beta^{-1}e(m) + hm\right) \tag{8.16a}$$

where a is defined in (8.15b) and

$$e(m) = -\frac{1+m}{2}\log\left[\frac{1+m}{2}\right] - \frac{1-m}{2}\log\left[\frac{1-m}{2}\right] \tag{8.16b}$$

Thus $e(m)$ is the entropy of a Bernoulli process (i.e. $-\beta$ times the free energy of the Ising model with no interactions) in $\{-1,1\}^{\mathbf{Z}^d}$ with average spin equal to m.

Remarks. The free energy $F(\beta,m)$ is defined as the Legendre transform of $p(\beta,h)$:

$$F(\beta,m) = \sup_h\{hm - p(\beta,h)\} \tag{8.17a}$$

and from (8.16) we get that

$$F(\beta,m) = CE\{-\frac{1}{2}am^2 - \beta^{-1}e(m)\} \tag{8.17b}$$

where $CE(f)$ denotes the convex envelope of f. If $\beta > \beta_c = 1/a$, F has a flat part in the interval $(-m_\beta, m_\beta)$ where m_β is the positive solution of

$$2\beta a m_\beta = \log\left[\frac{1+m_\beta}{1-m_\beta}\right] \tag{8.17c}$$

In Fig.8.4 (in the next page) we plot F versus m at β fixed. The lower curve in the figure on the right represents the convex envelop of the upper one, whose equation is $-\frac{1}{2}am^2 - \beta^{-1}e(m)$. At the points $\pm m^*_\beta$ the curvature changes.

The phase transition described in this way corresponds to the Van der Waals theory and the convex envelop to the Maxwell rule, metastability is also explained by the Van der Waals theory and in terms of the Lebowitz-Penrose limit, as we shall discuss in the sequel.

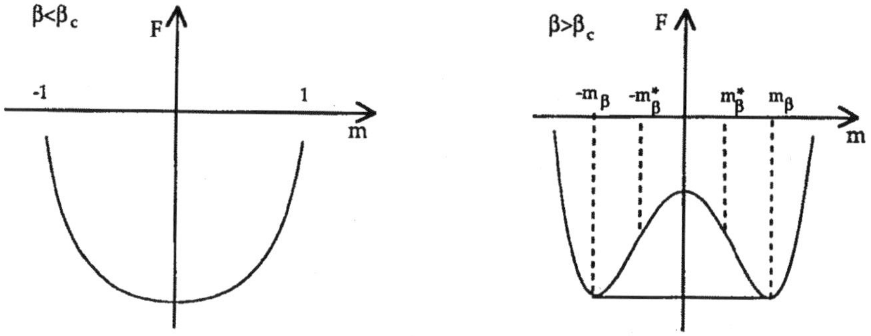

FIG. 8.4.

Sketch of the proof of Theorem 8.3.1. The proof is based on a block spin renormalization procedure. We divide $\mathbb{R}^d$ into squares of side $\gamma^{-\alpha}$, $0<\alpha<1$, and centers $i\gamma^{-\alpha}$, $i \in \mathbb{Z}^d$; we denote by $\mathcal{B}_i$ the elements in $\mathbb{Z}^d$ which are in the square of center $i\gamma^{-\alpha}$ and by $|\mathcal{B}_i|$ their number. Then the block spin renormalization is the transformation from the original spin configuration $S = \{S_x, x \in \mathbb{Z}^d\}$ to $\sigma = \{\sigma_i, i \in \mathbb{Z}^d\}$:

$$\sigma_i = \frac{1}{|\mathcal{B}_i|} \sum_{x \in \mathcal{B}_i} S_x \tag{8.18}$$

The new spins σ_i interact via a new "effective potential" obtained by summing the Gibbs factors over all the configurations S which give rise to the same spin configuration σ. This is an implicit and rather complicated expression, but, in the limit when $\gamma \to 0$ it can be worked out to give useful information.

With this in mind we begin the estimate of the partition function: we consider a square Λ with periodic boundary conditions [in the thermodynamic limit the pressure is independent of the boundary conditions for any $\gamma > 0$] and we assume that it is exactly paved by the squares $\mathcal{B}_i$, $i \in \tilde{\Lambda}$. We shall prove that in the limit when first the volume goes to infinity and then $\gamma \to 0$ an upper bound for the pressure is given by the right hand side of (8.16a). Since the same expression can be similarly proven for the lower bound this will complete the proof of the theorem. We use the notation:

$$1(S_\Lambda \to \sigma_{\tilde{\Lambda}}) = \prod_{i \in \tilde{\Lambda}} 1\Big(\sum_{x \in \mathcal{B}_i} S_x = |\mathcal{B}|\sigma_i\Big), \quad |\mathcal{B}| = |\mathcal{B}_i|$$

Then the partition function (in Λ with periodic condition) is

$$Z = \sum_{\sigma_{\tilde{\Lambda}}} \sum_{S_\Lambda} 1(S_\Lambda \to \sigma_{\tilde{\Lambda}}) \exp\left(\beta\gamma^d \frac{1}{2} \sum_{x \neq y} J(\gamma|x-y|) S_x S_y + \sum_x \beta h S_x\right)$$

We approximate $J(\gamma|x-y|)$ by $J(\gamma^{1-\alpha}|i-j|)$ for $x \in \mathcal{B}_i$ and $y \in \mathcal{B}_j$, more precisely, for a suitable constant c

$$\Big|\gamma^d \sum_{x \neq y} J(\gamma|x-y|) S_x S_y - \gamma^{-\alpha d} \sum_{i,j} \tilde{J}_\gamma(|i-j|)\sigma_i \sigma_j\Big| \leq c\gamma^{1-\alpha}|\Lambda|$$

where

$$\tilde{J}_\gamma(|i|) = \gamma^{(1-\alpha)d} J(|\gamma^{1-\alpha} i|) \tag{8.19a}$$

We therefore have that, for a suitable constant c,

$$Z \leq \sum_{\sigma_{\tilde{\Lambda}}} \sum_{S_\Lambda} 1(S_\Lambda \to \sigma_{\tilde{\Lambda}}) \exp\Big(c\gamma^{1-\alpha}|\Lambda| + \gamma^{-\alpha d}[\beta \frac{1}{2} \sum_{i,j} \tilde{J}_\gamma(|i-j|)\sigma_i \sigma_j + \sum_i \beta h \sigma_i]\Big) \tag{8.19b}$$

To estimate the sum over S_Λ, we consider the generic block $\mathcal{B}$ and denote by S the set of all the spins in that block and by $|S|$ their sum, (notice that $|S|$ may also be negative). Then

$$1(|S| = \sigma|\mathcal{B}|) = 1(|S| = \sigma|\mathcal{B}|) \exp\Big(|\mathcal{B}|[P^0_{h'} - h'\sigma]\Big) \prod_x \mu^0_{h'}(S_x) \tag{8.19c}$$

where

$$\mu^0_{h'}(S_x) = \frac{e^{h' S_x}}{e^{h'} + e^{-h'}}$$

is the probability that the spin at x has value S_x with respect to the Gibbs measure with $\beta = 1$, magnetic field h' and no interaction among the spins; $P^0_{h'}$ is the corresponding pressure

$$P^0_{h'} = \log[e^{h'} + e^{-h'}]$$

By choosing $h' = h(\sigma)$ as

$$P^0_{h'} - h'\sigma = \min_h [P^0_h - h\sigma] = e(\sigma)$$

we see that (8.19c) is bounded by $\exp(|\mathcal{B}| e(\sigma))$, hence

$$Z \leq e^{c\gamma^{1-\alpha}|\Lambda|} \sum_{\sigma_{\tilde{\Lambda}}} \exp\Big(\gamma^{-\alpha d}[\beta \frac{1}{2} \sum_{i,j} \tilde{J}_\gamma(|i-j|)\sigma_i\sigma_j + \sum_i (\beta h \sigma_i + e(\sigma_i))]\Big)$$

The second exponential on the right hand side is the Gibbs factor for a new system of spins which interact via the same potential as before the block spin renormalization, but with $\gamma \to \gamma^{1-\alpha}$. There are three more differences: i) the spins σ_i have $\gamma^{-\alpha d}$ values in $[-1, 1]$; ii) an entropy-term has been added to the one body potential; iii) the effective temperature of the new system vanishes like $\gamma^{\alpha d}$.

For the proof of the theorem it is convenient to rewrite the hamiltonian in a slightly different way. We use the identity

$$\sigma_i \sigma_j = -\frac{1}{2}(\sigma_i - \sigma_j)^2 + \frac{1}{2}\sigma_i^2 + \frac{1}{2}\sigma_j^2$$

which gives

$$Z \leq e^{c\gamma^{1-\alpha}|\Lambda|} \sum_{\sigma_{\tilde{\Lambda}}} \exp\Big(- \beta\gamma^{-\alpha d} \tilde{H}^h_\gamma(\sigma_{\tilde{\Lambda}})]\Big) \tag{8.20a}$$

where

$$\tilde{H}^h_\gamma(\sigma_\Lambda) = \sum_{i \in \tilde{\Lambda}} [f_\beta(\sigma_i) - h\sigma_i] + \frac{1}{4} \sum_{i \neq j \in \tilde{\Lambda}^2} \tilde{J}_\gamma(|i-j|)(\sigma_i - \sigma_j)^2 \tag{8.20b}$$

and $f_\beta(\sigma_i)$ is a "local free energy":

$$f_\beta(\sigma_i) = -\frac{1}{2}a\sigma_i^2 - \frac{1}{\beta}e(\sigma_i) \tag{8.20c}$$

In (8.20c) we have replaced $\sum_i \tilde{J}_\gamma(|i|)$ by a, see (8.15), since the error can be included in the first factor in (8.20a) by changing the value of the constant c. To complete the proof we notice that the interaction between different spins is positive, so that an upper bound for Z is simply obtained by dropping the interaction. We bound the one body energy by $p_0(\beta, h)$, see (8.16a); the contribution of the sum over i multiplied by $\gamma^{-\alpha d}$ produces a factor $\gamma^{-\alpha d}|\tilde{\Lambda}|$. By (8.18), the values of σ_i vary by $|\mathcal{B}_i|^{-1} \approx \gamma^{-\alpha d}$ so the sum over $\sigma_{\tilde{\Lambda}}$ is bounded by $\gamma^{-\alpha d|\tilde{\Lambda}|}$. We take the log of Z, divide it by $|\Lambda|$, let $|\Lambda| \to \infty$ and then $\gamma \to 0$ and obtain the desired upper bound.

The lower bound can be obtained similarly. The most delicate point is when we get at (8.19c). Given σ we choose, as before, $h' = h(\sigma)$, noticing that

$$\mathbb{E}_{\mu^0_{h(\sigma)}}(S_0) = \sigma$$

By the local central limit theorem there is a constant c so that

$$\sum_S 1(|S| = \sigma|\mathcal{B}|) \prod_x \mu^0_{h(\sigma)}(S_x) \geq \frac{c}{\sqrt{|\mathcal{B}|}} = c\gamma^{\alpha d/2}$$

There are $|\Lambda|/|\mathcal{B}|$ such factors, so that their contribution is negligible when we first take the limit as $|\Lambda| \to \infty$ and then $\gamma \to 0$. We omit the details. □

In the course of the proof of Theorem 8.3.1 we have introduced the new hamiltonian $\tilde{H}_\gamma$, which approximately describes the interaction between the block spins σ_i. It is conceivable and to some extent proven that such a hamiltonian is relevant in determining the structure of the typical configurations for the Gibbs measure. Since the effective temperature vanishes like $\gamma^{\alpha d}$ as $\gamma \to 0$, one expects them to be related to the profiles which minimize (maybe only locally) the hamiltonian (8.20), hence to its ground states. It is convenient to make one more approximation, namely to replace the sums in (8.20) by integrals and the spins configurations by trajectories $m(r)$ with values in the real interval $[-1, 1]$. This leads to the new hamiltonian

$$\hat{H}^{\beta,h}_{\hat{\Lambda}}(m) = \int_{\hat{\Lambda}} dr[f_\beta(m(r)) - hm(r)] + \int_{\hat{\Lambda}^2} dr\, dr' \frac{1}{4} J(|r - r'|)(m(r) - m(r'))^2 \tag{8.21}$$

The corresponding inverse temperature is then $\beta\gamma^{-1}$. $\hat{\Lambda}$ is the original region Λ contracted by a factor γ. Eisele and Ellis, [56], have studied the Gibbs measure with the interaction (8.15) in a torus $T(\gamma^{-1}L)$ of side $\gamma^{-1}L$ proving that the functional $\hat{H}^h_{\hat{\Lambda}}(m)$, $\hat{\Lambda} = T(L)$, is the rate function for its large deviations. In particular, therefore, the typical configurations are related to the profiles $m(r)$ which minimize (8.21). On a torus these are the constant $m(\beta, h)$; $m(\beta, 0)$, for $\beta > \beta_c$, takes the two values $\pm m_\beta$ corresponding to the two pure phases, see (8.17c).

In this setup a magnetization profile where the two pure phases are both present may only occur as a large deviation, see [30], it has therefore a vanishingly small probability. To observe it with non vanishing probability, we need to "forbid" the constant magnetization profiles with values $\pm m_\beta$: this can be done either by imposing suitable boundary conditions or, keep periodic boundary conditions, but imposing the constraint that the total magnetization has a fixed value $m \neq \pm m_\beta$

$$\int_{T(L)} dr\, m(r) = L^d m \tag{8.22}$$

At the spin level, this constraint amounts to studying the canonical Gibbs measure, for which Varadhan, (private communication) has proven that the functional (8.21) is the rate function for the large deviations. The analysis of the variational problem for (8.21) with the constraint (8.22) has been developed under a further approximation on the functional (8.21). This is based on the assumption that the range of J may be considered vanishingly small. If we keep $m(r)$ as a fixed C^1 function, then the second term in (8.21) behaves, when J approaches a delta function, as

$$\frac{1}{4}\int_{\hat{\Lambda}^2} dr\, dr' |\nabla m(r)|^2 J(|r-r'|)(r-r')^2$$

With this in mind, one introduces the new functional

$$H_\beta(m) = \int dr \left(\lambda^2 |\nabla m(r)|^2 + f_\beta(m(r)) \right) \tag{8.23a}$$

where λ is a parameter which eventually goes to zero. We have dropped the term with the magnetic field h since it becomes irrelevant when studying the minimum with the constraint (8.22). For notational simplicity, we have also dropped the reference to the region where the problem is studied.

Modica et al., [91], [98], have proven that in the limit when $\lambda \to 0$ the solutions (when $m = 0$ in (8.22) and $f_\beta(m)$ is a double well potential with minima $\pm m_\beta$) are the functions with support on $\pm m_\beta$ which minimize the interface between them, while the integral of the magnetization satisfies (8.22). Dobrushin, Kotecký and Shlosman, [51], have studied the same problem for the Ising model, finding that the solution is given in terms of the Wulff construction with a free energy functional which is determined by the Ising interaction. These are the only results, as far as we know, about the structure of the interface in equilibrium statistical mechanics.

We next study the dynamical problem.

8.4 Deterministic evolutions.

We have introduced in the previous subparagraph several hamiltonians which describe to various degree of approximation the same system. From each of them we can define an evolution and we start from the renormalized hamiltonians (after the block spin transformation). The corresponding effective temperature in the limit when $\gamma \to 0$ is 0 and the equilibrium states are the hamiltonian ground states. By taking a formal limit in (8.9) for $\beta \to \infty$, we see that the dynamics at 0-temperature becomes deterministic, and that the energy decreases on time, so that the hamiltonian is a Liapunov function. (One often does the reverse: given a function f a dynamics is introduced for which f is a Liapunov function, so that one has an algorithm for searching the minima of f. Such a procedure is also used in image reconstructions). We therefore expect that in the limiting dynamics, when the temperature goes to 0, H_β decreases. An evolution with such a property is the gradient evolution defined by the equation

$$\frac{\partial m(r,t)}{\partial t} = -\frac{\delta H_\beta((m(\cdot,t))}{\delta m(r,t)} \tag{8.23b}$$

The time derivative of the magnetization profile at (r,t) is in fact opposite to the variation of the functional with respect to $m(r,t)$, so that $H_\beta(m)$ decreases along the solutions of (8.23b). If we apply this to the functional defined by (8.21) we get the reaction diffusion equation

$$\frac{\partial m(r,t)}{\partial t} = \lambda^2 \Delta m(r,t) - [f'_\beta(m(r,t)) + h] \tag{8.24}$$

When $h = 0$, the reactive potential is the local free energy which, at $\beta > \beta_c$ has two equal minima, $\pm m_\beta$. This equation is capable of describing interface dynamics. For what said in the previous subparagraph, we should think of λ as a small, infinitesimal, parameter: in such a case it is known that if the initial state $m(r,0)$ is negative inside a regular region Γ and positive outside (some assumptions of regularity are also needed) then the solution $m_\lambda(r,\lambda^{-2}t)$ of (8.24) converges when $\lambda \to 0$ to a function which has value $\pm m_\beta$ outside, respectively inside, the region Γ_t. Furthermore, the boundary of Γ_t evolves by the motion by mean curvature: namely the velocity of any point of $\delta\Gamma_t$ equals a constant times the curvature of Γ_t at that point and it is directed toward the interior. We refer to the literature: De Mottoni-Schatzman, [45].

Of course an equation like (8.24) might not lead necessarily to the true minima of $H(m)$, but may be trapped in some local minimum. To avoid it, some noise is usually added to the equation: we can do it in a natural way by looking at the original spin process which is intrinsically stochastic. Before that, however, we discuss the deterministic evolution under the constraint (8.22). In such a case we modify (8.24) by requiring the magnetization to satisfy the continuity equation

$$\frac{\partial m(r,t)}{\partial t} = -\nabla j(r,t) \tag{8.25a}$$

where

$$j(r,t) = -\nabla \frac{\delta H_\beta((m(\cdot,t))}{\delta m(r,t)} \tag{8.25b}$$

If $H_\beta(m)$ is given by (8.23a), then (8.25) becomes the Cahn-Hilliard equation:

$$\frac{\partial m}{\partial t} = \Delta(- \Delta m + f'_\beta(m)) \tag{8.26}$$

The study of the evolution of the interface in this case is much more difficult, but some results have been obtained, see the bibliographical notes, §8.8.

8.5 The Ginzburg-Landau evolution.

Next we consider the evolution before taking the limit $\gamma \to 0$, the temperature is now different from zero and we have a stochastic evolution. We shall eventually go back to the original spin systems and introduce at that level the Kawasaki and the Glauber dynamics, in the usual way, but before doing it, we discuss an intermediate case corresponding to the hamiltonian (8.20b). It is convenient to rewrite it as

$$H(\sigma_\Lambda|\sigma_{\Lambda^c}) = \sum_{x\in\Lambda}[f(\sigma_x) - h\sigma_x] + \frac{1}{2}\sum_{x\neq y\in\Lambda} J_{x,y}(\sigma_x - \sigma_y)^2 + \sum_{\substack{x\in\Lambda\\ y\notin\Lambda}} J_{x,y}(\sigma_x - \sigma_y)^2 \tag{8.27}$$

with $\sigma_x \in \mathbf{R}$. To give sense to the Gibbs measure we assume superstability, namely that $f(\sigma) \geq A\sigma^2 - B$ with A suitably large and positive. The Gibbs measure is then

$$d\mu^{\beta}_{\Lambda,\sigma_{\Lambda^c}} = \frac{1}{Z_{\Lambda}(\sigma_{\Lambda^c})} e^{-\beta H(\sigma_\Lambda|\sigma_{\Lambda^c})} \prod_{x\in\Lambda} d\sigma_x \tag{8.28}$$

The Kawasaki evolution for this hamiltonian is the set of stochastic differential equations

$$d\sigma(x,t) = \sum_{\substack{|e|=1\\ x+e\in\Lambda}} \left([\frac{\partial H}{\partial \sigma(x+e,t)} - \frac{\partial H}{\partial \sigma(x,t)}]dt - dw(x,x+e,t) \right) \tag{8.29}$$

The first term is deterministic and it is the discretized version of (8.25), the second one represents the divergence of a current whose components are $j_i(x,t) = dw(x, x+e_i, t)$. The variables $w(x,x+e,t)$ are independent standard Brownian motions and one can verify that the measure (8.28) is invariant under (8.29).

The hydrodynamical behavior of the Ginzburg-Landau system has been extensively studied, we just state the following result

8.5.1 Theorem. *([117]) For each $\epsilon > 0$ such that $\epsilon^{-1}L$, $L > 0$, is an integer, consider (8.29) on a torus of side $\epsilon^{-1}L$. Assume [again, our assumptions are unnecessarily restrictive] that μ^ϵ is a product measure with continuous density and compact support such that the average of the spin σ_x equals $m_0(\epsilon x)$, where $m_0(r)$ is a C^∞ function on the torus of side L. Then for any $\delta > 0$*

$$\lim_{\epsilon\to 0} \mathbf{P}_{\mu^\epsilon}\left(|X^\epsilon_t(\phi) - \int drm(r,t)\phi(r)| > \delta \right) = 0 \tag{8.30}$$

where

$$X^\epsilon_t(\phi) = \epsilon^d \sum_x \phi(\epsilon x)\sigma(x, \epsilon^{-2}t) \tag{8.31}$$

$m(r,t)$ is the solution of

$$\frac{\partial m}{\partial t} = \nabla(D\nabla m), \qquad D_{\ell,\ell'} = \frac{1}{2\chi}\delta_{\ell,\ell'} \tag{8.32}$$

χ is the magnetic susceptibility

$$\chi(\beta,m) = \sum_x \mathbf{E}_{\mu_{\beta,m}}(\sigma_x\sigma_0 - m^2) \tag{8.33}$$

and $\mu_{\beta,m}$ is the infinite volume Gibbs measure at inverse temperature β and magnetization m.

Notice that this theorem covers cases where there are phase transitions: in the phase transition region the magnetic susceptibility is infinite so that the diffusion coefficient vanishes. The same conclusions were drawn for the nearest neighbor Ising model, but they were based on the assumption that the Green-Kubo theory could be applied, in this model the result is fully proven. The result in Theorem 8.5.1 proves also that $D = D^{(GK)}$, in fact, because of the gradient condition, satisfied in the Ginzburg-Landau model, the integral in (8.14a) vanishes. The gradient condition refers to the current, defined as in (8.14b), and it expresses the fact that it can be written as the discrete gradient of a local function. The sum over x in (8.14a) becomes then a telescopic sum and the integrand at any time vanishes.

The proof of the theorem is based on an extension of the Guo, Papanicolaou and Varadhan method presented in Chapter III. The difficulty to extend the proof to the nearest neighbor Ising model originates from the fact that this is a non gradient system, unlike the Ginzburg-Landau model.

8.6 The Kawasaki evolution in the Lebowitz-Penrose limit.

We restrict here to the one dimensional case. The evolution is defined by (8.9) when the spin interaction is given by (8.15a). Notice that the energy difference in (8.9b) is of the order of γ, so that, to leading order in γ, this model looks like the Ising model at infinite temperature, i.e. to the stirring process, which, as we said, gives rise, in the hydrodynamic limit, to the heat equation. This model has in fact been studied as a perturbation of the stirring process, which is relatively easy on a time scale of the order of γ^{-2} where it gives rise to a integro-differential macroscopic equation:

8.6.1 Theorem. *([87],[70]). Let ν_γ be the product measure on $\{-1,1\}^{\mathbf{Z}}$ such that the average of the spin at x equals $m_0(\gamma x)$, where m_0 is a C^∞ function with values in $[-1,1]$. For $n \geq 1$ denote by $\underline{x}$ any n-tuple of distinct sites $x_1, \ldots, x_n$ in $\mathbf{Z}$; then for any n and $t > 0$*

$$\lim_{\gamma\to 0} \sup_{\underline{x}} \left|\mathbf{E}_{\nu_\gamma}\Big(\prod_{i=1}^{n} \sigma(x_i, \gamma^{-2}t)\Big) - \prod_{i=1}^{n} m(\gamma x_i, t)\right| = 0 \tag{8.34}$$

where $m(r,t)$ solves the equation

$$\frac{\partial m(r,t)}{\partial t} = \frac{1}{2}\frac{\partial}{\partial r}\left(\frac{\partial m(r,t)}{\partial r} + \beta(1 - m(r,t)^2)\int_0^\infty dr' J'(r')[m(r+r',t) - m(r-r',t)]\right) \tag{8.35}$$

and $m(r,0) = m_0(r)$.

Notice that the functional $\hat{H}$ defined in (8.21) is a Liapunov functional for the evolution (8.35), so that (8.35) is one of the possible deterministic dynamics related to the free energy functional (8.21). In this sense we have therefore justified the approximation leading to $\hat{H}$ also in a dynamical frame.

The true hydrodynamical limit is obtained by considering an initial magnetization density which varies on a scale ϵ^{-1} much longer than γ^{-1}. The corresponding time scale is then $\epsilon^{-2}t$. Only partial results have been obtained for this case, [87], [70]:

8.6.2 Theorem. *Let $\epsilon = \gamma^{1+\alpha}$, $\alpha > 0$, and let μ^ϵ be the product measure on $\{-1,1\}^{\Lambda_\epsilon}$, such that the average of the spin at x equals $\tilde{m}_0(\epsilon x)$: Λ_ϵ is the interval centered at 0 of length ϵ^{-2} with identification of the extremes; $\tilde{m}_0(r)$ is a C^∞ function on the torus of side ϵ^{-1} which coincides for $|r| \leq \epsilon^{-1} - 1$ with $m_0(r)$, a fixed function on $\mathbf{R}$. It is assumed that each derivative of $\tilde{m}_0(r)$ is uniformly bounded. Assume finally that if $\beta > 1/a$ the range of m_0 does not intersect the interval $[-m_\beta^*, m_\beta^*]$ where*

$$m_\beta^* = \sqrt{1 - [\beta a]^{-1}} \tag{8.36}$$

Then, if α is small enough, [recall that $\epsilon = \gamma^{1+\alpha}$], for any $n \geq 1$ and $t > 0$

$$\lim_{\epsilon \to 0} \sup_{\underline{x}} \left| \mathbb{E}_{\mu^\epsilon}\Big(\prod_{i=1}^{n} \sigma(x_i, \epsilon^{-2}t)\Big) - \prod_{i=1}^{n} m(\epsilon x_i, t) \right| = 0 \tag{8.37}$$

where $m(r,t)$ solves the equation

$$\frac{\partial m(r,t)}{\partial t} = \frac{1}{2}\frac{\partial}{\partial r}\Big(D(m(r,t))\frac{\partial m(r,t)}{\partial r}\Big) \tag{8.38}$$

$$D(m) = 1 - \beta a(1 - m^2) \tag{8.39}$$

and $m(r,0) = m_0(r)$.

At first sight, at least, this is a surprising result, since it states that in the hydrodynamical limit the system behaves diffusively outside of the region

$$\{(\beta, m) : \beta > 1/a,\ m \in [-m^\star_\beta, m^\star_\beta]\} \tag{8.40}$$

which is not the phase transition region: the critical temperature is the same, $\beta_c = 1/a$, but the phase transitions interval $[-m_\beta, m_\beta]$, see (8.17c), is strictly larger than the interval in (8.40), for $\beta > 1/a$. In the Van der Waals theory the region (8.40) is thermodynamically unstable, its complement inside the phase transition region is the metastable region, see Fig. 8.4. Lebowitz and Penrose have given arguments which confirm this picture in the context of the model described in §8.3; Theorem 8.6.1 goes in the same direction, since it shows that at the hydrodynamical times the system in the metastable region does not feel the existence of a phase transition.

The different space-time scalings emploied in the Theorems 8.6.1 and 8.6.2 on the same spin system give rise to two different macroscopic equations, respectively (8.35) and (8.38). It is therefore natural to conjecture that these two equations are related by a space-time transformation. Let $\delta > 0$ be a parameter which goes eventually to 0, $\delta = \epsilon\gamma^{-1} = \gamma^\alpha$. Denote by $m^\delta(r,t)$ the solution to (8.35) with initial datum $m^\delta(r,0) = m_0(\delta r)$. Define $m_\delta(r,t) = m^\delta(\delta^{-1}r, \delta^{-2}t)$, then this is a candidate for being a solution of (8.38) when $\delta \to 0$. Indeed this is what happens for $\{\beta, m\}$ as in Theorem 8.6.2.

The next and more interesting question concerns the behavior of the system outside of this region where phase separation phenomena should take place. The only rigorous result in the above direction has been obtained by Dal Passo and de Mottoni, [32], who have proved the existence, below the critical temperature, of a stationary solution for (8.35) with different asymptotics to the right and to the left of the origin. This beautiful result allows to relate the statistical and the dynamical properties of the system. In fact Dal Passo and de Mottoni have shown that the asymptotic magnetizations in the stationary solution are just the values $\pm m_\beta$ given by (8.17c).

There are computer simulations, [70], on a modified version of the spin model we have described, which also confirm the occurrence of phase separation phenomena. We reproduce Fig.8.5, taken from [70], which shows that a profile inside the unstable (spinodal) region does not change on the hydrodynamic scale, in agreement with the fact that the diffusion vanishes. In the picture

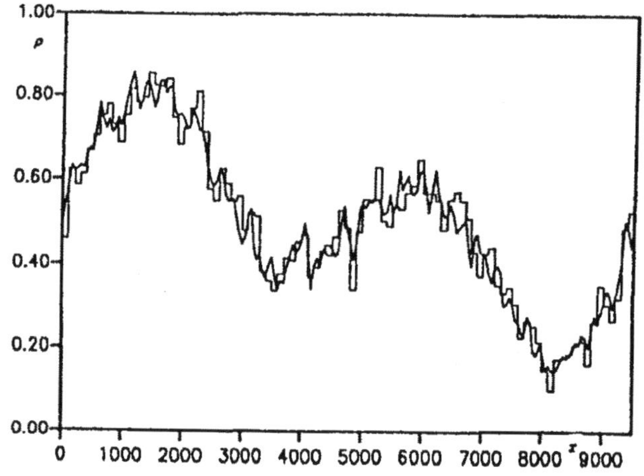

FIG. 8.5.

the initial spin configuration is averaged over intervals of 20 lattice sites, the whole lattice having 300 × 32 sites, the range of the potential being 32 and $\beta = 20$. The continuous line in Fig. 8.5 represent the density after 10^6 iterations averaged on intervals of size 100.

However if we look on a finer space scale, the picture changes dramatically and the phase separation is evident. In Fig. 8.6 it is reported the result of measurements of the magnetization at single sites, time averaged over a time interval of length 100:

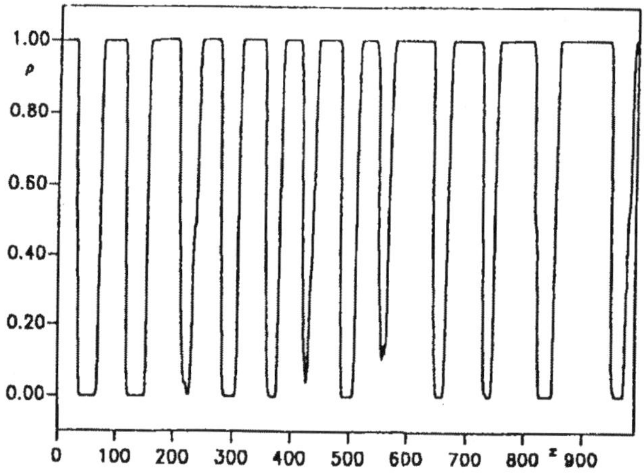

FIG. 8.6.

Other simulations presented in [70] show that initially the phases separate in agreement with he linearization of (8.35). Assume that initially $m_0(r)$ is a small perturbation of a constant, say = 0. One can then linearize (8.35) around 0 to see how it evolves. The Fourier analysis shows that elow the critical temperature there is an interval of frequencies whose corresponding eigenvalues are positive and one where the maximum is attained. Computer simulations prove that in the first stage of the escape the magnetic profile is approximately periodic with this frequency. At longer times there is some evidence that some of the clusters of the same phase cohalesce, but longer computer runs are needed to reach some reasonable evidence.

8.7 The Glauber evolution in the Lebowitz-Penrose limit.

We now consider the Glauber evolution in the Lebowitz-Penrose limit: its generator is given by (8.10), with potential as in (8.15). The analogue of Theorem 8.6.1 is:

8.7.1 Theorem. *Let ν_γ be as in Theorem 8.6.1. Then for any $\delta > 0$ and any $\phi \in S$*

$$\lim_{\gamma\to 0} \mathbf{P}_{\nu_\gamma}\left(\left|X_t^\gamma(\phi) - \int dr\phi(r)m(r,t)\right| > \delta\right) = 0 \tag{8.41a}$$

where

$$X_t^\gamma(\phi) = \gamma^d \sum_x \phi(\gamma x)S(x,t) \tag{8.41b}$$

and

$$\frac{\partial m}{\partial t} = -(1+m)\exp\left(-\beta(J\star m+h)\right) + (1-m)\exp\left(\beta(J\star m+h)\right) \tag{8.42a}$$

$$m(r.0) = m_0(r) \tag{8.42b}$$

$$(J\star m)(r,t) = \int dr' J(|r-r'|)m(r',t) \tag{8.42c}$$

The proof of this theorem can be obtained using the Guo, Papanicolaou and Varadhan method under the additional assumption that the system is confined to a torus of side $\gamma^{-1}L$, for any given $L > 0$, see also [30] and [31].

The functional (8.21) is a Liapunov function for the evolution defined by (8.42a), so that the Glauber evolution in the above scaling reproduces one of the deterministic dynamics associated to (8.21). The stationary equation which describes the interface is now (set $h = 0$)

$$m(r) = \tanh\left(\beta(J\star m)(r)\right) \tag{8.42d}$$

The constant solutions are $m(r) = \hat{m}$ where

$$\hat{m} = \tanh\left(\beta a\hat{m}\right)$$

which for $\beta a > 1$ have the non zero solutions $\pm m_\beta$, see (8.17c). The one dimensional interface profile is then the solution of (8.42d) with asymptotic values $\pm m_\beta$. This has the same solution found by dal Passo and de Mottoni for the steady state in (8.35).

Like in the Kawasaki evolution, the more interesting case, however, is when the system is studied on a longer space-time scale. No work on this aspect has been done, as far as we know. If we look for solutions of (8.42a) which are smooth, i.e. they have the form $m(r,t) = \tilde{m}(\delta r, t)$, with δ a small parameter, then by expanding the difference $m(r',t) - m(r,t)$ to second order and omitting the remainder, we obtain

$$\frac{\partial \tilde{m}(r,t)}{\partial t} = \delta^2 A(\tilde{m}(r,t))\Delta\tilde{m}(r,t) - V'(\tilde{m}(r,t)) \tag{8.43a}$$

where

$$A(x) = 2\cosh\left(x\beta(a+h)\right)\frac{1}{4d}\int_{\mathbb{R}^d} dr' J(|r'||)(r')^2 \tag{8.43b}$$

$$-V'(x) = 2\sinh\left(x\beta(a+h)\right) - 2x\cosh\left(x\beta(a+h)\right) \tag{8.43c}$$

This equation resembles (8.24), but we do not know whether it really shares the same qualitative behavior, particularly for what concerns phase-separation and interface-dynamics. These have been studied in the Glauber+Kawasaki process defined in Chapter VI. Its generator is $L_\epsilon = \epsilon^{-2}L_0 + L_G$, where L_0 is the generator of the stirring process (Kawasaki at infinite temperature), while L_G generates the Glauber evolution with nearest neighbor interactions. This process is in a sense similar to the Glauber evolution in the Lebowitz-Penrose limit, in fact the effect of the stirring is to delocalize the Glauber interaction: by exchanging the values of the spin the Glauber interaction might in fact involve spins previously quite far apart, typically at distances of the order of ϵ^{-1}. There is however a big difference between the two models, while in fact the Glauber evolution is reversible with respect to the Gibbs measure, for the Glauber+Kawasaki evolution the invariant measures are not known, in particular the ergodic problem is an open question for this system, even in one dimension.

In the limit $\epsilon \to 0$ the macroscopic equation for the Glauber+Kawasaki evolution is the reaction-diffusion equation (6.4), see Theorem 6.2.2. We consider hereafter the case $\gamma > 1/2$, see (6.8b), so that the reactive potential $V(m)$, see (6.11), is a double well potential with minima $\pm m^*$, where $m^* = (\alpha/\beta)^{-1/2}$, α and β, given in (6.11b) are positive. The values $\pm m^*$ are to be interpreted, in this context, as the values of the magnetization in the two pure phases.

With this choice of the parameters, the Glauber+Kawasaki process is a microscopic model for the evolution (8.24) and for the associated phase separation phenomena: these have been studied in the one dimensional case:

Theorem 8.7.2. *([40]) Let μ^ϵ be the product measure on $\{-1,1\}^{\Lambda_\epsilon}$, Λ_ϵ is $\mathbb{Z}$ modulo $\epsilon^{-1}|\ln\epsilon|$, $\epsilon^{-1}|\ln\epsilon|$ being an integer and let $\mathbb{E}_{\mu^\epsilon}(\sigma(x)) = 0$ for all $x \in \Lambda_\epsilon$. Then, setting $t_f = |\ln\epsilon|/(2\alpha) + |\ln\epsilon|^{1/3}$, see (6.11) for the definition of α,*

$$(i) \quad \lim_{\epsilon\to 0}\sup_{\underline{x}\in\mathcal{M}_n^\epsilon} |\mathbb{E}_{\mu^\epsilon}(\prod_{i=1}^n \sigma(x_i,\tau|\ln\epsilon|))| = 0 \qquad \text{if} \quad \tau \le 1/(2\alpha) \tag{8.44}$$

$$(ii) \quad \lim_{\epsilon\to 0}\sup_{\underline{x}\in\mathcal{M}_n^\epsilon} \left|\mathbb{E}_{\mu^\epsilon}(\prod_{i=1}^n \sigma(x_i,t_f)) - \tilde{E}\left(\prod_{i=1}^n \rho(\epsilon|\ln\epsilon|^{1/2}x_i)\right)\right| = 0 \tag{8.45}$$

where $\mathcal{M}_n^\epsilon$ is the set of all the n-tuples of distinct sites in Λ_ϵ, $\rho(r) = m^*$ sign $\tilde{X}(r)$; $(\tilde{X}(r))_{r\in\mathbb{R}}$ is a zero-average Gaussian process on some probability space $(\tilde{\Omega},\tilde{\mathcal{A}},\tilde{P})$ with $\tilde{E}(\tilde{X}(r)\tilde{X}(r')) = e^{-\alpha(r-r')^2/2}$, for $r, r' \in \mathbb{R}$.

We shall prove the theorem in the next two chapters for $|\Lambda_\epsilon| = \epsilon^{-1}$. In such a case we loose completely the space structure of the phenomenon, but we still have the possibility to observe the way the phases separate, as a consequence of the stochastic fluctuations which cause the growth of one phase from the initial "chaos".

An extension of Theorem 8.7.2 to more dimensions is at the moment in progress, G. Giacomin, the picture which emerges is similar to that in one dimensions. There are clusters of the different phases whose shape and positions are determined by a Gaussian process. The boundaries of the clusters are smooth and vary on distances $\epsilon^{-1}|\ln\epsilon|$, in lattice units. They then move on the times scale $|\ln\epsilon|$: a work in progress by L. Bonaventura shows that there is motion by mean curvature: such a result is based on the work of de Mottoni and Schatzman, [45], and at the moment suitable hypotheses are needed on the structure of the state at the boundaries between the clusters.

§8.8 Bibliographical notes.

For recent papers on the analysis of the growth of the clusters in the Ising model see [101], [93] and the short survey [92]. The equilibrium statistical mechanics for Kac potentials, to which we have referred in this chapter, is studied in [88], metastability is discussed in [110] and [27].

The hydrodynamic equation derived in [87] for Kac potentials, at the critical temperature becomes an equation for the porous media; the convergence in [87] however does not extend to the case when the initial density has compact support. For the derivation of equations for porous media when the interaction length is scaled as in Kac potentials see [103].

A discussion on the derivation of macroscopic equations for phase decomposition can be found for instance in [85], [64], [109],[108], [87], [106]. For numerical studies see also [102], [120], [57].

On interface dynamics, besides the papers of de Mottoni and Schatzman, we collect a (largely incomplete) list of references in [1].

CHAPTER IX

ESCAPE FROM AN UNSTABLE EQUILIBRIUM

In this and in the next chapters we study the Glauber+Kawasaki spin system of Chapter VI, when the initial measure is the product measure with zero magnetization, namely when all the spins have zero average. This corresponds, in the macroscopic limit, to a stationary, linearly unstable, state. We prove that the system escapes from such an unstable equilibrium at times of the order of $\log \epsilon^{-1}$.

§9.1 The model and the results.

The model is the Glauber+Kawasaki system of Chapter VI, but restricted to finite volumes. Namely for any $\epsilon > 0$ such that ϵ^{-1} is an integer, the configuration space is $\{-1,1\}^{\mathbb{Z}_\epsilon}$, $\mathbb{Z}_\epsilon = \mathbb{Z}$ modulo ϵ^{-1}. We then define a Markov process whose state space is the set of all the spin configurations and its generator is given by (6.6), (6.7) and (6.8) with $\gamma > 1/2$ in (6.8b). The limiting reaction diffusion equation is then

$$\frac{\partial m}{\partial t} = \frac{1}{2}\frac{\partial^2 m}{\partial r^2} + \alpha m - \beta m^3 \tag{9.1}$$

where $m = m(r,t)$ is a periodic function of $r \in [0,1]$; α and β, as defined in (6.11b), are both strictly positive. The finite volume assumption is here physically relevant: at infinite volumes there is an interesting spatial structure, as discussed in §8.1, and the analysis becomes more difficult and delicate, we just refer to [40].

The initial measure μ^ϵ is a product measure with zero magnetization, namely for all x

$$\mathbb{E}_{\mu^\epsilon}(\sigma(x)) = 0 \tag{9.2}$$

By Theorem 6.2.2 μ_t^ϵ (the law of the process at time t) converges, for each fixed t, when $\epsilon \to 0$, to the Bernoulli measure (product measure) with zero magnetization. We want to study the limit when $\epsilon \to 0$ and, at the same time, $t \to \infty$. Since the solution $m \equiv 0$ to (9.1) is linearly unstable ($\alpha > 0$) it is not difficult to guess that the relevant time scale unit is $\log \epsilon^{-1}$, as confirmed by the following theorem.

9.1.1 Theorem. *With the above notation and assumptions*

$$\lim_{\epsilon \to 0} \mu^\epsilon_{\tau \log \epsilon^{-1}} = \begin{cases} \nu_0 & \text{for } \tau < 1/(2\alpha) \\ \frac{1}{2}\nu_{m^*} + \frac{1}{2}\nu_{-m^*} & \text{for } \tau > 1/(2\alpha) \end{cases} \tag{9.3}$$

where the limit in (9.3) means that the expectation of the product of any fixed number of spins converges uniformly on their location. ν_m denotes the Bernoulli measure with spin average m and $\pm m^*$ are the non zero roots of $\alpha m - \beta m^3$.

Furthermore

$$\lim_{\epsilon \to 0} \mu^{\epsilon}_{1/(2\alpha)\log \epsilon^{-1}+t} = \int_{-m^*}^{m^*} dm \lambda_t(m) \nu_m \tag{9.4}$$

where $\lambda_t(m)$ is a continuous function, implicitly defined in §9.6, see (9.81) and (9.78).

9.1.2 Remarks.

A proof of this theorem given in [44] contains an error, see the Bibliographical Notes to this chapter, §9.8, which we correct here using an argument introduced in [40] to study infinite volumes.

There would be a lot to say about the physical motivations and the implications concerning Theorem 9.1.1, we refer to Chapter VIII for the relation with the theory of separation of phases and to the first two sections of [22], which we schematically summarize by the following four considerations. 1) Even though the leading mechanism for the escape (from the zero magnetization state) is random, yet the time when the escape occurs (in the relevant time scale $\log \epsilon^{-1}$) is deterministic, namely $1/(2\alpha)$. 2) However, if we consider a model where the potential $V(m)$ has a quartic maximum at $m = 0$, then the escape occurs at a random time in the new relevant time scale $\epsilon^{-1/2}$, cf. [22]. 3) At infinite volumes there is a non trivial space structure because the magnetizations in far away regions are independent, so there are several time scales which describe the evolution of the magnetization after the escape from its initial value. 4) The same behavior described by (9.3) and (9.4) is exhibited by the stochastic differential equation

$$dm = (\frac{1}{2}\frac{\partial^2}{\partial r^2} m + \alpha m - \beta m^3)dt + \sqrt{\epsilon} dw \tag{9.5a}$$

$$m(r,0) = 0 \tag{9.5b}$$

where dw denotes white noise in space and time. In this respect we are proving that the particle system is better approximated by (9.5) than (9.1).

We now turn to the proof of Theorem 9.1.1 which will take this and the next chapter. The general strategy is based on *the separation of time scales.* At an early stage the small fluctuations are dominant, these are well studied, even in a more general frame, see for instance [36], and we can exploit such a theory to analyse this first stage of the evolution. In this way we eventually reach values of the total magnetization still infinitesimal ($\approx \epsilon^a$, $a < 1/2$), but large enough for the drift to be the leading term, i.e. the linear instabilty in (9.1) takes over and the evolution becomes essentially deterministic. However, since the magnetization is still infinitesimal, the non linear terms can be neglected and this stage of the evolution is also accessible to our investigation: it essentially involves a linear theory. In this way, the magnetization becomes "almost finite", so that, going on, the linear approximation eventually fails and the non linear effects become important. This last time interval is however not too long and the behavior of the system can be studied by methods similar to those presented in Chapter V, to derive the Carleman equation. (The derivation of the reaction diffusion equation in Chapter VI did not need such a refined analysis). It is an useful exercise to try to implement the above ideas on the stochastic differential equation (9.5) and prove the analogue of Theorem 9.1.1.

As in Chapter V, we introduce a sort of truncated correlation functions, called the v-functions, which, for the Glauber+Kawasaki system, are defined as follows.

§9.2 The v-functions.

We give the definition for general Glauber interactions, where

$$(L_G f)(\sigma) = \sum_x \sum_{\underline{x} \in U(x)} c(\underline{x}, x)\sigma(\underline{x})[f(\sigma^x) - f(\sigma)] \tag{9.6a}$$

and, given x, $U(x)$ is a finite collection of finite subsets of $\mathbb{Z}$; $c(\underline{x}, x)$ should be consistent with the fact that the right hand side is the action of a generator on f. We consider translationally invariant interactions so that $c(\underline{x}, x) = c(\underline{x} - x, 0)$, here $\underline{x} - x = (x_1 - x, \cdots, x_n - x)$ if $\underline{x} = (x_1, \cdots, x_n)$. The maximal diameter of the subsets in $U(x)$ (with $c(\underline{x}, x) \neq 0$) is the range of the interaction. In (9.6a) and in the sequel we use the notation

$$\sigma(\underline{x}) = \prod_{x \in \underline{x}} \sigma(x), \qquad \sigma(\emptyset) = 1 \tag{9.6b}$$

In our specific case

$$c(\underline{x}, x) = \begin{cases} 1 & \underline{x} = \emptyset \\ -\gamma & \underline{x} = (x, x \pm 1) \\ \gamma^2 & \underline{x} = (x-1, x+1) \\ 0 & \text{otherwise} \end{cases} \tag{9.6c}$$

We write

$$U(x) = U_1(x) \cup U_2(x) \tag{9.7}$$

where $U_1(x)$ contains those $\underline{x}$ such that $x \notin \underline{x}$ and $U_2(x)$ the others.

We consider the generator $\epsilon^{-2}L_0 + L_G$ without distinguishing finite and infinite volumes, the actual meaning of the various terms will be easily recovered from the context. The reaction diffusion equation associated to the model is then, by Theorem 6.2.2,

$$\frac{\partial m}{\partial t} = \frac{1}{2}\frac{\partial^2 m}{\partial r^2} - 2 \sum_{\underline{x} \in U_1(0)} c(\underline{x}, 0) m^{|\underline{x}|+1} - 2 \sum_{\underline{x} \in U_2(0)} c(\underline{x}, 0) m^{|\underline{x}|-1} \tag{9.8}$$

As in Chapter V we consider a discretized version of this equation:

$$\begin{aligned} \frac{d}{dt} m^\epsilon(x,t) =& \frac{1}{2}\epsilon^{-2}[m^\epsilon(x-1,t) + m^\epsilon(x+1,t) - 2m^\epsilon(x,t)] \\ & - 2 \sum_{\underline{x} \in U_1(x)} c(\underline{x}, x) m^\epsilon(\underline{x} \cup x, t) - 2 \sum_{\underline{x} \in U_2(x)} c(\underline{x}, x) m^\epsilon(\underline{x} \backslash x, t) \end{aligned} \tag{9.9a}$$

where, cf. (9.6b),

$$m^\epsilon(\underline{x}, t) = \prod_{x \in \underline{x}} m^\epsilon(x, t) \tag{9.9b}$$

If needed, we add an extra argument in m^ϵ to specify the initial condition: when this is a spin configuration, say σ, then we write $m^\epsilon(x, t|\sigma)$.

In our case, by (9.6c), (9.9a) becomes [$m^\epsilon(x)$ below is a shorthand for $m^\epsilon(x, t|\sigma)$]:

$$\begin{aligned} \frac{d}{dt} m^\epsilon(x) =& \frac{1}{2}\epsilon^{-2}[m^\epsilon(x-1) + m^\epsilon(x+1) - 2m^\epsilon(x)] \\ & - 2m^\epsilon(x) + 2\gamma[m^\epsilon(x-1) + m^\epsilon(x+1)] - 2\gamma^2 m^\epsilon(x) m^\epsilon(x-1) m^\epsilon(x+1) \end{aligned} \tag{9.10a}$$

which, recalling (6.11), can be rewritten as

$$\begin{aligned}\frac{d}{dt}m^\epsilon(x) =& \frac{1}{2}(\epsilon^{-2}+4\gamma)[m^\epsilon(x-1)+m^\epsilon(x+1)-2m^\epsilon(x)] \\ &+\alpha m^\epsilon(x)-\beta m^\epsilon(x)m^\epsilon(x-1)m^\epsilon(x+1)\end{aligned} \tag{9.10b}$$

The v-functions are then defined as

$$v^\epsilon(\underline{x},t|\sigma) = \mathbb{E}^\epsilon_\sigma\Big(\prod_{x\in\underline{x}}\tilde{\sigma}(x,t)\Big) \tag{9.11a}$$

$$\tilde{\sigma}(x,t) = \sigma(x,t) - m^\epsilon(x,t|\sigma) \tag{9.11b}$$

As usual $\mathbb{E}^\epsilon_\sigma$ denotes the expectation with respect to the process with generator $\epsilon^{-2}L_0 + L_G$ starting from the configuration σ. We often write $v^\epsilon_n(\underline{x},t|\sigma)$ to underline the number of sites, n, in $\underline{x}$.

In the next chapter we prove:

9.2.1 Theorem. *There exist $\hat{a}>0$, $\delta>0$, $\beta^\star>0$ and a sequence c_n, $n\ge 1$, so that for all $\epsilon^{\beta^\star}\le t\le \hat{a}\log\epsilon^{-1}$, all $\underline{x}=(x_1,\cdots,x_n)$ and all configurations σ:*

$$|v^\epsilon_n(\underline{x},t|\sigma)| \le c_n\epsilon^{\delta n} \tag{9.12}$$

Furthermore if $m(r,t|\sigma)$ solves (9.1) with initial datum $m(r,0|\sigma)=\sigma([\epsilon^{-1}r])$ while $m^\epsilon(x,t|\sigma)$ solves (9.10) with initial datum σ, then there is a constant c such that

$$|m(\epsilon x,t|\sigma) - m^\epsilon(x,t|\sigma)| \le c\epsilon^\delta \tag{9.13}$$

for $\epsilon^{\beta^\star}\le t\le\hat{a}\log\epsilon^{-1}$.

§9.3 Equations for the v-functions.

We restrict henceforth to the specific case considered in Theorem 9.1.1. The initial measure is then μ^ϵ and, in the sequel, unless otherwise specified, $v^\epsilon_n(\underline{x},t)$ is defined by (9.11) with $\mathbb{E}_{\mu^\epsilon}$ in place of $\mathbb{E}_\sigma$ and with $\tilde{\sigma}\equiv\sigma$, since $m^\epsilon(x,t|\mu^\epsilon)=m^\epsilon(x,0|\mu^\epsilon)=\mathbb{E}_{\mu^\epsilon}(\sigma(x))=0$.

For all $n\ge 0$

$$v^\epsilon_{2n+1}(\underline{x},t)\equiv 0 \tag{9.14}$$

In fact by symmetry, at any time t, the probability of a configuration and of that obtained by flipping all the spins are the same, because this is true at time 0, by the symmetry of μ^ϵ, and because the evolution is symmetric under spin flip.

The equation for the v-functions is:

9.3.1 Lemma. *For any $t>0$ and $\underline{x}$*

$$v^\epsilon_n(\underline{x},t) = \int_0^t ds\; e^{\alpha n(t-s)}\mathbb{E}^\epsilon_{\underline{x}}(\mathcal{R}^\epsilon(\underline{x}(t-s),s) \tag{9.15}$$

where $\mathbb{E}^{\epsilon}_{\underline{x}}$ *is the expectation in the stirring process of intensity* $\epsilon^{-2}+4\gamma$ *starting from the configuration* $\underline{x}$*;* $\underline{x}(\tau)$ *denotes the positions of the particles at time* τ *and*

$$\begin{aligned}\mathcal{R}^{\epsilon}(\underline{x},t) = \sum_{x\in\underline{x}}\Big[&-2\gamma^2 v^{\epsilon}_{n+2}(\underline{x}+\delta_{x+1}+\delta_{x-1},t)1(x\pm 1\notin \underline{x})\\ &+2\gamma\sum_{b=\pm1} 1(x+b\in\underline{x})\{v^{\epsilon}_{n-2}(\underline{x}-\delta_x-\delta_{x+b},t)-v^{\epsilon}_n(\underline{x},t)\\ &-\gamma 1(x-b\notin\underline{x})v^{\epsilon}_n(\underline{x}-\delta_{x+b}+\delta_{x-b},t)-\gamma 1(x-b\in\underline{x})v^{\epsilon}_{n-2}(\underline{x}-\delta_{x+1}-\delta_{x-1},t)\}\Big]\end{aligned} \tag{9.16}$$

where $\underline{x}\pm\delta_y$ *is the configuration obtained from* $\underline{x}$ *by adding, respectively subtracting,* y *and* $1(\cdot)$ *is the characterstic function of* $(\cdot)$.

Proof. We have

$$\frac{d}{dt}v^{\epsilon}_n(\underline{x},t) = \epsilon^{-2}L_0 v^{\epsilon}_n(\underline{x},t)+\mathcal{R}^{\epsilon}_0(\underline{x},t) \tag{9.17a}$$

where

$$\begin{aligned}\mathcal{R}^{\epsilon}_0(\underline{x},t) = \sum_{x\in\underline{x}}\Big[2\gamma\sum_{b=\pm1}&\{v^{\epsilon}_n(\underline{x}-\delta_x+\delta_{x+b},t)1(x+b\notin\underline{x})\\ &+v^{\epsilon}_{n-2}(\underline{x}-\delta_x-\delta_{x+b},t)1(x+b\in\underline{x})\}-2v^{\epsilon}_n(\underline{x},t)\\ &-2\gamma^2\{v^{\epsilon}_{n+2}(\underline{x}+\delta_{x+1}+\delta_{x-1},t)1(x\pm1\notin\underline{x})\\ &+\sum_{b=\pm1}1(x+b\in\underline{x})1(x-b\notin\underline{x})v^{\epsilon}_n(\underline{x}-\delta_{x+b}+\delta_{x-b})\\ &+v^{\epsilon}_{n-2}(\underline{x}-\delta_{x+1}-\delta_{x-1},t)1(x\pm1\in\underline{x})\}\Big]\end{aligned} \tag{9.17b}$$

We add and subtract to $\mathcal{R}^{\epsilon}_0(\underline{x},t)$ the two terms $1(x\pm1\in\underline{x})v^{\epsilon}_n(\underline{x},t)$ and recalling that $\alpha=4\gamma-2$ we get

$$\frac{d}{dt}v^{\epsilon}_n(\underline{x},t) = (\epsilon^{-2}+4\gamma)L_0 v^{\epsilon}_n(\underline{x},t)+\alpha v^{\epsilon}_n(\underline{x},t)+\mathcal{R}^{\epsilon}(\underline{x},t) \tag{9.17c}$$

from which the Lemma follows. □

In the next paragraph we introduce the w-functions as solutions of a simplified version of (9.15). We then establish their asymptotic behavior as $\epsilon\to0$, at times $a\log\epsilon^{-1}$ with $a<1/(2\alpha)$. In the paragraph §9.5, we then prove that to leading order when $\epsilon\to0$, the v and w-functions are equal.

§9.4 The w-functions.

For any positive integer n, any time $t>0$ and any $\underline{x}$ we define

$$w^{\epsilon}_{2n}(\underline{x},t) = 2\gamma\int_0^t ds\, e^{\alpha 2n(t-s)}\mathbb{E}^{\epsilon}_{\underline{x}}\Big(\sum_{i,j}1(|x_i(t-s)-x_j(t-s)|=1)w^{\epsilon}_{2n-2}(\underline{x}^{i,j}(t-s),s)\Big) \tag{9.18}$$

where $\underline{x}^{i,j}(t-s)$ is the configuration obtained from $\underline{x}(t-s)$ by dropping $x_i(t-s)$ and $x_j(t-s)$. Notice that the w-functions are non negative.

9.4.1 Proposition. *There exists a sequence c_n such that for all t and ϵ*

$$\sup_{\underline{x}} |w^\epsilon_{2n}(\underline{x},t)| \le c_n e^{2n\alpha t}\epsilon^n \tag{9.19}$$

Proof. By iteration of (9.18) we get

$$w^\epsilon_{2n}(\underline{x},t) = (2\gamma)^n e^{2n\alpha t}\int_0^t ds_1 e^{-2\alpha s_1}\cdots\int_0^{s_{n-1}} ds_n e^{-2\alpha s_n}\sum_{(\underline{i},\underline{j})}\mathbb{E}^\epsilon_{\underline{x}}\Big(\prod_{\ell=1}^n \chi(s_\ell)\Big) \tag{9.20}$$

where the sum is over all the partitions $\{(i_1,j_1)\cdots,(i_n,j_n)\}$ of $\{1,\cdots,2n\}$ into disjoint sets of two elements. The expectation refers to the stirring process while

$$\chi(s_\ell) = 1(|x_{i_\ell}(t-s_\ell) - x_{j_\ell}(t-s_\ell)| = 1) \tag{9.21}$$

To bound the expectation in (9.20) we use an inequality proven in [90], see Proposition 1.7 and Corollary 1.9 in Chapter IX:

$$P^\epsilon_t(\underline{x}\to\underline{y}) \le \prod_{x\in\underline{x}}\sum_{y\in\underline{y}} P^\epsilon_t(x\to y) \tag{9.22}$$

We then have

$$\sum_{|y_1-y_2|=1} P^\epsilon_t((x_1,x_2)\to(y_1,y_2)) \le [P^\epsilon_t(x_1\to y_1)+P^\epsilon_t(x_1\to y_2)][P^\epsilon_t(x_2\to y_1)+P^\epsilon_t(x_2\to y_2)] \tag{9.23a}$$

Recalling that $P^\epsilon_t(x\to y)$ is the transition probability in a torus of side ϵ^{-1}, we have, using (4.34),

$$P^\epsilon_t(0\to x) = \sum_{n\in\mathbb{Z}} P^{\epsilon,\infty}_t(0\to x+\epsilon^{-1}n) \le \frac{c}{\sqrt{\epsilon^{-2}t}} + \sum_n G_{\epsilon^{-2}t}(x+\epsilon^{-1}n) \tag{9.23b}$$

where the probability in the second term refers to a symmetric random walk in the whole $\mathbb{Z}$ and with the same intensity ϵ^{-2}. G_t is the Gaussian kernel, (the Green function for the laplacian) given in (4.35). Hence

$$\sum_{|y_1-y_2|=1} P^\epsilon_t((x_1,x_2)\to(y_1,y_2)) \le c\epsilon(1+\frac{1}{\sqrt{t}}) \tag{9.23c}$$

Using (9.23c) we then have:

$$\begin{aligned}\mathbb{E}^\epsilon_{\underline{x}}\Big(\prod_{\ell=1}^n\chi(s_\ell)\Big) &= \mathbb{E}^\epsilon_{\underline{x}}\Big(\prod_{\ell=1}^{n-1}\chi(s_\ell)\mathbb{E}^\epsilon_{x_{i_n}(t-s_{n-1}),x_{j_n}(t-s_{n-1})}(\chi(s_n))\Big)\\ &\le \mathbb{E}^\epsilon_{\underline{x}}\Big(\prod_{\ell=1}^{n-1}\chi(s_\ell)c\epsilon(1+1/\sqrt{s_{n-1}-s_n})\Big)\end{aligned} \tag{9.23d}$$

and, by iteration,

$$|w^\epsilon_{2n}(\underline{x},t)| \le ce^{2n\alpha t}\int_0^t ds_1\, e^{-2\alpha s_1}\cdots\int_0^{s_{n-1}} ds_n\, e^{-2\alpha s_n}\pi(2n)\prod_{\ell=1}^n \epsilon[1+(\sqrt{s_{\ell-1}-s_\ell})^{-1}]$$

where $s_0 \equiv t$ and $\pi(2n)$ counts the number of ways the set $\{1, \cdots, 2n\}$ can be partitioned into n different pairs. From this the Proposition follows. □

Observe that at any time

$$t_a = a \log \epsilon^{-1} \tag{9.24}$$

with $a < 1/(2\alpha)$, the w-functions are infinitesimal as $\epsilon \to 0$, namely in the same time interval where the v-functions are, according to Theorem 9.1.1, infinitesimal. We shall prove more, namely that w_n^ϵ is the leading term in the expression for v_n^ϵ, but first we find the asymptotic behavior of w_n^ϵ.

The invariant measure for n stirring particles in $[1, N]$, $N = \epsilon^{-1}$, with periodic conditions is such that each configuration has the same weight. Since the number of configurations is

$$M(n) \equiv N(N-1)\cdots(N-n+1) \tag{9.25a}$$

the probability of a single configuration is

$$\Gamma(n) = M(n)^{-1} \tag{9.25b}$$

In the next Lemma we establish the rate of convergence to the invariant measure, which will then be used to estimates the transition probabilities in (9.18).

9.4.2 Lemma. *For any $b > 0$ there exist $\hat{\gamma} > 0$ and a sequence c_n so that for all $\underline{x}$ and $\underline{y}$*

$$|P_{t_b}^\epsilon(\underline{x} \to \underline{y}) - \Gamma(n)| \le c_n \Gamma(n) \epsilon^{\hat{\gamma}} \tag{9.25c}$$

Proof. We have

$$\begin{aligned} |P_t^\epsilon(\underline{x} \to \underline{y}) - \Gamma(n)| &= \Big|\sum_{\underline{x}'} [\Gamma(n) P_t^\epsilon(\underline{x} \to \underline{y}) - \Gamma(n) P_t^\epsilon(\underline{x}' \to \underline{y})]\Big| \\ &\le \sum_{\underline{x}'} \Gamma(n) |P_t^\epsilon(\underline{x} \to \underline{y}) - P_t^\epsilon(\underline{x}' \to \underline{y})| \end{aligned}$$

To prove (9.33) it is therefore sufficient to show that for any $b > 0$ there is $\hat{\gamma} > 0$ and a sequence c_n so that for all $\underline{x}$, $\underline{x}'$ and $\underline{y}$,

$$|P_{t_b}^\epsilon(\underline{x} \to \underline{y}) - P_{t_b}^\epsilon(\underline{x}' \to \underline{y})| \le c_n \epsilon^{\hat{\gamma}} \Gamma(n) \tag{9.26}$$

To prove (9.26) we write $\beta = b/2$ and

$$|P_{t_b}^\epsilon(\underline{x} \to \underline{y}) - P_{t_b}^\epsilon(\underline{x}' \to \underline{y})| \le \sum_{\underline{z}} |P_{t_\beta}^\epsilon(\underline{x} \to \underline{z}) - P_{t_\beta}^\epsilon(\underline{x}' \to \underline{z})| P_{t_\beta}^\epsilon(\underline{z} \to \underline{y}) \tag{9.27}$$

By (9.22)

$$P_{t_\beta}^\epsilon(\underline{z} \to \underline{y}) \le \prod_{i=1}^n \sum_{j=1}^n P_{t_\beta}^\epsilon(z_i \to y_j) \tag{9.28}$$

Hence from (9.23b), (9.28) and (9.27) we get (the value of the constant c is going to change from one line to the other)

$$|P^\epsilon_{t_b}(\underline{x} \to \underline{y}) - P^\epsilon_{t_b}(\underline{x}' \to \underline{y})| \le \sum_{\underline{z}} |P^\epsilon_{t_\beta}(\underline{x} \to \underline{z}) - P^\epsilon_{t_\beta}(\underline{x}' \to \underline{z})| c\,(\frac{n}{N})^n$$
$$\le c\Gamma(n) \sum_{\underline{z}} |P^\epsilon_{t_\beta}(\underline{x} \to \underline{z}) - P^\epsilon_{t_\beta}(\underline{x}' \to \underline{z})| \tag{9.29}$$

We now observe that for any $\bar{t} < t_\beta$ we can write

$$\sum_{\underline{z}} |P^\epsilon_{t_\beta}(\underline{x} \to \underline{z}) - P^\epsilon_{t_\beta}(\underline{x}' \to \underline{z})| = \|\lambda P^\epsilon_{\bar{t}} - \mu P^\epsilon_{\bar{t}}\| \tag{9.30}$$

where, for any probability ν,

$$\|\nu\| = \sum_{\underline{y}} \nu(\underline{y}) \tag{9.31}$$

and

$$\lambda(\underline{y}) = P^\epsilon_{t_\beta - \bar{t}}(\underline{x} \to \underline{y}), \qquad \mu(\underline{y}) = P^\epsilon_{t_\beta - \bar{t}}(\underline{x}' \to \underline{y}) \tag{9.32b}$$

Therefore

$$\lambda P^\epsilon_{\bar{t}}(\underline{y}) = \sum_{\underline{z}} \lambda(\underline{z}) P^\epsilon_{\bar{t}}(\underline{z} \to \underline{y}) = P^\epsilon_{t_\beta}(\underline{x} \to \underline{y}) \tag{9.32c}$$

Analogously $\mu P^\epsilon_{\bar{t}}(\underline{y}) = P^\epsilon_{t_\beta}(\underline{x}' \to \underline{y})$.

There are $d < 1$ and $t^\star > 0$ such that

$$\sup_{\|\underline{x}-\underline{x}'\| \le \epsilon^{-1}} \sum_{\underline{z}} |P^\epsilon_{t^\star}(\underline{x} \to \underline{z}) - P^\epsilon_{t^\star}(\underline{x}' \to \underline{z})| \le d < 1 \tag{9.33}$$

in fact the left hand side is bounded by the same expression but with $P^{\epsilon,\infty}$ (the transition probability in the whole $\mathbb{Z}$) replacing P^ϵ (see (9.23b)): then (9.33) follows from (6.26).

We thus have

$$\|\lambda P^\epsilon_{t^\star} - \mu P^\epsilon_{t^\star}\| \le d\|\lambda - \mu\| \tag{9.34}$$

Proof of (9.34).* We use the same argument we used when proving Proposition 3.1.4. Let

$$\nu(\underline{x}) = \min\{\lambda(\underline{x}), \mu(\underline{x})\}; \qquad \lambda'(\underline{x}) = \lambda(\underline{x}) - \nu(\underline{x}); \qquad \mu'(\underline{x}) = \mu(\underline{x}) - \nu(\underline{x})$$

then

$$\sum_{\underline{x}} [\lambda'(\underline{x}) + \mu'(\underline{x})] = \|\lambda - \mu\|; \quad \sum_{\underline{x}} [\lambda'(\underline{x}) - \mu'(\underline{x})] = 0$$

Fix any configuration $\underline{y}$, then

$$|\lambda P^\epsilon_{t^\star} - \mu P^\epsilon_{t^\star}| = \sum_{\underline{y}} |\sum_{\underline{z}} [\lambda(\underline{z}) - \mu(\underline{z})] P^\epsilon_{t^\star}(\underline{z} \to \underline{y})|$$
$$= \sum_{\underline{y}} |\sum_{\underline{z}} [\lambda'(\underline{z}) - \mu'(\underline{z})] P^\epsilon_{t^\star}(\underline{z} \to \underline{y})|$$
$$= \sum_{\underline{y}} |\sum_{\underline{z}} \{\lambda'(\underline{z})[P^\epsilon_{t^\star}(\underline{z} \to \underline{y}) - P^\epsilon_{t^\star}(\underline{y} \to \underline{y})] - \mu'(\underline{z})[P^\epsilon_{t^\star}(\underline{z} \to \underline{y}) - P^\epsilon_{t^\star}(\underline{y} \to \underline{y})]\}|$$
$$\le \sum_{\underline{y}} \sum_{\underline{z}} \{\lambda'(\underline{z})|P^\epsilon_{t^\star}(\underline{z} \to \underline{y}) - P^\epsilon_{t^\star}(\underline{y} \to \underline{y})| + \mu'(\underline{z})|P^\epsilon_{t^\star}(\underline{z} \to \underline{y}) - P^\epsilon_{t^\star}(\underline{y} \to \underline{y})|\}$$
$$\le \sum_{\underline{z}} [\lambda'(\underline{z}) + \mu'(\underline{z})] d \;\le d\|\lambda - \mu\|$$

(9.34) is therefore proven.

From (9.34) it follows that

$$\sum_{\underline{z}} |P^{\epsilon}_{t_\beta}(\underline{x} \to \underline{z}) - P^{\epsilon}_{t_\beta}(\underline{x}' \to \underline{z})| \le d^{t_\beta/t^*} \le c\epsilon^{\hat{\gamma}}, \quad \hat{\gamma} = \frac{\beta}{2t^*}|\log d| \tag{9.35}$$

which proves (9.26) and the Lemma. □

9.4.3 Proposition. *There are $\zeta > 0$ and a sequence c_n so that the following holds. For any $a^* < 1/(2\alpha)$ let k^* be the largest integer such that $a^* k^* < 1/(2\alpha)$. Then for all $k \le k^*$ and all $\underline{x}$*

$$|w^{\epsilon}_{2n}(\underline{x}, kt_{a^*}) - \frac{\pi(2n)}{n!}[\frac{2\gamma 2\epsilon}{2\alpha}]^n e^{2n\alpha t_a}| \le c_n[\epsilon^n e^{2n\alpha t_a}]\epsilon^{\zeta} \tag{9.36}$$

where $\pi(2n)$ counts the number of ways the set $\{1, \cdots, 2n\}$ can be partitioned into n different pairs.

Proof. We fix the value a^* and k for which we wish to prove (9.36) and we let $t = kt_{a^*}$. Then, given any $0 < a < a^*$, by (9.23c) and Proposition 9.4.1, the contribution to w^{ϵ}_{2n} in (9.18) of $\{s \ge t_a\}$ is bounded by

$$2\gamma \int_{t_a}^{t} ds e^{2n\alpha(t-s)} c_{n-1} e^{2(n-1)\alpha s} \epsilon^{n-1} c\epsilon \left(1 + 1/\sqrt{s_{n-1} - s_n}\right) \le c\epsilon^n e^{2n\alpha t} \epsilon^{2\alpha a} \tag{9.37}$$

Thus, choosing $\zeta < 2\alpha a$, we have

$$|w^{\epsilon}_{2n}(\underline{x}, kt_{a^*} - (2\gamma)^n e^{2n\alpha t} \int_0^{t_a} ds_1 e^{-2\alpha s_1} \ldots \int_0^{s_{n-1}} ds_n e^{-2\alpha s_n} \sum_{(\underline{i},\underline{j})} \mathbb{E}^{\epsilon}_{\underline{x}}\Big(\prod_{\ell=1}^{n} g_\ell(\tau_\ell)\Big)|$$
$$\le c[e^{2n\alpha t_a}\epsilon^n]\epsilon^{\zeta} \tag{9.38}$$

where, for $\ell = 1, \ldots, n$, we have set

$$\tau_\ell = kt_{a^*} - s_\ell, \qquad g_\ell(\tau_\ell) = 1(|x_{i_\ell}(\tau_\ell) - x_{j_\ell}(\tau_\ell)| = 1) \tag{9.39}$$

Let $b = ka^* - a$ so that $t_b \le \tau_1 < \cdots < \tau_n$. We are going to show that

$$|\mathbb{E}^{\epsilon}_{\underline{x}}[\prod_{\ell=1}^{n} g_\ell(\tau_\ell)] - (2\epsilon)^n| \le c\epsilon^{n+\hat{\gamma}} \tag{9.40}$$

where $\hat{\gamma}$ is related to b by Lemma 9.4.2. Using (9.40) in (9.38) and choosing $\zeta < \hat{\gamma}$, (9.36) easily follows. Therefore the Proposition will be proven once we show the validity of (9.40).

Proof of (9.40). Given a configuration $\underline{y}$, the pair $(\underline{i}, \underline{j})$ and $\ell \in [0, n]$, we define

$$\underline{y}^{(\ell)} = (y_{i_{\ell+1}}, y_{j_{\ell+1}}, \cdots, y_{i_n}, y_{j_n}), \qquad F_\ell(\underline{y}^{(\ell)}) = \mathbb{E}^{\epsilon}_{\underline{x}}\Big(\prod_{\bar{\ell}=\ell+1}^{n} g_{\bar{\ell}}(\tau_{\bar{\ell}}) \Big| \{\underline{x}^{(\ell)}(\tau_\ell) = \underline{y}^{(\ell)}\}\Big) \tag{9.41a}$$

where $\tau_0 = 0$ and $\underline{y}^{(0)} = \underline{x}$. In the proof of Proposition 9.4.1 we have seen that

$$|F_\ell(\underline{y}^{(\ell)})| \le \epsilon^{n-\ell} \tag{9.41b}$$

We write

$$\begin{aligned}
&|\sum_{\underline{y}} P^\epsilon_{\tau_1}(\underline{x} \to \underline{y}) 1(|y_{i_1} - y_{j_1}| = 1) F_1(\underline{y}^{(1)}) - 2\epsilon \sum_{\underline{y}^{(1)}} \Gamma(2n-2) F_1(\underline{y}^{(1)})| \\
&\quad \le \sum_{\underline{y}} |P^\epsilon_{\tau_1}(\underline{x} \to \underline{y}) - \Gamma(2n)| 1(|y_{i_1} - y_{j_1}| = 1) F_1(\underline{y}^{(1)}) \\
&\quad + |\sum_{\substack{y_{i_1}, y_{j_1} \\ y_{j_1} = y_{i_1} \pm 1}} \sum_{\underline{y}^{(1)}} \chi^*(\underline{y}) \Gamma(2n) F_1(\underline{y}^{(1)}) - 2\epsilon \sum_{\underline{y}^{(1)}} \Gamma(2n-2) F_1(\underline{y}^{(1)})|
\end{aligned} \tag{9.42}$$

where $\chi^*(\underline{y})$ is the characteristic function that the sites in $y^{(1)}$ are all different from y_{i_1}, y_{j_1}. Since

$$|\Gamma(2n) - \Gamma(2n-2)\Gamma(2)| \le c\epsilon \Gamma(2n-2)\Gamma(2) \tag{9.43}$$

by Lemma 9.4.2 and (9.41b), the right hand side of (9.42) is bounded by

$$\begin{aligned}
&c\{\epsilon^{\hat\gamma} \sum_{\underline{y}} \Gamma(2n) 1(|y_{i_1} - y_{j_1}| = 1) \epsilon^{n-1} + \sum_{\substack{y_{i_1}, y_{j_1} \\ y_{j_1} = y_{i_1} \pm 1}} \sum_{\underline{y}^{(1)}} \chi^*(\underline{y}) |\Gamma(2n) - \Gamma(2n-2)\Gamma(2)| \epsilon^{n-1} \\
&\quad + |\sum_{\substack{y_{i_1}, y_{j_1} \\ y_{j_1} = y_{i_1} \pm 1}} \sum_{\underline{y}^{(1)}} \chi^*(\underline{y}) \Gamma(2n-2)\Gamma(2) F_1(\underline{y}^{(1)}) - 2\epsilon \sum_{\underline{y}^{(1)}} \Gamma(2n-2) F_1(\underline{y}^{(1)})|\} \\
&\le c\{\epsilon^{\hat\gamma} \epsilon^{-2n+1} \epsilon^{2n} \epsilon^{n-1} + \epsilon \epsilon^{-2n+1} \epsilon^{2n} \epsilon^{n-1} + \sum_{\substack{y_{i_1}, y_{j_1} \\ y_{j_1} = y_{i_1} \pm 1}} \sum_{\underline{y}^{(1)}} (1 - \chi^*(\underline{y})) \Gamma(2n-2)\Gamma(2) \epsilon^{n-1}\} \\
&\le c(\epsilon^{\hat\gamma} + \epsilon) \epsilon^n
\end{aligned} \tag{9.44}$$

which is compatible with (9.40). We next estimate the term $2\epsilon \sum_{\underline{y}^{(1)}} \Gamma(2n-2) F_1(\underline{y}^{(1)})$. We do it iteratively and describe the ℓ-step. We use (9.43), and the invariance of the measure $\{\mu(\underline{y}^{(\ell)}) = \Gamma(2(n-\ell))\}$, to get

$$\begin{aligned}
&|\sum_{\underline{y}^{(\ell)}, \underline{z}} \Gamma(2(n-\ell)) P^\epsilon_{\tau_{\ell+1} - \tau_\ell}(\underline{y}^{(\ell)} \to \underline{z}) 1(|z_{i_{\ell+1}} - z_{j_{\ell+1}}| = 1) F_{\ell+1}(\underline{z}^{(\ell+1)}) \\
&\qquad - 2\epsilon \sum_{\underline{z}^{(\ell+1)}} \Gamma(2(n-\ell-1)) F_{\ell+1}(\underline{z}^{(\ell+1)})| \\
&\quad \le c\epsilon^2 \sum_{\underline{z}^{(\ell+1)}} \Gamma(2(n-\ell-1)) F_{\ell+1}(\underline{z}^{(\ell+1)})
\end{aligned} \tag{9.45}$$

Using (9.44) and (9.45) we conclude the proof of (9.40) and of the Proposition. □

§9.5 Closeness between the v and the w-functions.

In this paragraph we prove that for $t \le t_a$, $a < 1/(2\alpha)$, (see (9.24) for notation) the v and the w-functions are, to leading order, the same. We start by proving it at "short times":

9.5.1 Proposition. *There is $A > 0$, $\zeta > 0$ and a sequence c_n such that for all positive $a \le A$ and all n*

$$\sup_{\underline{x}} |v_{2n}^\epsilon(\underline{x}, t_a) - w_{2n}^\epsilon(\underline{x}, t_a)| \le c_n[e^{2n\alpha t_a}\epsilon^n]\epsilon^\zeta \tag{9.46}$$

Proof. Let $\hat{a}$, δ and β^* be as in Theorem 9.2.1. Of course we can always suppose that $\delta < 1/4$ and so small that the conditions stated in (9.71) below are satisfied. A is assumed $\le \hat{a}$, other requests on A will be specified later, while ζ will be chosen smaller than δ.

We write v_n^ϵ by means of (9.15). Since $\mathcal{R}^\epsilon$ is by (9.16) a sum of v-functions plus, possibly, (if $n = 2$ and $\underline{x} = (x_1, x_2)$ with $|x_1 - x_2| = 1$) a constant term, we can use again (9.15) to rewrite the v-functions in $\mathcal{R}^\epsilon$. We repeat this N times where N, for reasons which will become clear in the course of the proof, is chosen so that

$$N\delta > 2n \qquad N\beta^* > 4n \tag{9.47}$$

After N iterations, we have $v_n^\epsilon(\underline{x}, t)$ expressed as a huge but finite sum of terms. A first group is made of those where there is no v-function left, these will be treated later on.

Each of the remaining ones is an integral over the N times $s_1 \ge s_2 \ge \cdots \ge s_N$ and it has a v-function computed at time $t - s_N$. The integral is splitted into that with $s_N \ge \epsilon^{\beta^*}$ and that in the complement.

The case $s_N \ge \epsilon^{\beta^*}$. Let us fix one of the terms in this class, supposing that at the $i_1, \dots, i_k$ iterations, and only at these, no particle is created, i.e. the contribution does not come from the first term in (9.16). For $\ell = 1, \dots, k$, we call $\hat{\chi}(s_\ell)$ the characteristic function which specifies the labels of the particles which are at neighboring sites at s_{i_ℓ}, see (9.16). Which one occurs, is specified by the term under consideration. The v-function left at the N-th iteration has degree $2m$, where

$$2m \ge 2n + 2(N - k) - 2k \ge 0 \tag{9.48a}$$

By (9.12), the term we are considering is then bounded by

$$c \int_{\epsilon^{\beta^*}}^{t_a} ds_1\, e^{2n\alpha(t_a - s_1)} \dots \int_{\epsilon^{\beta^*}}^{s_{N-1}} ds_N\, e^{r_N\alpha(s_{N-1} - s_N)}\, \epsilon^{\delta 2m}\, \mathbb{E}_{\underline{x}}^\epsilon[\prod_{\ell=1}^{K} \hat{\chi}(s_{i_\ell})] \tag{9.48b}$$

where r_i denotes the number of particles present in the time interval (s_{i-1}, s_i) and t_a is defined in (9.24). We have bounded by 1 the remaining v-function. Finally the expectation in (9.48b) refers to the stirring process in the time intervals $(t - s_i, t - s_{i+1})$ and the relation between the configuration at $(t - s_i)^-$ and $(t - s_i)^+$ is determined by which one of the terms in (9.16) is taken.

By (9.22), the expression in (9.48b) is bounded by

$$c \int_0^{t_a} ds_1 \cdots \int_0^{s_{N-1}} ds_N\, e^{2(n+N-K)\alpha(t_a - s_N)}\, \epsilon^{\delta 2m} \prod_{\ell=1}^{K} \epsilon[1 + (\sqrt{s_{i_{\ell-1}} - s_{i_\ell}})^{-1}] \tag{9.49a}$$

which, using (9.48a), is in turns bounded by

$$ce^{2(n+N-K)\alpha t_a}\epsilon^{\delta(2n+2N)}\epsilon^{-4K\delta+K}(t_a)^N \le ce^{2n\alpha t_a}\epsilon^{\delta 2n}\epsilon^{\delta N}\epsilon^{-4K\delta+K}(t_a)^N$$

because $e^{2N\alpha t_a} \le \epsilon^{-2N\delta/2}$ if we take A so that $\alpha A \le \delta/2$.

Recalling that $\delta < 1/4$ we have that $\epsilon^{-4K\delta+K} \le 1$. By (9.47) $\epsilon^{\delta N} \le \epsilon^{2n}$, then, since $\epsilon^n (t_a)^N \le c\epsilon^\zeta$, $(n \ge 1,\ \zeta < 1)$ we get a bound compatible with (9.46), so that the analysis of this case is completed.

The case $s_N < \epsilon^{\beta^*}$ (when a v-function is still present). We apply again (9.15) and make N new iterations. Once again we find a huge but finite sum of terms. We first consider those where there is a v-function left. We bound it by 1, as well as all the characteristic functions which have been produced, then the generic term of this kind is bounded by

$$ce^{2(n+2N)\alpha t_a} \int_0^{t_a} ds_1 \cdots \int_0^{s_{N-1}} ds_N 1(s_N < \epsilon^{\beta^*}) \cdots \int_0^{s_{2N-1}} ds_{2N} \int_0^{\epsilon^{\beta^*}} ds_{M+1} \int_0^{s_{M+N-1}} ds_{M+N}$$
$$\le ce^{2(n+2N)\alpha t_a} \epsilon^{\beta^* N} (t_a)^N \tag{9.49b}$$

We take A so small that $4\alpha A < \beta^*/2$, then (9.49b) is bounded by $ce^{2n\alpha t_a} \epsilon^{\beta^* N/2} (t_a)^N$, which, by (9.47), is compatible with (9.46).

We are left with the terms where, at some of the iterations, say the H-th one, $H \le 2N$, no v-function is left. H is then the total number of time integrals in the term under consideration and let K be the number of events when no new particles are created. We first consider the case when there are only deaths, so that $H = K = n$.

There are two kinds of deaths, cf. (9.16): the first one is when only two particles are involved, the other needs three particles. All the terms when all the deaths are of the first type reconstruct w_{2n}^ϵ, so that the generic one among those which are left is bounded by

$$ce^{2n\alpha t} \int_0^t ds_1 e^{-2\alpha s_1} \cdots \int_0^{s_{n-1}} ds_n e^{-2\alpha s_n} \mathbf{E}_{\underline{x}}^\epsilon [\prod_{\ell=1}^n \hat{\chi}(s_\ell)]$$

with the condition that at least one of the $\hat{\chi}$'s involves three particles. By (9.22) and (9.23c), for any $b > 0$ there exists c so that for all t and $\underline{x}$

$$\sum_{\underline{y}} P_t^\epsilon(\underline{x} \to \underline{y}) 1(|y_1 - y_2| = 1,\ |y_1 - y_3| = 1) \le c\epsilon^{2-2b}[1 + 1/t^{1-b}]$$

hence the above term is bounded by

$$ce^{2n\alpha t} \epsilon^{n+1-2b}$$

which gives a contribution compatible with (9.46) for b small enough.

Each of the remaining terms is bounded by (9.49) with $m = 0$ and N replaced by H with $H > K$. We then get the bound

$$ce^{2n\alpha t_a} e^{2(H-K)\alpha t_a} \epsilon^K (t_a)^H$$

Call $0 \le K' < K$ the number of times when the particle number does not change, no birth nor death. Then since at the end no particle is left, we have $0 = 2n + 2(H-K) - 2(K-K')$, $K \le n + (H-K)$. Then recalling that $H > K$, we get the bound

$$c[e^{2n\alpha t_a} \epsilon^n][e^{2(H-K)\alpha t_a} \epsilon^{H-K} (t_a)^H]$$

By choosing $A < 1/(2\alpha)$, the second factor vanishes as some positive power of ϵ, hence the Proposition is proven. □

Let $a^\star < A/2$ (so that $a \equiv 2a^\star \le \hat{a}$, see Theorem 9.2.1) and such that $ka^\star \neq 1/(2\alpha)$ for all integers k. Let $k^\star$ be the largest integer such that $a^\star k^\star < 1/(2\alpha)$. We are going to see that (9.46) holds for all $t \le k^\star t_{a^\star}$, and this shows that the growth of the v-functions is essentially due to linear effects till any time t_a with $a < 1/(2\alpha)$.

9.5.2 Theorem. *Given $a^\star$ as above, there exist $\zeta > 0$ and a sequence c_n so that*

$$\sup_{\underline{x}} |v^\epsilon_{2n}(\underline{x}, kt_{a^\star}) - w^\epsilon_{2n}(\underline{x}, kt_{a^\star})| \le c_n[e^{2n\alpha kt_{a^\star}}\epsilon^n]\epsilon^\zeta \tag{9.50}$$

for all $k \le k^\star$.

Proof. We prove the theorem by induction on k. Since (9.50) has already been proven for $k = 1$, in Proposition 9.5.1, we only need to show that if it holds till $k \le k^\star - 1$, then it holds also at $k+1$. By (9.15) we have that for $kt_{a^\star} < t \le (k+1)t_{a^\star}$

$$v^\epsilon_{2n}(\underline{x}, t) = \mathbf{E}^\epsilon_{\underline{x}}\Big(e^{2n\alpha(t-kt_{a^\star})}v^\epsilon_{2n}(\underline{x}(t-kt_{a^\star}), kt_{a^\star}) + \int_{kt_{a^\star}}^t ds\, e^{2n\alpha(t-s)}\mathcal{R}^\epsilon(\underline{x}(t-s), s)\Big) \tag{9.51}$$

From (9.18), (9.19) and (9.23) with $n = 1$ we get

$$\begin{aligned}
&|w^\epsilon_{2n}(\underline{x}, t) - \mathbf{E}^\epsilon_{\underline{x}}[e^{2n\alpha(t-kt_{a^\star})}w^\epsilon_{2n}(\underline{x}(t-kt_{a^\star}), kt_{a^\star})]| \\
&\quad = 2\gamma\int_{kt_{a^\star}}^t ds e^{2n\alpha(t-s)}\mathbf{E}^\epsilon_{\underline{x}}\Big(\sum_{i,j} 1(|x_i(t-s) - x_j(t-s)| = 1)w^\epsilon_{2n-2}(\underline{x}^{i,j}(t-s), s)\Big) \\
&\quad \le ce^{-2\alpha kt_{a^\star}}e^{2n\alpha t}\epsilon^n \le c(e^{2n\alpha t}\epsilon^n)\epsilon^\zeta
\end{aligned} \tag{9.52}$$

choosing $\zeta < 2\alpha a^\star$. Therefore using (9.51) and (9.52), we get

$$\begin{aligned}
|v^\epsilon_{2n}(\underline{x}, kt_{a^\star}) - w^\epsilon_{2n}(\underline{x}, kt_{a^\star})| &\le e^{2n\alpha(t-kt_{a^\star})}\mathbf{E}^\epsilon_{\underline{x}}[|v^\epsilon_{2n}(\underline{x}(t-kt_{a^\star}), kt_{a^\star}) - w^\epsilon_{2n}(\underline{x}(t-kt_{a^\star}), kt_{a^\star})|] \\
&\quad + c_n(e^{2n\alpha t}\epsilon^n)\epsilon^\zeta + |\mathbf{E}^\epsilon_{\underline{x}}[\int_{kt_{a^\star}}^t ds e^{2n\alpha(t-s)}\mathcal{R}^\epsilon(\underline{x}(t-s), s)]|
\end{aligned} \tag{9.53}$$

The induction hypothesis and (9.53) prove (9.50) once we show the following inequality

$$|\mathbf{E}^\epsilon_{\underline{x}}(\int_{kt_{a^\star}}^t ds e^{2n\alpha(t-s)}\mathcal{R}^\epsilon(\underline{x}(t-s), s)]| \le c_n(e^{2n\alpha t}\epsilon^n)\epsilon^\zeta \tag{9.54}$$

In order to prove (9.54) we express $\mathcal{R}^\epsilon$ in terms of v-functions, we then use (9.51) for each of them and we repeat this N times, whenever a v-function at a time $t > kt_{a^\star}$ appears. We choose N so that

$$N\delta > 4n;\ N(1 - 2\alpha a^\star) > 16n;\ \alpha bN > 4n \tag{9.55a}$$

where

$$b \equiv 1/(2\alpha) - k^\star a^\star > 0 \tag{9.55b}$$

We first consider the case when the iteration at a certain step $h \le N$ hits $v^\epsilon_{2m}(\underline{x}, kt_{a*})$, for some m. From (9.46), (9.19) and the induction hypothesis

$$|v^\epsilon_{2m}(\underline{x}, kt_{a*})| \le c_m e^{2makt_{a*}} \epsilon^m \tag{9.56}$$

The generic term in this class is then bounded by

$$c \int_{kt_{a*}}^{t} ds_1 e^{2n\alpha[t-s_1]} \dots \int_{kt_{a*}}^{s_{h-1}} ds_h e^{r_h\alpha(s_{h-1}-s_h)} e^{2m\alpha(s_h-kt_{a*})} \mathbf{E}^\epsilon_{\underline{x}}(\prod_{\ell=1}^{u} \hat{\chi}(s_{i_\ell})) e^{2makt_{a*}} \epsilon^m \tag{9.57}$$

where $s_1 \cdots s_h$ are the times of iteration, $1 \le h \le N$, r_i are the number of particles in the time interval (s_{i-1}, s_i), $s_{i_1} \cdots s_{i_u}$ are the times when no particle is created and, as in (9.48a),

$$2m \ge 2n + 2(h-u) - 2u \tag{9.58a}$$

The term in (9.57) is then bounded by

$$ce^{2(n+(h-u))\alpha(t-kt_{a*})} e^{2makt_{a*}} \epsilon^m \int_0^{t-kt_{a*}} ds_1 \cdots \int_0^{s_{h-1}} ds_h \prod_{\ell=1}^{u} \epsilon[1 + (\sqrt{s_{i_{\ell-1}} - s_{i_\ell}})^{-1}]$$

$$\le ce^{2mat} \epsilon^{m+u} e^{2u\alpha(t-kt_{a*})} (t_{a*})^h \tag{9.58b}$$

By (9.58a) if $u = 0$, $m + u \ge n + 1$, (since $h \ge 1$) hence the right hand side of (9.58b) is bounded by

$$c[e^{2n\alpha t}\epsilon^n] e^{2h\alpha t} \epsilon^h (t_{a*})^N \le c[e^{2n\alpha t}\epsilon^n] \epsilon^{(1-2\alpha k^* a^*)h} (t_{a*})^N < c[e^{2n\alpha t}\epsilon^n] \epsilon^\zeta$$

if ζ is small enough, recall that $k^* a^* < 1/(2\alpha)$. We have therefore proven that the terms with $u = 0$ are bounded as in (9.54).

If $u > 0$, then by (9.58a) we have

$$\begin{aligned} c[e^{2n\alpha t}\epsilon^n] e^{-4u\alpha t} e^{2(h-u)\alpha t} \epsilon^{h-u} (t_{a*})^N &\le c[e^{2n\alpha t}\epsilon^n] \epsilon^{4\alpha a u} (t_{a*})^N \\ &\le c[e^{2n\alpha t}\epsilon^n] \epsilon^\zeta \end{aligned}$$

if ζ is small enough.

In the remaining terms there are N integrals between kt_{a*} and t. The generic one is bounded by

$$\begin{aligned} c \int_0^{t-kt_{a*}} ds_1 \cdots \int_0^{s_{N-1}} ds_N e^{2(n+N-u)\alpha(t-kt_{a*}-s_N)} \\ \times \mathbf{E}^\epsilon_{\underline{x}}\left(|v^\epsilon_{2m}(\underline{x}(t-kt_{a*}-s_N), kt_{a*}+s_N)| \prod_{\ell=1}^{u} \chi(s_{i_\ell}) \right) \end{aligned} \tag{9.59}$$

with the notation used so far.

To bound the v-function in (9.59) we condition on the values of all the spins at time $T \equiv (k-1)t_{a*}$. We first observe:

Claim 1. *For any $\gamma^* > 0$ there exists a sequence c_n so that for any $n \geq 1$*

$$\mathbb{E}^{\epsilon}_{\mu^{\epsilon}}(\chi_T) \geq 1 - c_n\epsilon^n, \quad \chi_T = 1(\|\sigma(\cdot,T)\| < S), \quad S = \epsilon^{-\gamma^*}\max\{(\epsilon^{-2+1/4})^{-1/4}, (e^{2\alpha T}\epsilon)^{1/2}\} \tag{9.60}$$

The seminorm $\|f\|$ is defined as

$$\|f\| = \sup_x |\sum_y P^{\epsilon}_{\epsilon^{\frac{1}{4}}}(x \to y) f(y)| \tag{9.61}$$

Proof of the Claim. Using the Chebitchev inequality with power $2n$ and the fact that

$$|v^{\epsilon}_{2n}(\underline{x}, T)| \leq c_n(e^{2\alpha T}\epsilon)^n$$

we get,

$$\begin{aligned}\mathbb{E}^{\epsilon}_{\mu^{\epsilon}}(1-\chi_T) &\leq S^{-2n}\sum_x E^{\epsilon}_{\mu^{\epsilon}}\Big(\sum_{y_1,\dots,y_{2n}}\prod_{i=1}^{2n} P^{\epsilon}_{\epsilon^{1/4}}(x \to y_i)\sigma(y_i,T)\Big) \\ &\leq S^{-2n}\sum_x\sum_{m=0}^{2n}\sum_{\underline{y}\in\mathcal{M}_m}\sum_{\underline{k}\in\mathbb{N}}\Big(\prod_{i=1}^{m} P^{\epsilon}_{\epsilon^{1/4}}(x \to y_i))^{k_i}\Big)|v^{\epsilon}(\underline{y}'(\underline{y},\underline{k}),T)|\end{aligned} \tag{9.62}$$

where $\mathcal{M}_m$ is the set of configurations with particles in m distinct sites, $\underline{y} = (y_1,\dots,y_m)$; $\underline{k} = (k_1,\dots,k_m)$, $k_i > 0$, and $\underline{y}'$ is the subset of $\underline{y}$ where $y_i \in \underline{y}'$ if and only if k_i is odd. Call $m_1 = m_1(m,\underline{k})$ the number of i's such that $k_i = 1$. Then (9.62) is bounded by

$$\begin{aligned}&cS^{-2n}\epsilon^{-1}\sum_{m=0}^{2n}\sum_{m_1=0}^{m}[\epsilon^{-2+1/4}]^{-(2n-m_1)/4}[e^{2\alpha T}\epsilon]^{m_1} \\ &\leq cS^{-2n}\epsilon^{-1}[\epsilon^{-2+1/4}]^{2n}\sum_{m=0}^{2n}\sum_{m_1=0}^{m}[\epsilon^{1/2-1/16}e^{-2\alpha T}\epsilon^{-1}]^{2n-m_1}\end{aligned}$$

which proves the claim.

Using Claim 1 we have that for any given n and for any t such that $(k+1)t_{a^*} \geq t \geq kt_{a^*}$,

$$|v^{\epsilon}_{2m}(\underline{x},t)| \leq \mathbb{E}^{\epsilon}_{\mu^{\epsilon}}\Big(\chi_T\mathbb{E}^{\epsilon}_{\sigma(\cdot,T)}\big(\prod_{i=1}^{2m}\sigma(x_i,t-T)\big)\Big) + c_n\epsilon^n \tag{9.63}$$

We add and subtract $m^{\epsilon}(x_i, t-T|\sigma(\cdot,T))$ to each spin in the last product. We expand and use (9.12) to estimate the v-functions; we add and subtract $m(\epsilon x_i, t-T|\sigma(\cdot,T))$ to each remaining factor $m^{\epsilon}(x_i,t-T|\sigma(\cdot,T))$ and use (9.13). It follows that

$$|v^{\epsilon}_{2m}(\underline{x},t)| \leq c_n\epsilon^n + c\max\{\epsilon^{2\delta m}, \|m(\cdot,t-T|\sigma(\cdot,T))\|^{2m}_{\infty}\} \tag{9.64}$$

We claim that the last term is bounded as follows:

Claim 2. *Let σ be such that*

$$\|\sigma\| \le S \tag{9.65}$$

where S is defined in (9.60). Then there exists c such that

$$|m(r,t-T|\sigma) - e^{\alpha(t-T)}\int dr' G_{t-T}(r\to r')\sigma([\epsilon^{-1}r'])| \le c(e^{\alpha(t-T)}\hat{S})^3 t_{a*} \tag{9.66a}$$

where

$$\hat{S} = \max\left(\epsilon^{1/12}, S\right)$$

and G_t is the Green function for the Laplace operator in [0,1] with periodic conditions. Furthermore

$$|m(r,t-T|\sigma)| \le ce^{\alpha(t-T)}\hat{S} \tag{9.66b}$$

where $T = (k-1)t_{a*} < T + \epsilon^{1/4} \le t \le (k+1)t_{a*} \le k^* t_{a*}$.

Proof of the Claim. The integral version of (9.1) gives

$$\begin{aligned} m(r,t|\sigma^{(0)}) &= \int dr' G_t(r\to r')\,\sigma^{(0)}([\epsilon^{-1}r'],0) \\ &+ \int_0^t ds \int dr' G_{t-s}(r\to r')[\alpha m(r',s|\sigma^{(0)}) - \beta m(r',s|\sigma^{(0)})^3] \end{aligned} \tag{9.67}$$

From (9.67) it follows that

$$|m(r,\epsilon^{1/4}|\sigma) - \int dr' G_{\epsilon^{1/4}}(r\to r')\sigma([\epsilon^{-1}r'])| \le c\epsilon^{1/4}$$

From (4.34) we have

$$\sum_{y\in Z_\epsilon} |\int_{y-1/2}^{y+1/2} \epsilon dz G_t(\epsilon x\to\epsilon z) - P_t^\epsilon(x\to y)| \le c\epsilon/\sqrt{t}, \quad t=\epsilon^{1/4} \tag{9.68}$$

hence

$$|m(r,\epsilon^{1/4}|\sigma)| \le \|\sigma\| + c[\epsilon^{1-1/8} + \epsilon^{1/4}] \tag{9.69}$$

The maximum principle holds for (9.1), namely if $\bar{m}(r,t)$ solves (9.1) and $m(r,0) \le \bar{m}(r,0)$, then $m(r,t) \le \bar{m}(r,t)$ for all $t > 0$. From this we have that $|m(r,t|\sigma)|$ is bounded for $t \ge \epsilon^{1/4}$ by $z(t-\epsilon^{1/4})$ where $z(t)$ solves

$$\begin{aligned} \frac{d}{dt}z &= \alpha z - \beta z^3 \\ z(0) &= c[\epsilon^{1-1/8} + \epsilon^{1/4}] + S \end{aligned}$$

Hence $z(t) \le ce^{\alpha t}z(0)$ for $t \le 2t_{a*} - \epsilon^{1/4}$. This proves (9.66b).

To prove (9.66a) we rewrite (9.1) as

$$\begin{aligned} m(r,t|\sigma) &= \int dr' e^{\alpha t} G_t(r\to r')\,\sigma([\epsilon^{-1}r'],0) \\ &- \int_0^t ds \int dr' e^{\alpha(t-s)} G_{t-s}(r\to r')\beta m(r',s|\sigma)^3 \end{aligned}$$

(9.66a) then follows using (9.66b), the claim is therefore proven.

We can now insert the estimate (9.64) in (9.59). By using (9.66b) we obtain the bound

$$c\int_0^{t-kt_{a^*}} ds_1 \cdots \int_0^{s_{N-1}} ds_N e^{2(n+N-u)\alpha(t-kt_{a^*}-s_N)}$$
$$\times \{\prod_{\ell=1}^{u} \epsilon[1+(\sqrt{s_{\ell-1}-s_\ell})^{-1}]\}[\epsilon^{-\gamma^* 2m} e^{2m\alpha(kt_{a^*}+s_N)}\epsilon^m + \epsilon^{2m\delta}] \tag{9.70}$$

To simplify notation we have assumed δ so small that

$$\epsilon^\delta \geq \epsilon^{-\gamma^*}(\epsilon^{-2+1/4})^{-1/4}) + \epsilon^{1/12} \tag{9.71}$$

(γ^*, see (9.60), will be chosen small enough). Using that $2n+2(N-u)-2u \leq 2m$, we bound the term in (9.70) by

$$ce^{2m\alpha t}\epsilon^m\epsilon^{-2m\gamma^*}e^{2u\alpha(t-kt_{a^*})}\epsilon^u(t_{a^*})^N + ce^{2(n+N-u)\alpha(t-kt_{a^*})}\epsilon^u\epsilon^{2m\delta}(t_{a^*})^N \tag{9.72}$$

The second term in (9.72) is bounded by

$$ce^{2(m+u)\alpha t_{a^*}}\epsilon^{u+2m\delta}(t_{a^*})^N \leq ce^{2u\alpha t_{a^*}}\epsilon^u\epsilon^{m\delta}(t_{a^*})^N \tag{9.73}$$

because $a^* \leq A$ and that $2\alpha A \leq \delta$, see the assumptions stated below (9.49). If $u \geq n$ (9.73) gives a bound compatible with (9.50). If $u < n$ then $m \geq N-n$ and $\epsilon^{m\delta} \leq \epsilon^{2n}$ by (9.55), so that we have in all these cases a bound compatible with (9.50).

For the first term in (9.72) we consider firstly the case when $u > N/4$. We choose γ^* in (9.60) so small that

$$e^{2\alpha k^* t_{a^*}}\epsilon^{1-2\gamma^*} \leq 1$$

hence the first term in (9.72) is bounded by

$$e^{2u\alpha t_{a^*}}\epsilon^u t_{a^*}^N$$

Since $u > N/4$ by (9.55) we then get the desired bound.

If $u \leq N/4$, the first term on the right hand side of (9.72) is bounded by

$$ce^{2m\alpha t}\epsilon^m\epsilon^{-2m\gamma^*}t_{a^*}^N \leq c[e^{2\alpha t}\epsilon^{1-2\gamma^*}]^{n+N/2}t_{a^*}^N$$

The square bracket term can be bounded by

$$\epsilon^{[1-2\gamma^*-2\alpha k^* a^*](n+N/2)} = \epsilon^{[2\alpha b-2\gamma^*](n+N/2)}$$

cf. (9.55b). By choosing $2\gamma^* < \alpha b$ and by using (9.55a) we find that the first term in (9.72) is bounded in a way compatible with (9.50), hence the proof of Theorem 9.5.2 is completed. □

§9.6 The magnetization field.

In the next Proposition we establish the asymptotic behavior of the renormalized total magnetization

$$M_k^\epsilon = [e^{\alpha k t_{a^*}}\epsilon^{1/2}]^{-1}\epsilon \sum_{x\in \mathbf{Z}_\epsilon} \sigma(x, kt_{a^*}) \tag{9.74}$$

9.6.1 Proposition. *For any $k \le k^*$, M_k^ϵ converges in distribution as $\epsilon \to 0$ to a Gaussian random variable M_k with 0 average and variance $2\gamma/\alpha$.*

Proof. We shall prove that the moments of M_k^ϵ converge to those of M_k. By symmetry, for all k, $\mathbf{E}^\epsilon_{\mu^\epsilon}(M^\epsilon_{2k+1}) = \mathbf{E}(M_{2k+1}) = 0$, so that we only have to consider the even values of k. By Proposition 9.4.3 and Theorem 9.5.2

$$\begin{aligned}\lim_{\epsilon\to 0} \mathbf{E}^\epsilon_{\mu^\epsilon}\left((M_k^\epsilon)^{2n}\right) &= \lim_{\epsilon\to 0}[e^{\alpha k t_{a^*}}\epsilon^{1/2}]^{-2n}\epsilon^{2n}\sum_{\underline{x}} w^\epsilon_{2n}(\underline{x}, kt_{a^*}) \\ &= \frac{\pi(2n)}{n!}[\frac{4\gamma}{2\alpha}]^n\end{aligned}$$

which is indeed the $2n$ moment of M_k. The first equality in the above equation was obtained by noticing that when expanding $(M_k^\epsilon)^{2n}$, the sum over the sites which are not all distinct gives a vanishing contribution. □

§9.7 The escape.

We have now all the elements to conclude the proof of Theorem 9.1.1. Let $k^* t_{a^*} \le t \le (k^*+1)t_{a^*}$ and fix any set $\underline{x}$ of $2n$ different sites. Then by Theorem 9.2.1 (recalling that $a = 2a^*$ fulfills the requests of the Theorem) we have (setting $T \equiv (k^*-1)t_{a^*}$)

$$\lim_{\epsilon\to 0}|\mathbf{E}^\epsilon_{\mu^\epsilon}(\prod_{i=1}^{2n}\sigma(x_i,t)) - \mathbf{E}^\epsilon_{\mu^\epsilon}[\prod_{i=1}^{2n} m(\epsilon x_i, t-T|\sigma(\cdot,T))]| = 0 \tag{9.75}$$

9.7.1 Lemma. *For any C and ϵ let σ be such that $\|\sigma(\cdot,T)\| \le S$, cf. (9.60), and such that $|M^\epsilon_{k^*-1}| \le C$. Then*

$$\lim_{\epsilon\to 0}\sup_x |m(\epsilon x, t-T|\sigma(\cdot,T)) - Z(t; M^\epsilon_{k^*-1})) = 0 \tag{9.76}$$

where letting $z(t|c)$ denote the solution at time t of

$$\frac{d}{dt}z = \alpha z - \beta z^3 \tag{9.77}$$

with value c at time 0, we have set

$$Z(t; M^\epsilon_{k^*-1}) \equiv z\left(t - k^* t_{a^*}|(e^{\alpha k^* t_{a^*}}\epsilon^{1/2})M^\epsilon_{k^*-1}\right) \tag{9.78}$$

Proof. By Claim 2 and because $|G_{t_{a^*}}(r\to r') - 1| \le c\epsilon^{\hat\gamma}$, it follows that

$$\left|m(\epsilon x, t_{a^*}|\sigma(\cdot,T)) - e^{\alpha k^* t_{a^*}}\epsilon^{1/2}M^\epsilon_{k^*-1}\right| \le c\epsilon^{-3\gamma^*}(e^{\alpha k^* t_{a^*}}\epsilon^{1/2})^3 t_{a^*} + c\epsilon^{\hat\gamma}e^{\alpha k^* t_{a^*}}\epsilon^{1/2}$$

Therefore, setting

$$f_{\pm} = (e^{\alpha k^* t_{a^*}} \epsilon^{1/2}) M^{\epsilon}_{k^*-1} \pm [(\epsilon^{-\gamma^*} e^{\alpha k^* t_{a^*}} \epsilon^{1/2})^3 t_{a^*} + c\epsilon^{\hat{\gamma}} e^{\alpha k^* t_{a^*}} \epsilon^{1/2}]$$

we have that

$$z(t - k^* t_{a^*} | f_-) \le m(\epsilon x, t - T | \sigma(\cdot, T)) \le z(t - k^* t_{a^*} | f_+) \tag{9.79}$$

as follows from the maximum principle applied to (9.1). On the other hand $z(t'|f_{\pm})$ can be explicitly computed and from this the lemma follows, having chosen γ^* and $\hat{\gamma}$ small enough. □

From (9.75) and (9.76) we then have

$$\lim_{\epsilon \to 0} \sup_{\underline{x}} |\mathbf{E}^{\epsilon}_{\mu^{\epsilon}}(\prod_{i=1}^{2n} \sigma(x_i, t) - Z(t; M^{\epsilon}_{k^*-1})^{2n}))| = 0 \tag{9.80}$$

as it follows from restricting the expectation to the set of $\sigma(\cdot, T)$ considered in Lemma 9.7.1 and using the fact that such a set has a measure which goes to 1 as $C \to \infty$ uniformly on ϵ, by Proposition 9.6.1.

From (9.80) and Proposition 9.6.1, denoting by $G(x)dx$ the distribution of M_k and by τ any fixed real,

$$\lim_{\epsilon \to 0} \sup_{\underline{x}} \left| \mathbf{E}^{\epsilon}_{\mu^{\epsilon}} \left(\prod_{i=1}^{2n} \sigma(x_i, t_{1/(2\alpha)} + \tau) \right) - \int dx G(x) [Z(t_{1/(2\alpha)} + \tau; x))]^{2n} \right| = 0 \tag{9.81a}$$

For $x > 0$ (by symmetry, same considerations apply as well to the case $x < 0$), we have that

$$\hat{Z}_{\epsilon}(\tau, x) \equiv Z(t_{1/(2\alpha)} + \tau; x)$$

is given by

$$\int_{x e^{\alpha k^* t_{a^*}}}^{\hat{Z}_{\epsilon}(\tau, x)} dz \frac{1}{\alpha z - \beta z^3} = (\frac{1}{2\alpha} - k^* a^*) \log \epsilon^{-1} + \tau$$

hence

$$\lim_{\epsilon \to 0} \hat{Z}_{\epsilon}(\tau, x) = \hat{Z}(\tau, x)$$

where $\hat{Z}(\tau, x)$ satisfies

$$\lim_{c \to 0} \left(\int_c^{\hat{Z}(\tau, x)} dz \frac{1}{\alpha z - \beta z^3} - \log c \right) = \tau + \log x \tag{9.81b}$$

(9.81a) and (9.81b) prove (9.4).

If we take a so that $1/(2\alpha) \log \epsilon^{-1} < t_a \le (k^* + 1) t_{a^*}$ then by analogous arguments we get (9.3) at such times. To reach times t_a with arbitrary a we exploit the fact that the magnetization is close to a stable point for (9.1). Using this and iterating Theorem 9.2.1 finitely many times we can indeed reach arbitrarily long times (in the time scale $\log \epsilon^{-1}$). We omit the details. Theorem 9.1.1 is therefore proven modulo Theorem 9.2.1, whose proof will be given in the next chapter. □

§9.8 Bibliographical notes.

The problem we have considered in this chapter has been first studied in [44]. There is, however, a gap in the proof given in [44]. The analysis in fact uses the inequality

$$\mathbb{E}_{\nu_0}(\sigma(x_1,t)\cdots\sigma(x_n,t)) \geq 0 \tag{9.82}$$

where the expectation is with respect to the Glauber process which starts from ν_0, a Bernoulli measure with spin average 0. The validity of the inequality is an open problem, as far as we know.

The inequality (9.82) was used in [44] to derive un upper bound on the v-functions (for the Glauber+Kawasaki process): by symmetry the odd v-functions are 0 and in the equation for the v-function of order n, those of order $n+2$ appear with a minus sign. In this way, if (9.82) holds, it is possible to find an upper bound for any given v function in terms of the solution of a finite hierarchy of equations. We have avoided this by using Theorem 9.2.1 and the iterative procedure described above.

The escape in the context of an ordinary stochastic differential equation was previously studied in [46], where the idea of exploiting the time scale separation was first used. The model in [46] is related to problems in laser physics, for such a connection see also [19].

If there are Glauber interactions also with next nearest neighbor sites, the reaction potential may have a quartic maximum, when the parameters are suitably chosen. This case has been studied in [22] in "finite volumes", in the same sense considered in this chapter. The escape when the potential has an exponentially flat maximum, is studied, for a stochastic differential equation, by

M.E. Vares, *A note on small random perturbations of dynamical systems*, Stoch. Proc. and their Appl. **35** (1990), 225–230.

Infinite volumes in one dimension are considered in [40] for the same model we have studied here, the extension to the many dimensional case has been obtained by G. Giacomin, paper in preparation.

CHAPTER X

ESTIMATES ON THE V–FUNCTIONS

In this chapter we prove Theorem 9.2.1 studying an integral equation for v_n^ϵ by an iterative procedure similar to that used to prove Theorem 5.4.3. Here however we have the extra difficulty of the exclusion interaction and also the case $L_G \equiv 0$ is non trivial. We proceed by steps considering first the stirring process alone and then the general case.

§10.1 The v-functions in the stirring process.

To have an easier comparison with the existing literature, we go back to the particles language, where the occupation numbers are $\eta(x) = 0, 1$. We realize the stirring process by means of the active-passive marks process (cf. §6.4.5). We set, for notational simplicity, $\epsilon = 1$: since the generator is $\epsilon^{-2}L_0$ the process depends on ϵ and t via $\epsilon^{-2}t$, hence the case $\epsilon < 1$ is immediately recovered. We denote by $\rho(x,t)$ the solution of

$$\frac{\partial\rho(x,t)}{\partial t} = \frac{1}{2}\left[\rho(x+1,t) + \rho(x-1,t) - 2\rho(x,t)\right], \qquad \rho(x,t) = \eta(x)$$

and, by (2.11),

$$\rho(x,t) = \mathbb{E}_\eta(\eta(x,t)) \tag{10.1}$$

For any integer n we set

$$\mathcal{M}_n = \{\underline{x} = (x_1, \cdots, x_n) : \ x_i \neq x_j \ \ \forall\, i \neq j\} \tag{10.2}$$

and for any $\underline{x} \in \mathcal{M}_n$

$$v_n(\underline{x}, t|\eta) = \mathbb{E}_\eta\left(\prod_{i=1}^n [\eta(x_i,t) - \rho(x_i,t)]\right), \qquad v_n(\underline{x}, 0) \equiv 0 \tag{10.3}$$

In the remaining of this paragraph we give a proof of the following theorem, firstly proven in [62],

10.1.1 Theorem. *There is a sequence c_n such that for all $\eta \in \{0,1\}^{\mathbb{Z}}$ and all $t > 0$*

$$\sup_{\underline{x}\in\mathcal{M}_n} |v_n(\underline{x}, t|\eta)| \leq c_n t^{-n/8} \tag{10.4}$$

In the sequel we shall simply write $v_n(\underline{x}, t)$ for $v_n(\underline{x}, t|\eta)$. We first establish an integral equation for the v-functions:

10.1.2 Lemma. *For any $\underline{x} \in \mathcal{M}_n$ and any $t \geq 0$,*

$$v_n(\underline{x},t) = \int_0^t ds \sum_{\underline{z}} P_s(\underline{x} \to \underline{z})(Av)(\underline{z}, t-s) \tag{10.5}$$

where, for any $t > 0$, $P_t(\underline{x} \to \underline{z})$ is the stirring process transition probability to go from $\underline{x}$ to $\underline{z}$ in a time t. (Av) is given by

$$\begin{aligned}(Av)(\underline{x},t) \equiv \sum_{h,k=1}^{n} 1(x_h = x_k + 1)\{[\rho(x_h,t) - \rho(x_k,t)][v_{n-1}(\underline{x}^{(h)},t) - v_{n-1}(\underline{x}^{(k)},t)] \\ - \frac{1}{2}[\rho(x_h,t) - \rho(x_k,t)]^2 v_{n-2}(\underline{x}^{(h,k)},t)\}\end{aligned} \tag{10.6}$$

where $\underline{x}^{(h)}(t)$ and $\underline{x}^{(k)}(t)$ are respectively equal to $\{x_i(t), i \neq h\}$, and $\{x_i(t), i \neq k\}$, while $\underline{x}^{(h,k)}(t) = \{x_i(t), i \neq h,k\}$.

Proof. We have

$$\frac{d}{dt} v_n(\underline{x},t) = \mathbb{E}_\eta\Big(L_0 \prod_{i=1}^{n}[\eta(x_i,t) - \rho(x_i,t)]\Big) - \sum_{i=1}^{n} \mathbb{E}_\eta\Big(\frac{\partial \rho(x_i,t)}{\partial t}\prod_{j\neq i}[\eta(x_i,t) - \rho(x_i,t)]\Big)$$

From this it is not difficult to check that

$$\frac{d}{dt} v_n(\underline{x},t) = L_0 v_n(\underline{x},t) + (Av)(\underline{x},t) \tag{10.7}$$

hence (10.5). □

Denoting by $\underline{x}(\cdot)$ the process of n stirring particles, we can rewrite (10.5) as

$$v_n(\underline{x},t) = \int_0^t ds\, \mathbb{E}_{\underline{x}}\big((Av)(\underline{x}(s), t-s)\big) \tag{10.8}$$

In the expression for (Av) there are terms which are small: by the local central limit theorem, see (4.34), in fact, for any $t > 0$, and z_i, $z_j \in \mathbb{Z}$

$$|\rho(z_i,t) - \rho(z_j,t)| = |\sum_z [P_t(z_i \to z) - P_t(z_j \to z)]\eta(z)| \leq c' \frac{|z_i - z_j|}{1 + t^{1/2}} \leq c \frac{|z_i - z_j|}{t^{-b+1/2}} \tag{10.9}$$

The last expression gives a worse estimate when $t \to \infty$, but it will allow for more compact notation in the sequel.

The bound (10.9) by itself is not sufficient for proving Theorem 10.1.1, we need to take into account also the characteristic function and the difference between the two v-functions both present in (10.6). We start by the latter.

10.1.3 Lemma. *For any $\underline{x} \in \mathcal{M}_n$, any $i \neq j$ in $\{1, \ldots, n\}$ and any $t > 0$:*

$$\begin{aligned}&v_{n-1}(\underline{x}^{(j)},t) - v_{n-1}(\underline{x}^{(i)},t) \\ &= \int_0^t ds\, \mathbb{E}_{\underline{x}}\Big(1(\tau_{i,j} > s)[(Av)(\underline{x}^{(j)}(s), t-s) - (Av)(\underline{x}^{(i)}(s), t-s)]\Big)\end{aligned} \tag{10.10}$$

where $\tau_{i,j}$ is defined in (6.32).

Proof. From Lemma 10.1.2 it follows that

$$v_{n-1}(\underline{x}^{(j)},t) - v_{n-1}(\underline{x}^{(i)},t) = \int_0^t ds \Big[\mathbf{E}_{\underline{x}^{(j)}}\Big((Av)(\underline{x}(s),t-s)\Big) - \mathbf{E}_{\underline{x}^{(i)}}\Big((Av)(\underline{x}(s),t-s)\Big)\Big]$$
$$= \int_0^t ds\, \mathbf{E}_{\underline{x}}\Big((Av)(\underline{x}^{(j)}(s),t-s) - (Av)(\underline{x}^{(i)}(s),t-s)\Big) \tag{10.11a}$$

where in the first equality we have used the simbol $\underline{x}(s)$ to denote the configuration of $n-1$ particles at time s starting from $\underline{x}^{(j)}$ and $\underline{x}^{(i)}$ respectively. In the last equality, instead $\underline{x}(s)$ is a configuration with n particles and $\underline{x}^{(j)}(s)$ the configuration with $n-1$ particles obtained from $\underline{x}(s)$ by dropping $x_j(s)$. $\underline{x}^{(i)}(s)$ is defined analogously. Finally $\underline{x}(0) = \underline{x}$.

The validity of the second equality in (10.11a) follows from the fact that it is the same to consider at any time t either the law of the process starting from $\underline{x}^{(j)}$, i.e. without the particle j, or the law of the process with all the particles, but then to take its marginal disregarding particle j at time t.

Recalling the definition of $\tau_{i,j}$ in (6.32), we have that

$$E_{\underline{x}}\Big(1(\tau_{i,j} \le s)[(Av)(\underline{x}^{(j)}(s),t-s) - (Av)(\underline{x}^{(i)}(s),t-s)]\Big) = 0 \tag{10.11b}$$

because the difference of the (Av)'s is antisymmetric under the exchange of $x_i(s)$ and $x_j(s)$. □

By iterating (10.5) and (10.10) we obtain an expression which can be conveniently interpreted in terms of a branching process. We start by the latter.

§*10.1.4 Definition: The stirring process with deaths.*

We fix $t > 0$ and $n > 1$. In this definition $t > 0$ and $\underline{x} \in \mathcal{M}_n$ are fixed and dependence on them is not explicited.

(1) $\mathcal{L}$ (the set of all the possible particles labels) is the collection of all the subsets of $\{1,\dots,n\}$.
(2) Ω (the set of all the labelled configurations) is the set of all $\{x_i,\ i \in L\}$, $L \in \mathcal{L}$, with $x_i \ne x_j$ if $i \ne j$. The "empty configuration" is denoted by $\emptyset$.
(3) Given $\underline{x} \in \Omega$, $L(\underline{x}) \in \mathcal{L}$ is the set of the labels of the particles in $\underline{x}$.
(4) Γ is the subset of all $\gamma = (\underline{\xi},\underline{\xi}') \in \Omega^2$ such that $\underline{\xi}$ has just two particles in the set $\{0,1\}$, while $\underline{\xi}'$ is either $\emptyset$ or it has two particles in $\{0,1\}$ and, in this latter case, $L(\underline{\xi}) = L(\underline{\xi}')$. We shall denote by $a = a(\gamma)$ the lowest label in $\underline{\xi}$ and by $d = d(\gamma)$ the highest one; finally $\delta = \delta(\gamma)$ is equal to 1 if $\underline{\xi}' = \emptyset$ and it is equal to 0 otherwise. The particles of $\underline{\xi}$ are called interacting particles.
(5) For any integer $m \in [n/2, n)$ we set

$$\mathcal{X}_m = D([0,t],\Omega) \times [0,t]^m \times \Gamma^m \tag{10.12a}$$

and

$$\mathcal{X} = \bigcup_m \mathcal{X}_m, \tag{10.12b}$$

writing $\underline{t} = (t_1, \ldots, t_m) \in [0,t]^m$, $t_0 = 0$, $t_{m+1} = t$; $\underline{\gamma} = (\gamma_1, \ldots, \gamma_m) \in \Gamma^m$. The t_i are called interacting times and we set $a_i(\underline{\gamma}) = a(\gamma_i)$, $d_i(\underline{\gamma}) = d(\gamma_i)$, $\delta_i(\underline{\gamma}) = \delta(\gamma_i)$.

We shall henceforth consider only elements of $\mathcal{X}$ such that $t_i < t_{i+1}$, for all i.

(6) In the sequel $\mathcal{P}$ will denote the measure on $\mathcal{X}$ supported by $\{\underline{x}(0) = \underline{x}\}$ and, in $\mathcal{X}_m$, for any m, by $\{\underline{x}(s) = \emptyset,\ s \geq t_m^+\}$. $\mathcal{P}$ is then completely defined by the integrals, for all m, of the bounded measurable functions f supported by $\mathcal{X}_m$:

$$\int d\mathcal{P} f = \int_0^t dt_1 \cdots \int_{t_{m-1}}^t dt_m \sum_{\underline{x}^1, \ldots, \underline{x}^{m-1}} \mathbb{E}_{\underline{x}, \ldots, \underline{x}^{m-1}} \left(f \prod_{i=1}^m B(\underline{x}(t_i^-) \to \underline{x}^i; \gamma_{i-1}, \gamma_i) \right) \tag{10.13}$$

In (10.13), $\mathbb{E}_{\underline{x}, \ldots, \underline{x}^{m-1}}$ is the expectation with respect to a process which, in any time interval $[t_{i-1}, t_i)$, $1 \leq i \leq m$, is the stirring process (realized in the active-passive mark process) with initial condition $\underline{x}^{i-1}$, $\underline{x}^0 = \underline{x}$. Each particle keeps its label during these time intervals. In (10.13) $\gamma_0 = \emptyset$ and the kernel B has values 0 and 1, according to the following rules.

(7) Given γ, γ', $\underline{x}$ there is at most one value of $\underline{y}$ for which $B(\underline{x} \to \underline{y}; \gamma, \gamma') = 1$. A first condition for the existence of such a $\underline{y}$ is:

$$\delta(\gamma) = 1 \quad \Longrightarrow \quad a(\gamma) \notin L(\underline{x}), d(\gamma) \notin L(\underline{x}) \tag{10.14a}$$

$$\delta(\gamma) = 0 \quad \Longrightarrow \quad a(\gamma) \in L(\underline{x}), d(\gamma) \in L(\underline{x}) \tag{10.14b}$$

Define in the first case $\underline{z} = \underline{x}$ and in the second one $\underline{z} = \underline{x} \setminus \{x_{d(\gamma)}\}$.
If the above condition is satisfied, then $B(\underline{x} \to \underline{y}; \gamma, \gamma') = 1$ if

$$a \in L(\underline{z}), d \in L(\underline{z}), \quad z_a - z_d = \xi_a - \xi_d \qquad \text{where} \quad \gamma' = (\xi, \xi'),\ L(\xi) = (a, d) \tag{10.14c}$$

Then if $\delta(\gamma') = 1$: $\underline{y} = \underline{z} \setminus \{z_a, z_d\}$.
If $\delta(\gamma') = 0$: $L(\underline{y}) = L(\underline{z})$, $y_j = z_j$ for all $j \in L(\underline{y}) \setminus \{a, d\}$, while $y_a = \xi'_a + z_a - \xi_a$ and $y_d = \xi'_d + z_d - \xi_d$.

10.1.5 Lemma. *Let $n > 1$, $\underline{x} \in \mathcal{M}_n$ and $t > 0$, and let $\mathcal{P}$ be as in 10.1.4. Then*

$$v_n(\underline{x}, t) = \sum_{m=n/2}^{n} \int_{\mathcal{X}_m} d\mathcal{P} [\prod_{i:\delta_i=1} \frac{-1}{2} \{\rho(x_{a_i}(t_i), t - t_i) - \rho(x_{d_i}(t_i), t - t_i)\}^2]$$
$$\times [\prod_{i:\delta_i=0} \{\rho(x_{d_i}(t_i), t - t_i) - \rho(x_{a_i}(t_i), t - t_i)\} 1(\tau_{a_i,d_i}(t_i) > t_{i+1})]) \tag{10.15}$$

where $a_i = a_i(\underline{\gamma})$ and $d_i = d_i(\underline{\gamma})$. Furthermore $\tau_{a_i,d_i}(t_i) = s$, with $t_i \leq s < t_{i+1}$, if s is the first time when there is a mark between $x_{a_i}(s)$ and $x_{d_i}(s)$; otherwise $\tau_{a_i,d_i}(t_i) = \infty$

Proof. We iterate (10.5). The first term on the right hand side of (10.6), by (10.10), gives rise to a term with $\delta = 0$, the second with $\delta = 1$. We then easily obtain (10.15). □

We next reduce (10.15) to an expression containing expectations with respect to an independent branching process with only deaths. This is defined by dropping the conditions in (7) of 10.1.4 on the positions of the particles $x_{a_i}(t_i^-)$ and $x_{d_i}(t_i^-)$ and by realizing the stirring process by the couplings introduced in Chapter VI.

10.1.6 Definition: The independent branching.

(1) The skeleton π of an element $(\underline{x}(\cdot), \underline{t}, \underline{\gamma}) \in \mathcal{X}$, is $\pi = (\underline{L}, \underline{\gamma})$, where $\underline{L} = \{L_i, 1 \le i \le m(\pi)\}$ and $m(\pi) = m$ if π is the skeleton of an element of $\mathcal{X}_m$; $L_i = L(\underline{x}(t_i^+))$ denotes the labels of the particles at t_i^+.

If $\pi = (\underline{L}, \underline{\gamma})$, we write $\delta_i(\pi)$ for $\delta_i(\underline{\gamma})$, analogously defined are $a_i(\pi)$ and $d_i(\pi)$.

(2) Given a skeleton π as above and $\underline{t} = (t_1, \dots, t_m)$, $0 < t_1 \cdots < t_m < t$, we define $\mathcal{P}^0_{\pi,\underline{t}}$ as the law of the process $\underline{x}^0(s), 0 \le s \le t$, which, in $[t_i, t_{i+1})$, is made of independent symmetric random walks of intensity 1, with labels in L_i and such that $x_j^0(t_i) = 0$ for all $j \in L_i$. At $t = 0$ the labels are $\{1, \dots, n\}$ and $x_i^0(0) = 0$ for all $1 \le i \le n$.

(3) On the same space we define the processes $\underline{z}^0(s)$ and $\hat{\underline{x}}(s)$, $0 \le s \le t$, as follows. We set

$$\underline{z}^0(s) = \underline{x}^0(s) + \underline{x}, \qquad 0 \le s < t_1 \tag{10.16a}$$

and we introduce the coupling defined in 6.6.1 between the stirring and the independent particles, with priorities given to the particles with lower labels. By this, the stirring configuration $\hat{\underline{x}}(s)$ becomes a function of $\underline{z}^0(s)$ for $s < t_1$, hence, by (10.16a), it is measurable with respect to $\underline{x}^0(\cdot)$.

Let $\gamma_1 = (\underline{\xi}, \underline{\xi}')$, $\delta_1 = \delta_1(\pi)$, $d_1 = d_1(\pi)$ and $a_1 = a_1(\pi)$. If $\delta_1 = 1$ we set $\hat{\underline{x}}(t_1^+) = \hat{\underline{x}}(t_1^-) \backslash \{\hat{x}_{a_1}(t_1^-), \hat{x}_{d_1}(t_1^-)\}$. If $\delta_1 = 0$ and $\underline{\xi} = \underline{\xi}'$ we set $\hat{\underline{x}}(t_1^+) = \hat{\underline{x}}(t_1^-)$. If $\underline{\xi} \ne \underline{\xi}'$ then we set

$$b = \xi_{d_1} - \xi_{a_1}, \quad u = \hat{x}_{a_1}(t_1^-), \quad w = u + b$$

and $\hat{\underline{x}}(t_1^+) = \hat{\underline{x}}(t_1^-)^{(u,w)}$, where, given $\underline{y}$ with $L(\underline{y}) = L(\hat{\underline{x}}(t_1^+))$ and $y_k = u$, then

$$y_k^{(u,w)} = \begin{cases} u & \text{if } y_j = w \text{ and } j < k \\ w & \text{otherwise} \end{cases} \tag{10.16b}$$

and, if, in the last alternative, there is $j > k$ such that $y_j = w$, then $y_j^{(u,w)} = u$, while the other particles are unchanged. In the first alternative in (10.16b), $\underline{y}^{(u,w)} = \underline{y}$.

[*Remarks.* In this way priority is given to lower labels. Notice also that $B(\hat{\underline{x}}(t_1^-) \to \hat{\underline{x}}(t_1^+); \emptyset, \gamma_1)$ may be 0: the probability restricted to the set where it equals 1 at all the interacting times, is the same as $\mathcal{P}$ restricted to the trajectories with skeleton π and conditioned on $\underline{t}$].

If $\delta_1 = 1$ we set $\underline{z}^0(t_1^+) = \underline{z}^0(t_1^-) \backslash \{z^0_{a_1}(t_1^-), z^0_{d_1}(t_1^-)\}$. If $\delta_1 = 0$ then $\underline{z}^0(t_1^+) = \underline{z}^0(t_1^-)$.

In the next time interval we define

$$\underline{z}^0(s) = \underline{x}^0(s) + \underline{z}^0(t_1^+), \qquad t_1 \le s < t_2 \tag{10.16c}$$

To construct $\hat{\underline{x}}(s)$ in (t_1, t_2), we use again the coupling 6.6.1, but choosing the priorities as follows. If $\delta_1 = 1$, as before higher priority is given to lower labels. If $\delta_1 = 0$, top priority is given to the label a_1, next comes d_1 and, after them, higher priorities go to lower labels. By this rule we define $\hat{\underline{x}}(s)$ till $s < t_2^-$ and at t_2^- we drop the particle with label d_1 both in $\hat{\underline{x}}$ and $\underline{z}^0$. We then apply the same rule used to go from t_1^- to t_1^+, with $\gamma_2(\pi)$ replacing $\gamma_1(\pi)$. Thus we determine $\hat{\underline{x}}(t_2^+)$ and $\underline{z}^0(t_2^+)$.

By iteration we construct the process till time t_m^-: at t_m^+ all the particles have disappeared and we have an empty configuration thereafter.

(4) For $i : \delta_i = 0$, let $\gamma_i = (\underline{\xi}, \underline{\xi}')$ and $\underline{\xi}' = (\xi'_{a_i}, \xi'_{d_i})$. We define

$$\tau^0_{a_i,d_i}(t_i) = \begin{cases} s & \text{if } t_i \le s < t_{i+1} \text{ is the first time when } x^0_{a_i}(s) + \xi'_{a_i} = x^0_{d_i}(s) + \xi'_{d_i} \\ \infty & \text{otherwise} \end{cases} \tag{10.17}$$

Notice that the coupling allows to reconstruct the marks which involve the sites $\hat{\underline{x}}(s)$ so that in this space we can define the variables $\hat{\tau}_{a_i,d_i}(t_i)$, for $\delta_i = 0$, which have the values of the first time $s \in [t_i, t_{i+1})$ when there is a mark between $\hat{x}_{a_i}(s)$ and $\hat{x}_{d_i}(s)$; otherwise we set $\hat{\tau}_{a_i,d_i}(t_i) = \infty$.

If $\delta_i = 0$, $\gamma_i = (\underline{\xi}, \underline{\xi}')$ and $\hat{x}_{a_i}(t_i^+) - \hat{x}_{d_i}(t_i^+) = \xi'_{a_i} - \xi'_{d_i}$, then $\hat{\tau}_{a_i,d_i}(t_i)$ has the same law as $\tau_{a_i,d_i}(t_i)$, as defined in Lemma 10.1.5. In the next lemma we relate $\hat{\tau}$ to τ^0, so that it will be possible to relate τ and τ^0

10.1.7 Lemma. *With the same notation as in 10.1.6, if $\delta_i = 0$ and $\gamma_i = (\underline{\xi}, \underline{\xi}')$, then*

$$\mathcal{P}^0_{\pi,\underline{t}}\Big(\{\tau^0_{a_i,d_i}(t_i) = \hat{\tau}_{a_i,d_i}(t_i)\}\Big|\{\hat{x}_{a_i}(t_i^+) - \hat{x}_{d_i}(t_i^+) = \xi'_{a_i} - \xi'_{d_i}\}\Big) = 1 \tag{10.18}$$

The proof of the lemma is immediate, once the definition of the coupling in 6.6.1 is recalled, we therefore omit it.

As a Corollary of Lemma 10.1.7 we have, from (10.15):

$$\begin{aligned} v_n(\underline{x}, t) = \sum_\pi \int_0^t dt_1 \cdots \int_{t_{m-1}}^t dt_m \mathcal{E}^0_{\pi,\underline{t}}\Big([\prod_{i=1}^m 1(\hat{x}_{a_i}(t_i^-) - \hat{x}_{d_i}(t_i^-) = \xi^{(i)}_{a_i} - \xi^{(i)}_{d_i})] \\ \times [\prod_{i:\delta_i=1} \frac{-1}{2}\{\rho(\hat{x}_{a_i}(t_i^-), t - t_i) - \rho(\hat{x}_{d_i}(t_i^-), t - t_i)\}^2] \\ \times [\prod_{i:\delta_i=0} \{\rho(\hat{x}_{d_i}(t_i), t - t_i) - \rho(\hat{x}_{a_i}(t_i), t - t_i)\} 1(\tau^0_{a_i,d_i}(t_i) > t_{i+1})]\Big) \end{aligned} \tag{10.19}$$

In (10.19) we have written $m = m(\pi)$, $(\underline{\xi}^{(i)}, \underline{\xi}'^{(i)}) = \gamma_i(\pi)$, $a_i = a_i(\pi)$ and $d_i = d_i(\pi)$.

Since the sum over π contains finitely many terms, it will be enough to bound the generic one. To this purpose we use (10.9) and Proposition 6.6.3.

10.1.8 Corollary. *For any $d < 1$, a and k there is c so that*

$$|v_n(\underline{x}, t)| \le c[t^{-k} + \sum_\pi w_\pi(\underline{x}, t)] \tag{10.20}$$

where, if π is the skeleton of an element in $\mathcal{X}_m$,

$$\begin{aligned} w_\pi(\underline{x}, t) = \int_0^t dt_1 \cdots \int_{t_{m-1}}^t dt_m [\prod_{\delta_i=1} (t - t_i)^{2b-1}][\prod_{\delta_i=0} (t - t_i)^{b-1/2}] \\ \mathcal{P}^0_{\pi,\underline{t}}\Big([\bigcap_{\delta_i=0} \{\tau^0_{a_i,d_i}(t_i) > t_{i+1}\}] \bigcap_{t_i > t^d} \{|z^0_{a_i}(t_i^-) - z^0_{d_i}(t_i^-)| \le t_i^{1/4+a}\}\Big) \end{aligned} \tag{10.21}$$

Proof. (10.21) follows from (10.19) by using (10.9) and Proposition 6.6.3, whose validity extends trivially to the present case, where different time intervals may have different priorities. In this way we obtain an expression similar to that in the right hand side of (10.21), where the last intersection is replaced by one involving the sets

$$\{|z^0_{a_i}(t_i^-) - z^0_{d_i}(t_i^-)| \le 2t_i^{1/4+a'} + 1 + D\}, \quad a' > 0$$

D is determined by π and it takes into account the displacements of the particles in $\hat{\underline{x}}$ which may occur at the interaction times t_i and which are not matched in $\underline{z}^0$, where no displacement occurs at the t_i's. Since D is finite for any π, we have that

$$2t_i^{1/4+a'} + 1 + D \le t_i^{1/4+2a'}$$

if t is large enogh (recall that $t_i > t^d$ in the case that we are considering). We have thus proven (10.21) with $a = 2a'$ and for the above values of t; by a suitable choice of c in (10.20) we then obtain the proof of the Corollary at all t's. □

We shall bound (10.21) by an iterative procedure, we first need some notation.

10.1.9 Notation.

Given a skeleton π we define

$$\Delta_i = \begin{cases} \frac{1}{2}[t_{i+1} - t_i] & \text{if } \delta_i = 0 \\ 0 & \text{otherwise} \end{cases} \tag{10.22}$$

and call $\mathcal{G}$ the σ-algebra generated by

$$\{\underline{x}^0(s), \quad t_i \le s \le t_i + \Delta_i, \quad \text{for all } i : \delta_i = 0\} \tag{10.23a}$$

For $1 < k \le n$ we define

$$\mathcal{F}_k = \text{ the } \sigma\text{-algebra generated by } \{x_j^0(s),\ 0 \le s \le t;\ j < k\} \tag{10.23b}$$

The probability in (10.21) is then bounded by

$$\mathcal{E}^0_{\pi,\underline{t}}\Big(\prod_{\delta_i=0} 1(\tau^0_{a_i,d_i}(t_i) > t_i + \Delta_i)] \mathcal{P}^0_{\pi,\underline{t}}\big(\bigcap_{t_i > t^d} \{|z^0_{a_i}(t_i^-) - z^0_{d_i}(t_i^-)| \le t_i^{1/4+a}\} | \mathcal{G}\big)\Big) \tag{10.24}$$

We shall bound the conditional probability in (10.24) iteratively, starting by the contribution of the last particle, particle n.

10.1.10 Lemma. *Let i be such that $d_i = n$ and suppose that $t_i > t^d$. Then for any $a > 0$ there is c so that*

$$\mathcal{P}^0_{\pi,\underline{t}}\Big(\bigcap_{t_j > t^d} \{|z^0_{a_j}(t_j^-) - z^0_{d_j}(t_j^-)| \le t_j^{1/4+a}\} \Big| \mathcal{G}\Big)$$

$$\le ct^{-1/4+a} \mathcal{P}^0_{\pi,\underline{t}}\Big(\bigcap_{j \ne i: t_j > t^d} \{|z^0_{a_j}(t_j^-) - z^0_{d_j}(t_j^-)| \le t_j^{1/4+a}\} \Big| \mathcal{G}\Big) \tag{10.25}$$

Proof. By conditioning on $\mathcal{F}_n$, the left hand side of (10.25) equals

$$\mathcal{E}^0_{\pi,\underline{t}}\Big(\prod_{j\neq i: t_j > t^d} 1(|z^0_{a_j}(t_j^-) - z^0_{d_j}(t_j^-)| \le t_j^{1/4+a}) \mathcal{P}^0_{\pi,\underline{t}}(\{|z^0_{a_i}(t_i^-) - z^0_n(t_i^-)| \le t_i^{1/4+a}\} | \mathcal{G} \vee \mathcal{F}_n) \Big| \mathcal{G}\Big) \tag{10.26}$$

where $\mathcal{F}_n \vee \mathcal{G}$ is the minimal σ-algebra which contains both $\mathcal{F}_n$ and $\mathcal{G}$. (We have used that $a_j \neq n$, $d_j \neq n$ for all $j \neq i$).

The last probability in (10.26) equals

$$\mathcal{P}(\{|x(s) - y| \le t_i^{1/4+a}\}) \tag{10.27}$$

where $\mathcal{P}$ is the law of a symmetric random walk starting at the origin and jumping with intensity 1;

$$s = t_i - \sum_{j<i} \Delta_j$$

and y in (10.27) is a suitable constant. Since $s > t_i/2$, because of the definition of the Δ_j, by using the local central limit theorem, see (4.34), we get

$$\mathcal{P}(\{|x(s) - y| \le t_i^{1/4+a}\}) \le ct_i^{1/4+a-1/2}$$

and the Lemma is proven. □

The characteristic function involving the particle with label n does not appear anymore on the right hand side of (10.25), we can therefore repeat what done in Lemma 10.1.10 with $n-1$ replacing n. By iteration we finally get

$$\mathcal{P}^0_{\pi,\underline{t}}\Big(\bigcap_{t_j > t^d} \{|z^0_{a_j}(t_j^-) - z^0_{d_j}(t_j^-)| \le t_j^{1/4+a}\} \Big| \mathcal{G}\Big) \le c \prod_{i: t_i > t^d} (t_i)^{a-1/4} \tag{10.28}$$

After using in (10.21) the bound (10.28), which is uniform on the conditioning $\mathcal{G}$, we obtain an intersection of independent events. Each of them is bounded in terms of the probability that the first hit to the origin of a random walk is bigger than Δ_i. Thus we have

$$\mathcal{E}^0_{\pi,\underline{t}}\Big(\bigcap_{i:\delta_i=0} \{\tau^0_{a_i,d_i}(t_i) > t_i + \Delta_i\}\Big) \le c \prod_{i:\delta_i=0} (t_{i+1} - t_i)^{-1/2} \tag{10.29}$$

By (10.28) and (10.29) we get

$$\begin{aligned} w_\pi(\underline{x}, t) \le c \int_0^t dt_1 \cdots \int_{t_{m-1}}^t dt_m [\prod_{t_i > t^d} t_i^{a-1/4}][\prod_{i:\delta_i=1} (t - t_i)^{2b-1}] \\ \times [\prod_{i:\delta_i=0} (t - t_i)^{b-1/2} (t_{i+1} - t_i)^{-1/2}] \end{aligned} \tag{10.30}$$

If $t_i \le t^d$:

$$\frac{t^{d(1/4-a)}}{t_i^{1/4-a}} \ge 1 \tag{10.31}$$

so that we can bound (10.30) as

$$w_\pi(\underline{x},t) \le ct^{d(1/4-a)m} \int_0^t dt_1 \cdots \int_{t_{m-1}}^t dt_m$$
$$\times [\prod_{i=1}^m (t_i)^{a-1/4}][\prod_{\delta_i=1}(t-t_i)^{2b-1}][\prod_{\delta_i=0}(t-t_i)^{b-1/2}(t_{i+1}-t_i)^{-1/2}] \qquad (10.32)$$

We change variables, writing $t_i' = t_i/t$ and $s_i = 1 - t_i'$ so that

$$w_\pi(\underline{x},t) \le ct^m (t^d)^{(1/4-a)m} (t^{a-1/4})^m (t^{2b-1})^p (t^{b-1})^q$$
$$\times \int_0^1 ds_1 \cdots \int_0^{s_{m-1}} ds_m [\prod_{i=1}^m (1-s_i)^{a-1/4}][\prod_{\delta_i=1} s_i^{2b-1}]$$
$$\times [\prod_{\delta_i=0} s_i^{b-1/2}(s_i - s_{i+1})^{-1/2}] \qquad (10.33)$$

where p [q] stands for the number of indices i such that $\delta_i = 1$ [$= 0$]. We shall prove in 10.1.11 below that the integral in (10.33) is finite so that the dependence on t in (10.33) is fully explicited. Noticing that $p+q=m$ we get

$$w_\pi(\underline{x},t) \le ct^{m[d(1/4-a)+2b]} t^{(a-1/4)m} \qquad (10.34)$$

We first consider the cases where $m > n/2$, the case $m = n/2$ is studied in 10.1.12 below. The right hand side of (10.34) is then bounded by

$$ct^{-n/8}\, t^{m[a+d(1/4-a)+2b]} t^{-1/4(m-n/2)} \le c't^{-n/8} \qquad (10.35)$$

if a,d and b are small enough.

10.1.11 Proof that the integral in (10.33) is finite.

Recall that $\delta_1, \cdots, \delta_m$ are specified by π and fixed in (10.33). For $0 \le k \le m$ we introduce the set

$$A_k = \{(s_1, \cdots, s_m) : s_k > \frac{1}{2} \ge s_{k+1}\}$$

and we consider the integral over this set. Then we bound by a constant all the factors s_i^u, $u = b - 1/2$ or $u = 2b-1$, whenever $i \le k$, and the factors $(1-s_i)^{1/4-a}$ for $i > k$. To estimate the integral after having used such bounds, we drop the condition that $\{s_1, \cdots, s_m\} \in A_k$ and bound the terms of the form $(1-s_i)^{a-1/4}$, if present, by $(s_{i-1}-s_i)^{a-1/4}$. We use the identity

$$\int_0^t ds (t-s)^{-q} s^{-p} = c_{p,q} t^{1-p-q} \qquad (10.36)$$

where $c_{p,q}$ is finite if both p and q are strictly less than 1. We integrate over s_m and the result, by (10.36), has again the same structure, so that we can iterate the computation and perform all the integrals. Notice in fact that at each integration we have an expression like that in (10.36) with $p+q<1$ and $q \ge 0$, $p > 0$: this is a consequence of the bounds obtained by restricting the integration to A_k. By summing over $k \in \{0, \ldots, m\}$ we prove that the integral in (10.33) is finite.

From (10.20) it follows then that for a, d and b small enough

$$|v_n(\underline{x},t)| \le ct^{-n/8} + c \sum_{\pi:\delta_i(\pi)\equiv 1} \int_0^t dt_1 \cdots \int_{t_{m-1}}^t dt_m [\prod_{\delta_i=1} (t-t_i)^{2b-1}]$$
$$\times \mathcal{P}^0_{\pi,\underline{t}}(\bigcap_{i=1}^m \{|\hat{x}_{a_i}(t_i) - \hat{x}_{d_i}(t_i)| = 1\}) \tag{10.37}$$

The first term on the right hand side, by (10.34) and 10.1.11, bounds the contribution of all π having some $\delta_i = 0$. The others, present only if n is even, contribute to the second term in (10.37).

10.1.12 Bound on the right hand side of (10.37).

We consider the coupling 6.6.1 in the time interval $[0, t_1)$ which gives higher priority to a_1 and d_1. By Proposition 6.6.3 we obtain that for any k there is c so that the probability in (10.37) is bounded by

$$ct^{-k} + \mathcal{P}^0_{\pi,\underline{t}}\Big(\{|\hat{x}_{a_1}(t_1) - \hat{x}_{d_1}(t_1)| = 1\} \bigcap_{i>1:t_i>t^d} \{|z^0_{a_i}(t_i^-) - z^0_{d_i}(t_i^-)| \le t_i^{1/4+a}\}\Big) \tag{10.38}$$

By Proposition 6.6.3, $\hat{x}_{a_1}$ and $\hat{x}_{d_1}$ are measurable with respect to $x^0_{a_1}$ and $x^0_{d_1}$, so that the events in (10.38) are independent. We then have a bound as in (10.28) for the latter intersection, while, using Liggett's inequality, see (9.22), we get that the probability of the first event is bounded by $ct_1^{-1/2}$. Hence, overall, from (10.37) we get

$$|v^\epsilon_n(\underline{x},t)| \le c\big[t^{-n/8} + t^{[2b+a+d(1/4-a)](n-1)+2b-1/2}\big] \tag{10.39}$$

Since n is fixed for a, d and b small enough we get that the right hand side of (10.39) is bounded by $ct^{-n/8}$, so that Theorem 10.1.1 is proven. □

§10.2 Estimates at small macroscopic times.

In this paragraph we study the full Glauber+Kawasaki process proving the same bound obtained in §10.1, but only at infinitesimally small macroscopic times:

10.2.1 Theorem. *For any $\beta > 0$ and any n there is c so that for all σ and all $\underline{x} \in \mathcal{M}_n$, see (10.2):*

$$|v^\epsilon_n(\underline{x},t|\sigma)| \le c\,(\epsilon^{-2}t)^{-n/8}, \quad t \le \epsilon^\beta \tag{10.40}$$

We shall drop henceforth the dependence on the initial configuration σ and, sometimes, the subscript n from v^ϵ_n. Our considerations apply as well to more general Glauber interactions, namely whenever

$$\frac{d}{dt}v^\epsilon_n(\underline{x},t) = \epsilon^{-2}\big[L_0 v^\epsilon_n(\underline{x},t) + (A_\epsilon v^\epsilon)(\underline{x},t)\big] + (C_\epsilon v^\epsilon)(\underline{x},t) \tag{10.41}$$

in general, however, we only obtain the bound $c(\epsilon^{-2}t)^{-\delta n}$, with $\delta > 0$ independent of β, which is however sufficient for proving Theorem 9.2.1.

In (10.41), $(A_\epsilon v^\epsilon)$ is given by (10.6) with m^ϵ replacing ρ. $(C_\epsilon v^\epsilon)$, which arises exclusively from the Glauber generator, shorthands the following expression

$$(C_\epsilon v^\epsilon)(\underline{x},t) = \sum_{x,\underline{\xi},\underline{\xi}'} c_\epsilon(x,\underline{\xi},\underline{\xi}',\underline{x},t)v^\epsilon(\underline{x}',t), \quad \underline{x}' = \underline{x}'(x,\underline{\xi},\underline{\xi}',\underline{x}) \tag{10.42}$$

All configurations are unlabelled, labels will be superimposed later; $\underline{\xi}$ and $\underline{\xi}'$ are configurations in $\{0,1\}^\Lambda$, where Λ is a finite set in $\mathbf{Z}$. The functions c_ϵ vanish unless:

$$(\underline{x} - x) \cap \Lambda = \underline{\xi}, \qquad \text{where } (\underline{x} - x)_i = x_i - x \tag{10.43}$$

In such a case

$$\underline{x}'(x,\underline{\xi},\underline{\xi}',\underline{x}) = (\underline{\xi}' + x) \cup (\underline{x} \cap (x+\Lambda)^c) \tag{10.44}$$

In words, the transformations $\underline{x} \to \underline{x}'$ which are allowed are those where $\underline{x}$ and $\underline{x}'$ in $x+\Lambda$ is like $\underline{\xi}$, respectively $\underline{\xi}'$, in Λ; while they are equal outside of $(x+\Lambda)$, i.e. in $(x+\Lambda)^c$.

Our proof relies on a further property of the functions c_ϵ:

10.2.2 Assumptions on c_ϵ.

There is c so that

$$\sup_{x,\underline{\xi},\underline{\xi}',\underline{x},t} |c_\epsilon(x,\underline{\xi},\underline{\xi}',\underline{x},t)| \le c \tag{10.45a}$$

and

$$c_\epsilon(x,\emptyset,\underline{\xi}',\underline{x}) = 0, \qquad c_\epsilon(x,\underline{\xi},\emptyset,\underline{x}) = 0 \text{ if } |\underline{\xi}| = 1 \tag{10.45b}$$

where $|\underline{\xi}|$ is the number of particles in $\underline{\xi}$.

Some simple computations which are omitted, allow to prove the following lemma:

10.2.3 Lemma. *If L_G is given by (9.6), then (10.45) holds, and Λ consists of an interval of three consecutive sites.*

We study by iteration the integral version of (10.43). As in §10.1, we interpret it in terms of a branching process: due to the presence of $(C_\epsilon v^\epsilon)$, this has a more complex structure with births and not only deaths. The Definition 10.1.4, to which the reader is addressed, is then modified as follows.

10.2.4 Definition: The interacting branching.

(1) $\mathcal{L}$ is now the collection of all the finite subsets of $\mathbb{N}_+$ and Ω the set of all the labelled configurations with finitely many particles. Γ is still a collection of subset in Ω^2, but its elements, denoted by $\gamma = (\underline{\xi},\underline{\xi}')$, range in a larger set, in order to describe both the stirring and the Glauber interactions.

(2) For any integer m

$$\mathcal{X}_m \subset D([0,t],\Omega) \times [0,t]^m \times \Gamma^m \times \{0,1,2,3,4\}^m \tag{10.46}$$

In $\mathcal{X}_m$ $\underline{\delta} = (\delta_1,\dots,\delta_m) \in \{0,1,2,3,4\}^m$ and $\underline{\gamma} = (\gamma_1,\dots,\gamma_m) \in \Gamma^m$ are related: if $\delta_i = 0,1$, then $\gamma_i \equiv (\underline{\xi}^{(i)},\underline{\xi}'^{(i)})$ is like in the stirring case. If $\delta_i = 2$, then $|\underline{\xi}^{(i)}| = |\underline{\xi}'^{(i)}|$; if $\delta_i = 3$,

$|\underline{\xi}'^{(i)}| > |\underline{\xi}^{(i)}|$; if $\delta_i = 4$, $|\underline{\xi}'^{(i)}| < |\underline{\xi}^{(i)}|$. Furthermore $\underline{\xi}^{(i)}$ and $\underline{\xi}'^{(i)}$ fulfill the condition in (10.45b): $\underline{\xi}^{(i)} \neq \emptyset$ and if $|\underline{\xi}^{(i)}| = 1$ then $\underline{\xi}'^{(i)} \neq \emptyset$.

The sequences in $\mathcal{X}_m$ have also constraints concerning the particles labels. Let L_i be the set of the labels of the particles at time t_i^+, setting $L_0 = \{1, \ldots, n\}$. Then the labels are kept in the time intervals (t_{i-1}, t_i) and may only change at the "interacting times" t_i: the relation between L_{i+1} and L_i is given by $L_{i+1} = [L_i \backslash L(\underline{\xi}^{(i)})] \cup L(\underline{\xi}'^{(i)})$. Furthermore the labelling of the $\underline{\xi}$ is such that if a label disappears, it never appear again (the particle dies), while, for the new labels:

$$L(\underline{\xi}'^{(i)}) = L(\underline{\xi}^{(i)}) \cup \{N+1, .., N + |\underline{\xi}'^{(i)}| - |\underline{\xi}^{(i)}|\}, \text{ where } N = \max\{\ell \in L_j : j \leq i\} \qquad (10.47)$$

(3) The interacting time t_i is called a stirring or Glauber time if $\delta_i \leq 1$, or, respectively, if $\delta_i > 1$.

(4) $\mathcal{X}$ is defined as in (10.12b) and $\mathcal{P}^\epsilon$ is a measure on $\mathcal{X}$ supported by $\{\underline{x}(0) = \underline{x}\}$. [The condition that eventually the configuration becomes empty is now dropped, because there are births which could prevent this from happening]. An expression like (10.13) still holds, with the expectations having the same meaning, but the stirring process in the time intervals (t_i, t_{i+1}) has intensity ϵ^{-2}; B has a new expression, as we are going to say.

(5) The transitions which are allowed, $(B = 1)$, are those described by 10.1.4 if δ_i and δ_{i-1} have values in $\{0, 1\}$. If $\delta_i \in \{0, 1\}$ and $\delta_{i-1} > 1$, then we set $\underline{z} = \underline{x}(t_i^-)$ and apply (10.14c). For all the other cases the allowed transitions are only those for which there is x so that:

$$(\underline{x}(t_i^-) - x) \cap \Lambda = \underline{\xi}, \quad (\underline{x}(t_i^+) - x) \cap \Lambda = \underline{\xi}', \quad (\underline{x}(t_i^+) - x) \cap \Lambda^c = (\underline{x}(t_i^-) - x) \cap \Lambda^c \qquad (10.48)$$

(as labelled configurations). There is a restriction on the possible labelling of $\gamma = (\underline{\xi}, \underline{\xi}')$ which comes from imposing that there is no exchange when $\delta_i \geq 2$, more precisely that

$$\xi'_j \neq \xi_i, \quad \text{for all } i \neq j \in L(\underline{\xi}) \cap L(\underline{\xi}') \qquad (10.49)$$

With these notation, using (10.9), we have that for any $N \geq n/2$ there is c so that

$$\begin{aligned}
|v_n^\epsilon(\underline{x}, t)| \leq c \int_{\mathcal{X}_N \cap \{x(t_N^+) \neq \emptyset\}} & d\mathcal{P}^\epsilon \, [\prod_{\delta_i = 1} \epsilon^{-2} (\epsilon^{-2}(t - t_i))^{2b-1}] \\
& \times [\prod_{\delta_i = 0} \epsilon^{-2} (\epsilon^{-2}(t - t_i))^{b-1/2} 1(\tau_{a_i, d_i}(t_i) > t_{i+1})] \\
+ \sum_{m=n/2}^{N} c \int_{\mathcal{X}_m \cap \{x(t_m^+) = \emptyset\}} & d\mathcal{P}^\epsilon \, [\prod_{\delta_i = 1} \epsilon^{-2} (\epsilon^{-2}(t - t_i))^{2b-1}] \\
& \times [\prod_{\delta_i = 0} \epsilon^{-2} (\epsilon^{-2}(t - t_i))^{b-1/2} 1(\tau_{a_i, d_i}(t_i) > t_{i+1})])
\end{aligned} \qquad (10.50)$$

The first term, called the remainder term, is new, it is due to the births originating from the Glauber interactions. In such a term there was originally a v-function computed at time $(t - t_N^+)$, but we have bounded it by 1, so that it does not appear in the final expression. By using the assumption that $t \le \epsilon^\beta$ it will be easy to show that such a remainder term is suitably small, if N is large.

As in §10.1, our next step is to rewrite the integrals in (10.50) in terms of an independent process.

10.2.5 Definition: The independent branching.

The skeleton π of an element of $\mathcal{X}$ is $(\underline{L}, \underline{\gamma}, \underline{\delta})$, see 10.2.4. The independent process $\underline{x}^0(\cdot)$, whose law will be denoted by $\mathcal{P}^{0\epsilon}_{\pi,\underline{t}}$, is defined in the intervals (t_i, t_{i+1}) as the process of independent symmetric random walks with jump intensity ϵ^{-2} and initial condition $\underline{x}^0(t_i) = 0$, just as in 10.1.6. Also $\underline{z}^0(\cdot)$ and $\hat{\underline{x}}(\cdot)$ are defined as in 10.1.6, but we need to specify how to construct $\hat{\underline{x}}(t_i^+)$ from $\hat{\underline{x}}(t_i^-)$.

Same rules as in item (3) of 10.1.6 apply if t_i is a stirring time. If instead $\delta_i = 2, 3$ we define for $j \in L(\underline{\xi})$, $b_j = \xi'_j - \xi_j$, $\gamma_i = (\underline{\xi}, \underline{\xi}')$. We then displace successively, for increasing j, $\hat{x}_j(t_i^-)$ by b_j, with the priority rule defined in Chapter VI: a displacement is allowed if and only if it takes a particle to either an empty site or to one occupied by a particle with higher label (lower priority). In such a case the particle with higher label moves back to where the particle with higher label was.

We leave the positions of the other particles in $L(\underline{\xi}(t_i^-))$ unchanged and we call $\underline{z}$, $L(\underline{z}) = L(\hat{\underline{x}}(t_i^-))$, the configuration obtained in this way. We then set $\hat{x}_\ell(t_i^+) = z_\ell$, for all $\ell \in L(\hat{\underline{x}}(t_i^-))$. This completes the definition of $\hat{\underline{x}}(t_i^+)$ if $\delta_i = 2$. If $\delta_i = 3$, let $b_j = \xi'_j - \xi'_\ell$, $j = L(\underline{\xi}')\backslash L(\underline{\xi})$, $\ell = \min\{h \in L(\underline{\xi})\}$. Then starting from the smallest j we set $\hat{x}_j(t_i^+) = b_j + \hat{x}_\ell(t_i^+)$, if none of the particles of $\hat{\underline{x}}(t_i^+)$ already defined is there, otherwise we set $\hat{x}_j(t_i^+)$ equal to the first empty site to the right of $b_j + \hat{x}_\ell(t_i^+)$. Proceeding iteratively, we then define the whole $\hat{\underline{x}}(t_i^+)$.
For $\delta_i = 4$, we drop the particles in $\hat{\underline{x}}(t_i^-)$ with labels in $L(\underline{\xi})\backslash L(\underline{\xi}')$ and then proceed as for $\delta_i = 2$.

Finally $\underline{z}^0(t_i^+)$ is defined in terms of $\underline{z}^0(t_i^-)$ and π, but, as a difference from the stirring case, also in terms of $\hat{\underline{x}}(t_i^+)$. In fact we set: $z_j^0(t_i^+) = z_j^0(t_i^-)$ for all $j \in L_{i-1} \cap L_i$ and $z_j^0(t_i^+) = \hat{x}_j(t_i^+)$ for all $j \in L_i \backslash L_{i-1}$.

Given a skeleton π and $\underline{t}$, for $1 \le i \le m(\pi)$ we define

$$\chi_i = 1\left(|\hat{x}_\ell(t_i^-) - \hat{x}_h(t_i^-)| \le \text{diam}\,\Lambda;\ \text{for all } h, \ell \in L(\underline{\xi}^{(i)})\right) \tag{10.51a}$$

Then, from (10.50),

$$|v_n^\epsilon(\underline{x}, t)| \le c \sum_{m(\pi) \le N} w_\pi^\epsilon(\underline{x}, t) \tag{10.51b}$$

where, writing $m = m(\pi)$,

$$w_\pi^\epsilon(\underline{x}, t) = \int_0^t dt_1 \ldots \int_{t_{m-1}}^t dt_m [\prod_{\delta_i=1} \epsilon^{-2}(\epsilon^{-2}(t-t_i))^{2b-1}][\prod_{\delta_i=0} \epsilon^{-2}(\epsilon^{-2}(t-t_i))^{b-1/2}]$$
$$\times \mathcal{E}^{0,\epsilon}_{\pi,\underline{t}}\left([\prod_{i=1}^N \chi_i][\prod_{i:\delta_i=0} 1(\tau^0_{a_i,d_i}(t_i) > t_i + \Delta_i)]\right) \tag{10.51c}$$

where τ^0 is defined in (10.17) and Δ_i in (10.22).

We call

$$R_\epsilon(\underline{x},t)=\sum_{\substack{m(\pi)=N\\ L_N(\pi)\neq\emptyset}} w^\epsilon_\pi(\underline{x},t) \tag{10.51d}$$

and, using (10.29), we have

$$R_\epsilon(\underline{x},t)\le c\sum_{\substack{m(\pi)=N\\ L_N(\pi)\neq\emptyset}}\int_0^t dt_1\cdots\int_{t_{N-1}}^t dt_N\Big[\prod_{\delta_i=1}\frac{\epsilon^{-4b}}{(t-t_i)^{1-2b}}\Big]$$
$$\times\Big[\prod_{\delta_i=0}\frac{\epsilon^{-2b}}{(t-t_i)^{1/2-b}(t_{i+1}-t_i)^{1/2}}\Big]$$

We bound the right hand side by replacing $(t-t_i)$ by $(t_{i+1}-t_i)$, so that, if $\ell=\ell(\pi)$ is the number of Glauber times in π, we get

$$R_\epsilon(\underline{x},t)\le c\sum_{\substack{m(\pi)=N\\ L_N(\pi)\neq\emptyset}}\epsilon^{-4b(N-\ell)}t^\ell$$

We multiply and divide by $\epsilon^{2n/8}$, then, since $\epsilon^{-1}\le t^{-1/\beta}$,

$$R_\epsilon(\underline{x},t)\le c\sum_{\substack{m(\pi)=N\\ L_N(\pi)\neq\emptyset}}\epsilon^{2n/8}t^{\ell-[n/4+4b(N-\ell)]/\beta}$$

We have that $n+\ell(|\Lambda|-1)>N-\ell$, because (1) at the Glauber interactions there are at most $|\Lambda|-1$ new particles, by (10.44) and (10.45); (2) at the stirring interactions at least one particle dies and (3) $L_N(\pi)\neq\emptyset$. We choose N so that

$$\ell>\frac{N-n}{|\Lambda|}\ge\frac{n}{8}+1+\frac{n}{4\beta}$$

and given such a N we take b so small that $4bN<\beta$, so that R_ϵ is bounded by $c(\epsilon^{-2}t)^{n/8}$.

Next we bound the generic $w^\epsilon_\pi(\underline{x},t)$ with $m(\pi)\equiv m\le N$ and $L_m(\pi)=\emptyset$. For reasons which will become clear below it is not convenient to use Proposition 6.6.3 to replace $\hat{\underline{x}}$ by $\underline{z}^0$, as in §10.1, if in π there are births. We shall rather use an iterative procedure, whose first step is reported below.

10.2.6 First step for bounding $w^\epsilon_\pi(\underline{x},t)$.

We call k the largest label in π, assuming that $k>n$. Let $s_k\equiv t_{i(k)}$ and $s'_k\equiv t_{i'(k)}$ be the times when k is born and when it dies. Denoting by $\Sigma(\cdot)$ the σ-algebra generated by $(\cdot)$, and by $\mathcal{A}\vee\mathcal{B}$ the minimal σ-algebra which contains $\mathcal{A}$ and $\mathcal{B}$, we define

$$\mathcal{H}_k=\Sigma(\{\underline{x}^0(s),\ s\le s_k\})\vee\Sigma(\{x^0_j(s),\ 0\le s\le t,\ j<k\})\vee\Sigma(\{\underline{x}^0(s),\ s>s'_k,\ j\ge k\}) \tag{10.52}$$

(the definition applies also when k is not the largest label). Let finally $\chi'_{i'(k)}$ be the characteristic function defined by (10.51a), with h and ℓ different from k. [In particular if $|\underline{\xi}^{(i'(k))}|=2$, $\chi'_{i'(k)}\equiv 1$].

We are now ready for bounding (10.51c). Suppose first that $\epsilon^{-2}(s'_k - s_k) \leq (\epsilon^{-2}t)^d$, where, as in §10.1, $d > 0$ is a suitably small parameter whose value will be determined below. Then, in this case, we simply replace $\chi_{i'(k)}$ by $\chi'_{i'(k)}$. If, on the other hand, $\epsilon^{-2}(s'_k - s_k) > (\epsilon^{-2}t)^d$, we use Proposition 6.6.3 and, recalling the definition 10.1.9 of $\mathcal{G}$, we have that for any M there is c so that the expectation in (10.51c) is bounded by

$$c(\epsilon^{-2}t)^M + c\mathcal{E}^{0,\epsilon}_{\pi,\underline{t}}\Big([\chi'_{i'(k)} \prod_{i\neq i'(k)} \chi_i][\prod_{i:\delta_i=0} 1(\tau^0_{a_i,d_i}(t_i) > t_i + \Delta_i)]$$
$$\times \mathcal{P}^{0,\epsilon}_{\pi,\underline{t}}(\{|z^0_k(s'^-_k) - \hat{x}_j(s'^-_k)| \leq u^\epsilon_k|\mathcal{G} \vee \mathcal{H}_k)\Big) \tag{10.53a}$$

where j in (10.53a) is the lowest label in the configuration $\underline{\xi}^{(i'(k))}$ and

$$u^\epsilon_k = (\epsilon^{-2}|s'_k - s_k|)^{1/4+a}$$

To derive (10.53a) we have used that the conditioning fixes the values of all the $\hat{x}_\ell(\cdot)$ with $\ell < k$. In fact, by the last statement in Proposition 6.6.3, a stirring particle at a time s and with a label j only depends on the σ-algebra generated by all $x^0_h(s')$ with $s' \leq s$, h having priority on j. Since k is the largest label, $\hat{x}_j$ does not depend on z^0_k.

Notice however that a particle with higher label may have priority, if it is one of the two particles involved in a stirring interaction with $\delta = 0$, but this does not apply to the present case because at the times t_i with $i < i'(k)$ the particle k cannot be involved in an interaction with deaths, because, being the largest label, by our convention, it would die. One should also take care of the displacements which occur at the interaction times, but the way we have defined them is such that they respect the priority of lower labelling, see 10.2.5.

Since $\hat{x}_j(s'^-_k)$ is fixed by the conditioning, the probability on the right hand side of (10.53a) has the form (10.27), therefore, using (10.31), we have that for all values of $(s'_k - s_k)$

$$w^\epsilon_\pi(\underline{x}, t) \leq c(\epsilon^{-2}t)^M + c(\epsilon^{-2}t)^{d(1/4-a)} \int_0^t dt_1 \cdots \int_{t_{N-1}}^t dt_N [\prod_{\delta_i=1} \epsilon^{-2}(\epsilon^{-2}(t-t_i))^{2b-1}]$$
$$\times [\prod_{\delta_i=0} \epsilon^{-2}(\epsilon^{-2}(t-t_i))^{b-1/2}][(\epsilon^{-2}(s'_k - s_k)]^{-1/4+a}$$
$$\times \mathcal{E}^{0,\epsilon}_{\pi,\underline{t}}\Big([\chi'_{i(k)} \prod_{i\neq i(k)} \chi_i][\prod_{i:\delta_i=0} 1(\tau^0_{a_i,d_i}(t_i) > t_i + \Delta_i)]\Big) \tag{10.53b}$$

which completes this first step.

Remarks. Observe that if we had used Proposition 6.6.3 to replace all $\hat{x}_h$ by z^0_h we would have had u^ϵ_k in (10.53a) replaced by

$$[(\epsilon^{-2}(s'_k - s_k)]^{1/4+a} + [(\epsilon^{-2}(s'_k - s_j)]^{1/4+a}$$

where s_j is the time when j is born, $s_j = 0$ if $j \leq n$. The bound for the corresponding probability would then be the above expression divided by $[(\epsilon^{-2}(s'_k - s_k)]^{1/2}$, which is not necessarily small, if the particle k dies "too early".

We shall next apply the above procedure to the other labels which are larger than n. We first drop in (10.53b) the characteristic functions χ_i corresponding to $\delta_i = 2, 3$ and call $\hat{\chi}_i = \chi_i$ if $\delta_i \neq 2, 3$ and $\hat{\chi}_i \equiv 1$ otherwise. In this way the particle with the largest label is no longer present in any of the $\hat{\chi}_i$; it might still appear in the characteristic function of the event $\{\tau^0_{a_i,d_i} > t_i + \Delta_i\}$, with $i = i'(k)$, but this is a constant when we condition on $\mathcal{G}$.

Even so, we cannot repeat exactly what done in 10.2.6 for the next largest label, call it $k-1$. In fact it might be that $\delta_{i'(k)} = 0$ ($i'(k)$ as in 10.2.6) and that $k - 1 = a_{i'(k)}$, so that the particle $k-1$ survives at $t_{i'(k)}$ and in the time interval $[t_{i'(k)}, t_{i'(k)+1}]$ it has top priority. Then if j is the lowest label in $L(\underline{\xi}^{(i'(k-1))})$, i.e. when particle $k-1$ dies, it is no longer true that $\hat{x}_j(s'_{k-1})$ is fixed by conditioning on $\mathcal{G} \vee H_{k-1}$. This problem can be avoided by changing the definition of $\mathcal{H}_h$ when h is not the largest label, but we shall then obtain a weaker bound than that proven in 10.2.6 for the largest label. We set:

10.2.7 Notation.

Let π and $\underline{t}$ be fixed, assume that the largest label in $\underline{L}(\pi)$ is larger than n. Then:

(1) Given any label k in $\underline{L}(\pi)$, we call s'_k the time $t_{i'(k)}$ when k dies. If $k > n$, s_k denotes the time when k is born if for no j: $\delta_j = 0$ and $k = a_j$. If, on the other hand, there is j: $\delta_j = 0$ with $k = a_j$, and j is the largest index for which this holds, then $s_k = t_{j+1}$. With this notation, the definition of $\mathcal{H}_k$ given in (10.52) extends to all $k > n$.

(2) We denote by K, $[J]$, the set of labels $k > n$, $[k \le n]$, such that there is $j < k$ in $L(\underline{\xi}^{(i'(k))})$.

We can now apply the same arguments as in 10.2.6 to the particle with next largest label, say $h > n$. The analysis is the same because of our definition of the time s_h, which makes still valid the measurability condition used in 10.2.6. In fact, before s_h everything is fixed by the conditioning on $\mathcal{H}_h$, while, after s_h, the particle h dies at the first time, s'_h, when it appears in one of the characteristic functions $\hat{\chi}_i$. Recall that also the displacements at the interacting times obey the rule which gives priority to lower labels.

So we bound, one after the other, all the contributions coming from the labels $j > n$ and after that we can proceed as in §10.1, by using first Proposition 6.6.3 to replace the remaining $\hat{x}_j$ by the z^0_j and then using (10.28) and (10.29). At the end we get

$$w^\epsilon_\pi(\underline{x}, t) \le c(\epsilon^{-2}t)^M + c(\epsilon^{-2}t)^{d(1/4-a)m} \int_0^t dt_1 \cdots \int_{t_{N-1}}^t dt_N [\prod_{\delta_i=1} \epsilon^{-2}(\epsilon^{-2}(t-t_i))^{2b-1}]$$
$$\times [\prod_{\delta_i=0} \epsilon^{-2}(\epsilon^{-2}(t-t_i))^{b-1}][\prod_{k\in K}\{1+\epsilon^{-2}(s'_k - s_k)\}^{-1/4+a}][\prod_{j\in J}\{1+\epsilon^{-2}s'_j\}^{-1/4+a}] \tag{10.54a}$$

To bound the integral in (10.54a) we need a few more notation. We call

$$\mathcal{B} = \{i = i(k), k \in K, \text{ such that } s_k \text{ is the time when } k \text{ is born }\}$$

and n_i the number of $k \in K$ with $i(k) = i$ and s_k as above, if such number is ≤ 3; otherwise we set n_i equal to 3. [In our case $|\Lambda| = 3$ so that the last alternative is not present, but it might occur for general Glauber interactions]. We then have

$$\prod_{k\in K}\{1+\epsilon^{-2}(s'_k - s_k)\}^{-1/4+a}\} \le \prod_{i\in\mathcal{B}} \left(\epsilon^{-2}(t_{i+1}-t_i)\right)^{(-1/4+a)n_i}$$

We start by considering the smallest $i \in \mathcal{B}$ and enlarge the range of integration to the whole $t_i \leq t_{i+1}$. We then integrate over t_i, keeping all the other times fixed, so that the only term which varies is

$$(\epsilon^{-2}(t_{i+1}-t_i))^{(-1/4+a)n_i}$$

whose integral is bounded by

$$\frac{1}{1-(-1/4+a)n_i}t\,(\epsilon^{-2}t)^{(-1/4+a)n_i}$$

because $n_i \leq 3$ and a is chosen small enough. The dependence on t_i is thus lost and we can proceed similarly and integrate all the t_j with $j \in \mathcal{B}$.

We then call G_4, $[G_0, G_1]$, the set of indices i such that $\{t_i = s'_k,\ k \in J;$ and $\delta_i = 4\}$, [respectively $\delta_i = 0, 1$], and by m_i either the number of such k's which give rise to the same i or, if such number is larger than 3, we set m_i equal to 3. By enlarging the range of integration of t_i, $i \in G_4$, to $\{t_i \leq t\}$, we can perform the integral over all such times. We are then left with one of the expressions that have been considered in §10.1, so that, in conclusion we have proven that

$$\begin{aligned} w^{\epsilon}_{\pi}(\underline{x},t) \leq c(\epsilon^{-2}t)^M &+ c(\epsilon^{-2}t)^{d(1/4-a)m}(\epsilon^{-2}t)^{2bp+bq}[\prod_{i\in\mathcal{B}} t(\epsilon^{-2}t)^{(-1/4+a)n_i}] \\ &\times [\prod_{i\in G_4} t(\epsilon^{-2}t)^{(-1/4+a)m_i}][\prod_{i\in G_0\cup G_1}(\epsilon^{-2}t)^{(-1/4+a)m_i}] \end{aligned} \tag{10.54b}$$

where, as in (10.33), p and q are the number of i such that $\delta_i = 1$, respectvely 0.

Call ℓ the number of indices $j \leq n$ such that when the particle j dies, the other particles which "interact" with j have all labels larger than n. Then

$$\sum_i m_i \geq \frac{1}{2}(n-\ell) \tag{10.54c}$$

and the inequality is strict if some $m_i > 1$. After some thought, one can convince himself that $|\mathcal{B}| \geq \ell$, so that if $\ell > 0$ and/or G_0 is non empty, we obtain the desired estimate (10.40), after choosing a, b and d small enough. If G_4 is non empty we use (10.54c) and we exploit the presence of the factor $t \leq \epsilon^{\beta}$ to prove the bound (10.40). Therefore (10.54b) gives (10.40) unless π is such that $\delta_i = 1$ for all i, but this is the case considered in 10.1.12, so that the proof of the Theorem 10.2.1 is completed. □

§10.3 Proof of Theorem 9.2.1.

In this paragraph we shall prove that if we suitably weaken the bound (10.40), we can extend it to longer times, in this way we shall prove Theorem 9.2.1.

We turn back to finite volumes, i.e. to the lattice $\mathbb{Z}_\epsilon = [0, \epsilon^{-1}]$ (ϵ^{-1} being a positive integer) with periodic boundary conditions. The proof is essentially the same in $\mathbb{Z}_\epsilon = [0, \epsilon^{-k}]$, with any $k > 1$. We shall proceed as in Chapter V, §5.5.

10.3.1 Definition: the seminorm $\|\cdot\|$.

For any function f on $\mathbb{Z}_\epsilon$ we define

$$\|f\| = \sup_x |\sum_y P^{\epsilon}_{\epsilon^{1/4}}(x \to y) f(y)| \tag{10.55}$$

here P_t^ϵ is the transition probability for a symmetric random walk on $[0, \epsilon^{-1}]$ which jumps on the nearest neighbor sites with intensity ϵ^{-2}. For more general Glauber interaction where we have a weaker bound than that in (10.40), we should change $\epsilon^{1/4}$ into a suitably smaller time.

For any given positive ϵ, $\beta^\star$, $r \in [1,2]$ and a we define

$$\mathcal{N}_\epsilon \equiv \{k \in \mathbb{N} : kr \leq \epsilon^{-\beta^\star} a \log \epsilon^{-1}\} \tag{10.56a}$$

and

$$t(k) = kr\epsilon^{-\beta^\star} \qquad\qquad k \in \mathcal{N}_\epsilon \tag{10.56b}$$

Denoting by $\sigma_{t(k)}$ a configuration at time $t(k)$, we introduce the *Good set of trajectories* H^ϵ: given any $\zeta > 0$, we define

$$H^\epsilon = \{\sigma^{(k)} \equiv \sigma_{t(k)},\ k \in \mathcal{N}_\epsilon : \sup_{k \in \mathcal{N}_\epsilon} \|\sigma^{(k)} - m^\epsilon(\cdot, r\epsilon^{\beta^\star} | \sigma^{(k-1)})\| \leq \epsilon^\zeta\} \tag{10.56c}$$

where, as usual, $m^\epsilon(\cdot, t|\sigma)$ denotes the solution to (9.10) with initial condition σ.

We first prove that H^ϵ has large probability:

10.3.2 Lemma. *For any $\zeta < 3/16$ and $\beta^\star < 1/4 - \zeta$ positive, there exists a sequence c_n such that for all $a > 0$, (H^ϵ depends on a)*

$$\mathbb{P}^\epsilon_{\sigma^{(0)}}(H^\epsilon) \geq 1 - c_n \epsilon^n \tag{10.57a}$$

where $\mathbb{P}^\epsilon_{\sigma^{(0)}}$ denotes the law of the Glauber + Kawasaki process starting from $\sigma^{(0)}$.

Proof. We first prove a bound on $m^\epsilon(x, t|\sigma)$ for $t \leq \epsilon^{\beta^\star}$, uniform on σ, which will be used in several instances in the sequel.

The bound is obtained recalling the equation for m^ϵ, which we report in integral form in (10.66) below. Denoting by $\|m_t^\epsilon\|_\infty$ the sup norm of $|m^\epsilon(x, t|\sigma)|$, we get

$$\|m_t^\epsilon\|_\infty \leq 1 + \int_0^t ds \big(\alpha \|m_s^\epsilon\|_\infty + \beta \|m_s^\epsilon\|_\infty^3\big)$$

Let

$$T = \max\Big(\sup\{t \leq \epsilon^{\beta^\star} : \|m_t^\epsilon\|_\infty \leq 2\}, \epsilon^{\beta^\star}\Big)$$

Then for $t \leq T$ and ϵ small enough,

$$\|m_t^\epsilon\|_\infty \leq 1 + \int_0^t ds (\alpha + 4\beta) \|m_s^\epsilon\| \leq e^{t(\alpha+4\beta)} \leq 2 \tag{10.57b}$$

which shows that $T = \epsilon^{\beta^\star}$, hence that (10.57b) holds for all $t \leq \epsilon^{\beta^\star}$.

We now come back to the proof of (10.57a). We have

$$\begin{aligned}\mathbb{P}^\epsilon_{\sigma^{(0)}}(H^\epsilon) &\geq 1 - \epsilon^{-\beta^\star} a \log \epsilon^{-1} \sup_\sigma \mathbb{P}^\epsilon_\sigma (\|\sigma^{(1)} - m^\epsilon(\cdot, t(1)|\sigma^{(0)})\| > \epsilon^\zeta) \\ &\geq 1 - \epsilon^{-\beta^\star} a (\log \epsilon^{-1}) \epsilon^{-1} \sup_\sigma \mathbb{P}^\epsilon_\sigma (|\sum_y P^\epsilon_{\epsilon^{1/4}}(0 \to y)[\sigma(y, t(1)) - m^\epsilon(y, t(1)|\sigma)]| > \epsilon^\zeta)\end{aligned} \tag{10.58}$$

Using the Chebyshev inequality with power $2h$ we then get

$$\mathbb{P}^{\epsilon}_{\sigma}\Big(\Big|\sum_{y} P^{\epsilon}_{\epsilon^{1/4}}(0\to y)[\sigma(y,t(1)) - m^{\epsilon}(y,t(1)|\sigma)]\Big| > \epsilon^{\zeta}\Big)$$

$$\le \epsilon^{-2h\zeta} \sum_{y_1,\dots,y_{2h}} [\prod_{i=1}^{2h} P^{\epsilon}_{\epsilon^{1/4}}(0\to y_i)][\mathbb{E}^{\epsilon}_{\sigma}(\prod_{i=1}^{2h}\{\sigma(y_i,t(1)) - m^{\epsilon}(y_i,t(1)|\sigma)\})] \tag{10.59}$$

By Theorem 10.2.1,

$$\sum_{y_1\neq\cdots\neq y_{2h}} [\prod_{i=1}^{2h} P^{\epsilon}_{\epsilon^{\frac{1}{4}}}(0\to y_i)][\mathbb{E}^{\epsilon}_{\sigma}(\prod_{i=1}^{2h}(\sigma(y_i,t(1)) - m^{\epsilon}(y_i,t(1)|\sigma))]$$

$$\le c \sup_{\underline{y}\in\mathcal{M}_{2h}} |v^{\epsilon}_{2h}(\underline{y},t(1)|\sigma)| \le c\,(\epsilon^{-2+\beta^*})^{-2h/8} \tag{10.60a}$$

The term on the right hand side of (10.59) with the y_i's pairwise equal is bounded by (10.60b) below, by using (10.57b) and the local central limit theorem to estimate the probability that two random walks are at same site after a time $\epsilon^{1/4}$, if they jump with intensity ϵ^{-2}:

$$\sum_{y_1\neq\cdots\neq y_h} [\prod_{i=1}^{h} P^{\epsilon}_{\epsilon^{1/4}}(0\to y_i)^2][\mathbb{E}^{\epsilon}_{\sigma}\Big(\prod_{i=1}^{h}(\sigma(y_i,t(1)) - m^{\epsilon}(y_i,t(1)|\sigma))^2\Big)] \le 3^{2h}(\epsilon^{-2+1/4})^{-h/2} \tag{10.60b}$$

From (10.59) and (10.60) it follows that there is c so that

$$\mathbb{P}^{\epsilon}_{\sigma}\Big(\Big|\sum_{y} P^{\epsilon}_{\epsilon^{1/4}}(0\to y)[\sigma(y,t(1)) - m^{\epsilon}(y,t(1)|\sigma)]\Big| > \epsilon^{\zeta}\Big)$$

$$\le c\epsilon^{-2h\zeta}\max\{(\epsilon^{-2+1/4})^{-h/2},(\epsilon^{-2+\beta^*})^{-2h/8}\} \tag{10.61}$$

Since all the other cases which arise when we expand the product in (10.54) lead to terms bounded as in (10.61), we omit the details.

By choosing ζ so small that

$$-2\zeta - \frac{1}{2}(-2+1/4) > 0$$

we make infinitesimal the contribution arising from the first term in the max. Given such a ζ which is also chosen smaller than 1/4, we take β^* so small that

$$-\zeta - (-2+\beta^*)/8 = \frac{1}{4} - (\zeta+\beta^*) > 0$$

Hence also the second term in the max is infinitesimal. The right hand side of (10.61) then vanishes as fast as any given power of ϵ if h is large enough and from this and (10.58) Lemma 10.3.2 follows. Notice that the right hand side of (10.61) can be made smaller than any positive power of ϵ so that the extension of this lemma to processes defined in interval of size ϵ^{-k} is trivial. □

Next we prove the following

10.3.3 Lemma. *Let $\beta^* < \zeta$ be as in Lemma 10.3.2. Let $m(r,t|\sigma^{(0)})$ be the solution to (9.1) with initial condition $\sigma^{(0)}$, i.e. $m(r,0|\sigma^{(0)}) = \sigma^{(0)}([\epsilon^{-1}r])$, $r \in [0,1]$, and periodic boundary condition. Then for a small enough there is $\delta' > 0$ such that for any sequence $(\sigma^{(k)})_{k\in\mathcal{N}_\epsilon} \in H^\epsilon$ and all $k \in \mathcal{N}_\epsilon$*

$$|m^\epsilon(x,t(1)|\sigma^{(k-1)}) - m(\epsilon x, t(k)|\sigma^{(0)})| \le c\epsilon^{\delta'} \tag{10.62}$$

Proof. The integral version of (9.1) is

$$\begin{aligned} m(r,t|\sigma^{(0)}) = &\int dr' G_t(r-r')\,\sigma^{(0)}([\epsilon^{-1}r'],0) \\ &+ \int_0^t ds \int dr' G_{t-s}(r-r')[\alpha m(r',s|\sigma^{(0)}) - \beta m(r',s|\sigma^{(0)})^3] \end{aligned} \tag{10.63}$$

where G_t, $[G_t^\infty]$, is the Green function for the Laplace operator in [0,1] with periodic conditions, [respectively in $\mathbb{R}$]. We have that for a suitable constant c

$$\int dx |G_t(r-x) - G_t(r'-x)| \le \int dx |G_t^\infty(r-x) - G_t^\infty(r'-x)| \le c\frac{|r-r'|}{\sqrt{t}}$$

Then, since m is bounded, we have that, uniformly on $\sigma^{(0)}$,

$$\sup_{|r-r'|\le\epsilon} |m(r,t|\sigma^{(0)}) - m(r',t|\sigma^{(0)})| \le c\frac{\epsilon}{\sqrt{t}} \tag{10.64a}$$

for a suitable value of c. We also have, by the local central limit theorem, see (4.34), that for $t \ge \epsilon^{1/4}$

$$\sum_{y\in\mathbb{Z}_\epsilon} \Big| \int_{y-1/2}^{y+1/2} \epsilon dz G_t(\epsilon(x-z)) - P_t^\epsilon(x \to y)\Big| \le c\epsilon/\sqrt{t} \tag{10.64b}$$

By (10.63), for $t(k) + \epsilon^{1/4} \le t \le t(k+1)$

$$\begin{aligned} &|m(\epsilon x,t|\sigma^{(0)}) - \sum_y P_{t-t(k)}^\epsilon(x \to y) m(\epsilon y, t(k)|\sigma^{(0)}) \\ &- \int_{t(k)+\epsilon^{1/4}}^t ds \sum_y P_{t-s}^\epsilon(x \to y)[\alpha m(\epsilon y,s|\sigma^{(0)}) - \beta m(\epsilon y,s|\sigma^{(0)})^3]| \le c\epsilon^{1/4} \end{aligned} \tag{10.65}$$

To compare m above with m^ϵ, it is convenient to rewrite (10.65) by replacing

(1) $m(\epsilon y,s|\sigma^{(0)})^3$ by $m(\epsilon(y-1),s|\sigma^{(0)})m(\epsilon y,s|\sigma^{(0)})m(\epsilon(y+1),s|\sigma^{(0)})$

(2) P_t^ϵ by $\hat{P}_t^\epsilon$, where $\hat{P}_t^\epsilon$ is the transition probability of a symmetric random walk having intensity $\epsilon^{-2} + 4\gamma$. It is easy to see that after these changes the same estimate (10.65) remains valid, maybe with a different constant c, we omit the details.

From (9.10b) it follows that

$$m^\epsilon(x,t|\sigma) = \sum_y \hat{P}_t^\epsilon(x \to y)\sigma(y) + \int_0^t ds \sum_y \hat{P}_{t-s}^\epsilon(x \to y)[\alpha m^\epsilon(y,s|\sigma) - \beta \prod_{b=-1}^{1} m^\epsilon(y+b,s|\sigma)] \tag{10.66}$$

For $t(k) < t \le t(k+1)$ we set

$$h^\epsilon(x,t) = |m(\epsilon x, t|\sigma^{(0)}) - m^\epsilon(x, t - t(k)|\sigma^{(k)})| \tag{10.67a}$$

$$h_t^\epsilon = \sup_{x \in \mathbf{Z}_\epsilon} h^\epsilon(x,t) \tag{10.67b}$$

Both $|m^\epsilon|$ and $|m|$ are uniformly bounded, the former by (10.57b), the latter by using the maximum principle for (9.1). We then have

$$\sup_{t(k)<t\le t(k+1)} h_t^\epsilon \le \bar{c} \tag{10.68}$$

Using (10.65), (10.66), (10.68) and the assumption on $(\sigma^{(k)})_{k\in\mathcal{N}_\epsilon}$, we have for all t: $t(k)+\epsilon^{1/4} \le t \le t(k+1)$

$$h_t^\epsilon \le c(\epsilon^{1/4} + \epsilon^\zeta) + h_{t(k)}^\epsilon + \int_{t(k)+\epsilon^{1/4}}^t ds\, c'\, h_s^\epsilon \tag{10.69}$$

We rewrite $h_{t(k)}^\epsilon$ in (10.69) by using again (10.69) and we keep doing this obtaining in the end an integral inequality which yields for all $k \in \mathcal{N}_\epsilon$ and for all $t \in (t(k), t(k+1)]$,

$$h_t^\epsilon \le e^{c't} c(\epsilon^{1/4} + \epsilon^\zeta) k \le e^{c'\, a\, \log \epsilon^{-1}} \epsilon^{-\beta^*} a\, \log \epsilon^{-1} \tag{10.70}$$

We now choose a and β^* so small that the right hand side of (10.70) is infinitesimal, thus h_t^ϵ is also infinitesimal for $t \le a \log \epsilon^{-1}$ and the proof of the Lemma is completed. [Since we have always used sup norms, the same proof applies to the case when the volume has size ϵ^{-k}]. □

10.3.4 Proof of Theorem 9.2.1.

Proof of (9.13). We define

$$\hat{h}^\epsilon(x,t) = \begin{cases} |m^\epsilon(x,t|\sigma^{(0)}) - m^\epsilon(x,t|\sigma^{(k)})| & \text{if } t(k)+\epsilon^{1/4} < t \le t(k+1) \\ 0 & \text{if } t(k) < t < t(k)+\epsilon^{1/4} \end{cases}$$

We also define

$$\hat{h}_t^\epsilon = \sup_{x\in\mathbf{Z}_\epsilon} \hat{h}^\epsilon(x,t)$$

and

$$t_\epsilon = \min\{a \log \epsilon^{-1}, \sup(t : \hat{h}_t^\epsilon \le 2), \sup(t : \sup_x |m^\epsilon(x,t|\sigma^{(0)})| \le 2)\}$$

By (10.57b) we then have for $t \le t_\epsilon$

$$\begin{aligned} \hat{h}^\epsilon(x,t) &\le c(\epsilon^{1/4} + \epsilon^\zeta) + \hat{h}_{t(k)}^\epsilon + \int_{t(k)}^t ds\, (\alpha + 28\beta)\hat{h}_s^\epsilon \\ &\le e^{(\alpha+28\beta)t} c(\epsilon^{1/4} + \epsilon^\zeta)\epsilon^{-\beta^*} a \log \epsilon^{-1} \end{aligned}$$

By choosing the parameter $\hat{a}$ in the statement of Theorem 9.2.1 small enough, we make the right hand side of the last inequality infinitesimal as $\epsilon \to 0$ for all $t \le \hat{a} \log \epsilon^{-1}$, which proves that $t_\epsilon = \hat{a} \log \epsilon^{-1}$. Using this and Lemma 10.3.3 we then prove (9.13). □

Proof of (9.12). Given a value of t in (9.12), we choose r in $[1,2]$ so that there is k^* which satisfies the relation $(k^*+1)r\epsilon^{\beta^*}=t$. We have

$$v_n^\epsilon(\underline{x},t|\sigma^{(0)}) = \mathbf{E}_{\sigma^{(0)}}^\epsilon\left(\mathbf{E}_{\sigma^{(k^*)}}^\epsilon\left(\prod_{i=1}^n[\sigma(x_i,r\epsilon^{\beta^*})-m^\epsilon(x_i,t|\sigma^{(0)})]\right)\right) \tag{10.71}$$

We consider the set H^ϵ as in (10.56) but with $k \leq k^*$. By Lemma 10.3.2 the probability of the complement of H^ϵ is smaller than any given power of ϵ. By (9.13), $m^\epsilon(x,t|\sigma^{(0)})$ is infinitesimally close to $m(\epsilon x,t|\sigma^{(0)})$, then its absolute value is uniformly bounded: therefore the contribution to (10.71) of configurations which are not in H^ϵ is compatible with the estimate (9.12). For the others we add and subtract $m^\epsilon(x_i,t|\sigma^{(k^*)})$. For the difference $m^\epsilon(x_i,t|\sigma^{(k^*)})-m^\epsilon(x_i,t|\sigma^{(0)})$ we use (9.13) and (10.62), see also 10.3.4. The other factors reconstruct a v-function for which we can use Theorem 10.2.1. The proof of Theorem 9.2.1 is thus completed. □

§10.4 Bibliographical notes.

The estimate on the v-functions in the stirring process has been proven in [62a], the short time estimate when also the Glauber interaction is present is proven in the Appendix of [22]. Theorem 9.2.1 in a volume of size ϵ^{-k} is used in [40], for the proof this Chapter X is quoted.

In the weakly asymmetric simple exclusion process the algebra of the v-functions is more complex, the case is studied in [43], where it is also proven that if the initial measure is smooth, as in Chapter VI, then the bound for the v-function can be improved: it is shown that v_{2n}^ϵ behaves as ϵ^n and such an estimate is used for studying the fluctuation fields.

In [86], estimates on the v-functions have been obtained and used to derive the Burgers equation, for the Boghosian Levermore model with weak asymmetry. The v-functions have also been used to study some of the stochastic models discussed in Chapter VII. Estimates for v-functions in the low density HPP automaton are proven in [35].

References

[1] N. Alikakos, P.W. Bates, G. Fusco, *Slow motion for the Chan-Hilliard equation in one space dimension*, J.Diff.Eqns. **90** (1991), 81–135.

S. Allen, J. Cahn, *A microscopic theory for antiphase boundary motion and its application to antiphase domain coarsening*, Acta Metall. **27** (1979), 1084–1095.

P.W. Bates, P.C. Fife, *Spectral comparison principles for the Cahn-Hilliard and phase-field equations and time scales for coarsening*, Physica D **43** (1990), 335–348.

L. Bronsard, R.V. Kohn, *Motion by mean curvature as the singular limit of Ginzburg Landau dynamics*, J.Diff.Eqns..

G. Caginalp, P.C. Fife, *Dynamics of layered interfaces arising from phase boundaries*, SIAM J. Appl. Math. **48** (1988), 506–518.

G. Fusco, *A geometric approach to the dynamics of $u_t = \epsilon^2 u_{xx} + f(u)$, for small ϵ*, Lecture Notes in Physics **359** (1990), 53–73.

G. Fusco, J.K. Hale, *Slow motion manifolds, dormant instability and singular perturbations*, Dynamics and Diff. Equations **1** (1989), 75–94.

P. Fife,J.B. McLeod, *The approach of solutions of nonlinear difusion equations to traveling front solutions*, Arch. Rat. Mech. Anal. **65** (1977), 335–361.

R.Pego, *Front migration in the non linear Cahn-Hilliard equation*, Proc.Royal Soc. London **A 422** (1989), 261–278.

J. Rubinstein, P. Sternberg, J. Keller, *Fast reaction, slow diffusion and curve shortening*, SIAM J. Appl. Math. **49** (1989), 116–133.

[2] E.D. Andjel, *A correlation inequality for the symmetric exclusion process*, Annals of Probability **16** (1988), 717–721.

[3] E. Andjel, M. Bramson, T.M. Liggett, *Shocks in the asymmetric exclusion process*, Th. Rel. Fields **78** (1988), 231–247.

[4] E. Andjel, C. Kipnis, *Derivation of the hydrodynamical equation for the zero range interaction process*, Annals of Probability **12** (1984), 325–334.

[5] E. Andjel, M.E. Vares, *Hydrodynamic equations for attractive particle systems on Z*, J. Stat. Phys. **47** (1987), 265–288.

[6] L. Arnold, *On the consistency of the mathematical models of chemical reactions*, Proc. Int. Symp. Synergetics, Bielefeld (H.Haken, eds.), Springer-Verlag, New York, 1980.

[7] L. Arnold, M. Theodosopulu, *Deterministic limit of the stochastic model of chemical reactions with diffusion*, Adv. Appl. Prob. **12** (1980), 367–379.

[8] D. Aronson, H. Weinberger, *Non linear diffusion in population genetics, combustion and nerve propagation*, Lectures Notes in Mathematics n.446, Springer-Verlag, New York, 1975, pp. 5–49.

[9] R. Arratia, *Symmetric exclusion processes: a comparison inequality and a large deviation result*, Annals of Probability **13** (1985), 53–61.

[10] A. Benassi, J.P. Fouque, *Hydrodynamic limit for the asymmetric simple exclusion process*, Annals of Probability **15** (1987), 546–560.

[11] A. Benassi, J.P. Fouque, E. Saada, M.E. Vares, *Asymmetric attractive particle systems on $\mathbb{Z}$: hydrodynamic limit for monotone initial profiles*, J. Stat. Phys. (to appear).

[12] F. Bertein, A. Galves, *Asymptotic behaviour of two simple random walks interacting by simple exclusion*, Z. Wahrsch. Verw. Gebiete **41** (1977), 73–85.

[13] P. Billingsley, *Convergence of probability measures*, Wiley, New York, 1968.

[14] B.M. Boghosian, C.D. Levermore, *A cellular automaton for Burgers equation*, Complex Systems **1** (1987), 17–30.

[15] C. Boldrighini, G. Cosimi, S. Frigio, M. Grasso-Nunes, *Computer simulations of shock waves in the completely asymmetric simple exclusion*, J. Stat. Phys. **55** (1989), 611–624.

[16] C. Boldrighini, A. De Masi, A. Pellegrinotti, *Non equilibrium fluctuations in systems modelling diffusion reaction equations*, Stochastic Processes and their Applications (to appear).

[17] C. Boldrighini, A. De Masi, A. Pellegrinotti, E. Presutti, *Collective phenomena in interacting particle systems*, Stochastic Processes and their Applications **25** (1987), 137–152.

[18] M. Bramsom, P. Calderoni, A. De Masi, P. Ferrari, J.L. Lebowitz, R. Schonmann, *Microscopic selection principle for Reaction Diffusion equation*, J. Stat. Phys. **45** (1986), 905–920.

[19] G. Broggi, L.A. Lugiato, A. Colombo, Phys. Rev. A **32** (1985), 2803-2812.

[20] R. Caflisch, *The fluid dynamic limit of the non linear Boltzmann equation*, Commun. Pure Appl. Math. **33** (1980), 651–666; *Fluid dynamics and the Boltzmann equation* (1983), 193–224, Studies in Statistical Mechanics, Vol.X. Non Equilibrium Phenomena I, The Boltzmann Equation (J.L. Lebowitz and E.W. Montroll publ North Holland, eds.), Amsterdam - New York.

C. Bardos, R. Caflisch, B. Nicolaenko, *Thermal layer solutions of the Boltzmann equation.*

[21] R.E. Caflisch, G. Papanicolaou, *The fluid dynamical limit of a non linear model Boltzmann equation*, Commun. Pure Appl. Math. **32** (1979), 589–616.

[22] P. Calderoni, A. Pellegrinotti, E. Presutti, M.E. Vares, *Transient bimodality in interacting particle systems*, J. Stat. Phys. **55** (1989), 523–578.

[23] P. Calderoni, M. Pulvirenti, *Propagation of chaos for Burger's equation*, Ann. Inst. H. Poincaré **39** (1983), 85–97.

[24] C. Cammarota, P.A. Ferrari, *Invariance principle for the edge of the branching exclusion process*, CARR reports in Math. Phys. **n.17/89** (1989).

[25] S. Caprino, A. De Masi, E. Presutti, M. Pulvirenti, *A stochastic particle system modelling the Carleman equation*, J. Stat. Phys. **55** (1989), 625–638; *Addendum*, J. Stat. Phys. **59** (1989), 535–537.

[26] ———, *A derivation of the Broadwell equation*, Commun. Math. Phys. **135** (1991), 443–465.

[27] M. Cassandro, A. Galves, E. Olivieri, M.E. Vares, *Metastable behaviour of stochastic dynamics: a pathwise approach*, J. Stat. Phys. **35** (1984), 603–634.

[28] C. Cercignani, *Theory and applications of the Boltzmann equation*, Scottish Academic Press, Edinburgh, 1975.

[29] P.A. Z. Cheng, J.L. Lebowitz, E.R. Speer, *Microscopic shock structure in model particle systems: the Boghosian Levermore revisited*, preprint (1990).

[30] F. Comets, *Nucleation for a long range magnetic model*, Ann. Inst. H. Poincaré **23** (1987), 135–178.

[31] F. Comets, Th. Eisele, *Asymptotic dynamics, noncritical and critical fluctuations for a geometric long range interacting model*, Commun. Math. Phys. **118** (1988), 531–568.

[32] R. Dal Passo, P. de Mottoni, *The heat equation with a nonlocal density dependent advection term*, preprint, 1991.

[33] A. De Masi, R. Esposito, J.L.Lebowitz, *Incompressible Navier-Stokes and Euler limits of the Boltzmann equation*, Commun. Pure Appl. Math. **42** (1989), 1189–1214.

[34] A. De Masi, R. Esposito, J.L. Lebowitz, E. Presutti, *Hydrodynamics of stochastic cellular automata*, Commun. Math. Phys. **125** (1989), 127–145.

[35] A. De Masi, R. Esposito, E. Presutti, *Kinetic limits of the HPP cellular automaton*, J. Stat. Phys. (1992).

[36] A. De Masi, P.A. Ferrari, J.L. Lebowitz, *Rigorous derivation of Reaction Diffusion equations with fluctuations*, Phys. Rev. Lett. **55** (1985), 1947–1949; *Reaction diffusion equations for interacting particle systems*, J. Stat. Phys. **44** (1986), 589–644.

[37] A. De Masi, C. Kipnis, E. Presutti, E. Saada, *Microscopic structure at the shock in the asymmetric simple exclusion*, Stochastics **27** (1989), 151–165.

[38] A. De Masi, N. Ianiro, A. Pellegrinotti, E. Presutti, *A survey of the hydrodynamical behaviour of many particle systems*, Studies in Statistical Mechanics, Vol.XI (E.W. Montroll and J.L.Lebowitz, eds.), North Holland, Amsterdam - New York, 1984, pp. 123–294.

[39] A. De Masi, N. Ianiro, E. Presutti, *Small deviations from local equilibrium for a process which exhibits hydrodynamical behaviour*, J. Stat. Phys. **29** (1982), 57–79.

[40] A. De Masi, A. Pellegrinotti, E. Presutti, M.E. Vares, *Spatial patterns when phases separate in an interacting particle system*, CARR reports in Math. Phys. n.1/91, 1991.

[41] A. De Masi, E. Presutti, *Probability estimates for simple exclusion random walks*, Ann. Inst. H. Poincaré A **19** (1983), 71–85.

[42] ———, *Lectures on the collective behavior of particle systems*, CARR preprint n.5/89 (1989).

[43] A. De Masi, E. Presutti, E. Scacciatelli, *The weakly asymmetric simple exclusion process*, Ann. Instit. H. Poincaré A **25** (1989), 1–38.

[44] A. De Masi, E. Presutti, M.E.Vares, *Escape from the unstable equilibrium in a random process with infinitely many interacting particles*, J. Stat. Phys. **44** (1986), 645–694.

[45] P. de Mottoni, M. Schatzman, *Geometrical evolution of developped interfaces*, preprint,, 1989; *Development of interfaces in n-dimensional space*, Proc. Royal Soc. Edinburgh (to appear).

[46] F. de Pasquale, P. Tartaglia, P. Tombesi, *Early stage domain formation and growth in one-dimensional systems*, Phys. Rev. A **31** (1985), 2447–2453.

[47] P. Dittrich, *A stochastic model of a chemical reaction with diffusion*, Prob. Theory Rel. Fields **79** (1988), 115–128.

[48] ———, *A stochastic particle system: fluctuations around a non linear reaction diffusion equation*, Stoch. Proc. and their Appl. **30** (1988), 149–164.

[49] ———, *Traveling waves and long time behaviour of the weakly asymmetric exclusion process*, Probab. Th. Rel. Fields **86** (1990), 443–455; *Long time behaviour of the weakly asymmetric exclusion process and the Burgers equation without viscosity*, preprint P-Math-32/90 (1990).

[50] P.A. P. Dittrich, J. Gärtner, *A central limit theorem for the weakly asymmetric simple exclusion process*, Math. Nachr. (to appear).

[51] R.L. Dobrushin, R. Kotecký, S. Shlosman, *Wulff construction: a global shape from local interactions*, AMS (to appear).

[52] R.L. Dobrushin,A. Pellegrinotti,Yu.M. Suhov,L. Triolo, *One-dimensional harmonic lattice caricature of hydrodynamics: second approximation*, J. Stat. Phys. **52** (1988).

[53] J. Doob, *Stochastic processes*, Wiley, New York, 1953.

[54] N. Dunford, J.T. Schwartz, *Linear operators. Part* I: general theory, Interscience Publishers, Inc., New York, 1957.

[55] R. Durrett, *Lecture notes on particle systems and percolation*, Wadsworth and Brooks/Cole Advanced Books, Software, Pacific Grove, California, 1988.

[56] P.A. T. Eisele, R.S. Ellis, *Symmetry breaking and random walks for magnetic systems on a circle*, Z. Wahr. verw. Gebiete **63** (1983), 297–348.

[57] C.M. Elliott, D.A. French, *Numerical studies of the Cahn-Hilliard equation for phase separation*, IMA J. Appl. Math. **38** (1987), 97–128.

[58] W. Feller, *An introduction to probability theory and its applications Vol.*I, Wiley, Inc., New York, 1957; *Vol.* II, 1966.

[59] P.A. Ferrari, *Shock fluctuations in asymmetric simple exclusion*, Probab. Theory and Rel. Fields (to appear).

[60] P.A. Ferrari, A. Galves, E. Presutti, *Closeness between a symmetric simple exclusion process and a system of independent random walks. Applications*, Nota Interna dell'Università dell'Aquila, 1981.

[61] P.A. Ferrari, C. Kipnis. E. Saada, *Microscopic structure of travelling waves in the asymmetric simple exclusion process*, Ann. of Probability **18** (1991).

[62a] P.A. Ferrari, E. Presutti, E. Scacciatelli, M.E. Vares, *The symmetric simple exclusion process.* I probability estimates, Stoch. Proc. and their Appl. (to appear).

[62b] P.A. Ferrari, E. Presutti, M.E. Vares, *Non equilibrium fluctuations for a zero range process*, Ann. Inst. H. Poincaré **24** (1988), 237-268.

[63] P. A. Ferrari, K. Ravishankar, *Shocks in the asymmetric exclusion automata*, CARR Reports in Math. Phys. n.2/91 (1991).

[64] P. Fife, *Dynamical aspects of the Cahn-Hilliard equations*, Barrett Lectures, 1990.

[65] U. Frisch, B. Hasslacher, Y. Pomeau, *Lattice gas automata for navier-Stokes equations*, Phys. Rev. Letters **56** (1986), 1505.
D. d'Humieres, P. Lallemand, U. Frisch, *Lattice gas models for 3d hydrodynamics*, Europhys. Lett. **2** (1986), 1991.
U. Frisch, D. d'Humieres, B. Hasslacher, P. Lallemand, Y. Pomeau, *Lattice gas hydrodynamics in two and three dimensions*, Complex Systems (1987).
S. Wolfram, *Cellular automaton fluids: basic theory*, J. Stat. Phys. **45** (1986), 471.

[66] J. Fritz, *On the hydrodynamic limit of a scalar Ginzburg Landau lattice model*, IMA Volume 9, Springer-Verlag (1987).

[67] A. Galves, E. Presutti, *Edge fluctuations for the one dimensional supercritical contact process*, Annals of Probability **15** (1987), 1131–1145; *Traveling wave structure of the one dimensional contact process*, Stoch. Proc. and their Appl. **25** (1987), 153–163.

[68] J. Gärtner, *Convergence towards Burgers' equation and propagation of chaos for weakly asymmetric simple exclusion processes*, Stoch. Proc. and their Appl. **27** (1988), 233–260.

[69] J. Gärtner, E. Presutti, *Shock fluctuations in a particle system*, Ann. Inst. H. Poincaré B **53** (1990), 1–14.

[70] G. Giacomin, *Van der Waals limit and phase separation in a particle model with Kawasaki dynamics*, J. Stat. Phys. **65** (1991), 217–234.

[71] M.Z. Guo, G.C. Papanicolaou, S.R.S. Varadhan, *Non linear diffusion limit for a system with nearest neighbor interactions*, Commun. Math. Phys. **118** (1988), 31–59.

[72] J. Hardy, O. de Pazzis, Y. Pomeau, *Molecular dynamics of classical lattice gas: transport properties and time correlation functions*, Phys. Rev. **A.13** (1976), 1949.

[73] R. Holley, D.W. Stroock, *Generalized Ornstein Uhlenbeck processes and infinite branching Brownian motions*, Kyoto Univ. RIMSA **14** (1978), 741.

[74] ———, *Rescaling short range interacting stochastic processes in higher dimensions*, Colloquia Mathematica Societatis Janos Bolyai **27** (1979).

[75] ———, *Central limit phenomena for various interacting systems*, Ann. Math. **110** (1979), 333.

[76] R. Illner, *Global existence for two velocity models of the Boltzmann equation*, Math. Meth. in the Apl. Sci. **1** (1979), 187–193.

[77] R. Illner, M. Pulvirenti, *Global validity of the Boltzmann equation for a two and three dimensional rare gas in vacuum: erratum and improved results*, Commun. Math. Phys. **121** (1989), 143–146.

[78] J. Jacob, J. Memin, M. Metivier, *On tightness and stopping times*, Stoch. Proc. and their Appl. **14** (1983), 109.

[79] C. Kipnis, S. Olla, S.R.S. Varadhan, *Hydrodynamics and large deviations for simple exclusion processes*, Commun. Pure Appl. Math. **42** (1989), 115–137.

[80] M. Kolb, T. Gobron, J.F. Gouyet, B. Sapoval, *Spinodal decomposition in a concentration gradient*, Europhys. Letter **11** (1990), 601–606.

[81] T. Kurtz, *Convergence of sequences of semigroups of non linear operators with an application to gas kinetics*, Trans. Am. Math. Soc. **186** (1973), 259–272.

[82] C. Landim, *Hydrodynamical equation for attractive particle systems on* $\mathbb{Z}^d$, Annals of Probability (to appear); *Hydrodynamical limit for asymmetric attractive particle systems on* $\mathbb{Z}^d$, preprint (1990).

[83] O.E. Lanford, *Time evolution of large classical systems*, Lecture Notes in Physics, vol. 38, Springer-Verlag, New York, 1975, pp. 1.

[84] R. Lang, N.X. Xanh, *Smoluchowski's theory of coagulation of colloids holds rigorously in the Boltzmann-Grad limit*, Z. Wahrsch. Verw. Geb. **54** (1980), 227–280.

[85] J.S. Langer, *Theory of spinodal decomposition in alloys*, Annals of Physics **65** (1971), 53.

[86] J.L. Lebowitz, E. Orlandi, E. Presutti, *Convergence of stochastic cellular automaton to Burgers' equation: fluctuations and stability*, Physica D **33** (1988), 165–188.

[87] ———, *A particle model for spinodal decomposition*, J. Stat. Phys. **63** (1991), 933–974.

[88] J. Lebowitz, O. Penrose, *Rigorous treatment of the Van der Waals Maxwell theory of the liquid vapour transition*, J. Math. Phys. **7** (1966), 98.

[89] J.L. Lebowitz, E. Presutti, H. Spohn, *Microscopic models of hydrodynamic behavior*, J. Stat. Phys. **51** (1988), 841–962.

[90] T.M. Liggett, *Interacting particle systems*, Springer-Verlag, New York, 1985.

[91] S. Luckhaus, L. Modica, *The Gibbs-Thompson relation within the gradient theory of phase transitions*, Archiv. Rational Mech. Anal. **107** (1989), 71–83.

[92] F. Martinelli, *Low temperature stochastic spin dynamics: metastability, convergence to equilibrium and phase segragation*, Proceedings of the 1991 $M \cap \Phi$ congress (1991).

[93] F. Martinelli, E. Olivieri, E. Scoppola, *Small random perturbation of finite and infinite dimensional systems: unpredictability of exit times*, J.Stat.Phys. **55** (1989), 477; *Rigorous analysis of low temperature stochastic Ising models: metastability and exponential approach to equilibrium*, Europhys Lett. **12** (90), 223–228; *Metastability and exponential approach to equilibrium for low temperature stochastic Ising models*, J. Stat. Phys. **61 1105** (1990), 1105; *On the Swendsen and wang dynamics*. I. Exponential convergence to equilibrium, J. Stat. Phys. **62** (1991), 117; *On the Swendsen and wang dynamics*. II. Critical droplets and homogeneous nucleation at low temperature for the two dimensional Ising model, J. Stat. Phys. **62** (1991), 135–159.

[94] ———, *Sufficient conditions for tightness and weak convergence of a sequence of processes*, Preprint, University of Minnesota (1980), Preprint, Ècole Polytechnique, Palaiseau.

[95] ———, *Convergence faible et principe d'invariance pour des martingales a valeurs dans des espaces de Sobolev*, Ann. Inst. H. Poincaré **20** (1984), 329.

[96] H. McKean, *Propagation of chaos for a class of parabolic equations*, Lectures series in Diff. Eq. (Catholic Univ.), 1967, pp. 41-57.

[97] I. Mitoma, *Tightness of probabilities on* $C([0,1],S')$ *and* $D([0,1],S')$, Annals Probability **11** (1983), 989.

[98] L. Modica, S. Mortola, *Un esempio di* Γ*-convergenza*, Boll. Un. Mat. Ital. **14b** (1977), 285–299.

[99] C.B. Morrey, *On the derivation of the equations of hydrodynamics from statistical mechanics*, Commun. Pure Appl. Math. **8** (1955), 279.

[100] G. Nappo, E. Orlandi, H. Rost, *A reaction-diffusion model for moderately interacting particles*, J. Stat. Phys. **55** (1989), 579–600.

[101] E.J. Neves, R.H. Schonmann, *Critical droplets and metastability for a Glauber dynamics at very low temperatures*, Preprint (1989); *Behavior of droplets for a class of Glauber dynamics at very low temperature*, Preprint (1990).

[102] P. Nielaba, *Dynamical block analysis in a non equilibrium system*, Physica A **611** (1990).

[103] K. Oelschläger, *Large systems of interacting particles and the porous medium equation*, J. Diff. Eq. (to appear).

[104] ———, *A law of large numbers for moderately interacting diffusion processes*, Z. Wahr. verw. gebiete **69** (1985), 279–322; *A fluctuation theorem for moderately interacting diffusion processes*, Probab. Th. Rel. Fields **74** (1987), 591–616; *On the derivation of reaction diffusion equations as limit dynamics of systems of moderately interacting stochastic processes*, Probab. Th. Rel. Fields **82** (1989), 565-586.

[105] S. Olla, S.R.S. Varadhan, *Scaling limit for interacting Ornstein-Uhlenbeck processes*, Commun. Math. Phys. **135** (1991), 355–378.

[106] E.Orlandi, E.Presutti, *Some rigorous results on phase segregation for stochastic cellular automata*, Discrete models of fluid dynamics (A.S. Alves, eds.), World Scientific, Singapore, 1991, pp. 45–59.

[107] K.R. Parthasarathy, *Probability measures on metric spaces*, Academic Press, New York, San Francisco, London, 1967.

[108] O. Penrose, *A mean field equation of motion for the dynamic Ising model*, J. Stat. Phys. **63** (1991), 975–986.

[109] O. Penrose, P.C. Fife, *Thermodynamically consistent models of phase-field type for the kinetics of phase transitions*, Physica D **43** (1990), 44–62.

[110] O. Penrose, J. Lebowitz, *Rigorous treatment of metastable states in the Van der Waals Maxwell theory*, J. Stat. Phys. **3** (1971), 211–236.

[111] E.Presutti, *Collective behaviour in interacting particle systems*, Invited talk at the I World Congress of the Bernoulli Society, Tashkent, 1986; *Collective phenomena in stochastic particle systems*, Lecture Notes in Math. n.1250, Springer-Verlag, New York, 1987, pp. 195–232.

[112] E. Presutti, D.W. Wick, *Macroscopic fluctuations in a one dimensional mechanical system*, J. Stat. Phys. **52** (1988), 497–502.

[113] J. Quastel, *Diffusion of colors in the simple exclusion process*, Preprint (1990).

[114] K.Ravishankar, *Interface fluctuations in the two dimensional weakly asymmetric simple exclusion*, Stoch. Proc. and their Appl. (to appear).

[115] R. Rebolledo, *Sur l'existence de solutions à certains problèmes de semimartingales*, C.R. Acad. Sci. Paris **290** (1980), 843.

116] D. Revuz, *Markov chains*, North Holland, Amsterdam - New York, 1975.

117] F. Rezakhanlou, *Hydrodynamic limit for a system with finite range interactions*, Commun. Math. Phys. **129** (1990), 445-480.

118] ———, *Hydrodynamical limit for asymmetric attractive particle systems on $\mathbb{Z}^d$*, preprint (1990).

119] H. Rost, *Non equilibrium behavior of many particle process: dnsity profiles and local equilibria*, Z. Wahrsch. Verw. Gebiete **58** (1981), 41–53.

120] M. Rovere, D.W. Heermann, K. Binder, *Gas liquid transition of the two dimensional Lennard-Jones fluid*, J. Physics **C2** (1990), 7009.

121] R. Schonmann, *An approach to carachterize metastability and critical droplets in stochastic Ising models*, Preprint.

122] Ya.G.Sinai, *Dynamics of Local equilibrium Gibbs distributions and Euler equations. The one dimensional case*, Selecta Mathematica Sovietica **7** (1988).

123] J.Smoller, *Shock Waves and Reaction Diffusion Equations*, Springer Verlag, New York, 1983.

124] H.Spohn, *Kinetic equations for Hamiltonian systems: Markovian limits*, Rev. Mod. Phys. **53** (1980), 569.

125] ———, *The dynamics of systems with many particles*, Springer-Verlag, New York, 1991.

126] ———, *Hydrodynamical theory for equilibrium time correlation functions of hard rods*, Annals of Physics **141** (1982), 353–364.

127] A. Sznitmann, *A propagation of chaos result for Burgers equation*, Prob. Th. Rel. Fields **71** (1988), 581–613.

128] ———, *Topics in Propagation of Chaos*, Ecole d'été de Probabilités des Saint Flour, 1989.

129] K. Uchiyama, *On the Boltzmann-Grad limit for the Broadwell model of the Boltzmann equation*, J. Stat. Phys. **52** (1988), 331–355.

130] S.R.S. Varadhan, *Scaling limits for interacting diffusions*, Commun. Math. Phys. (1991) 313–353, (to appear).

[131] D.W. Wick, *A dynamical phase transition in an infinite particle system*, J. Stat. Phys. Vol **38** (1985), 1005–1025.

[132] ———, *Entropy arguments in the study of the hydrodynamical limit*, CARR reports in Math. Phys. n. 3/88 (1988).

[133] ———, *Hydrodynamic limit of a non gradient interacting particle process*, J. Stat. Phys. **54** (1989), 873.

[134] H.T. Yau, *Relative entropy and hydrodynamics of Ginzburg-Landau models*, Letters Math. Phys. **22** (1991), 63–80.

SUBJECT INDEX